Digital Image and Signal Processing for Measurement Systems

RIVER PUBLISHERS SERIES IN INFORMATION SCIENCE AND TECHNOLOGY

Information science and technology enables 21st century into an Internet and multimedia era. Multimedia means the theory and application of filtering, coding, estimating, analyzing, detecting and recognizing, synthesizing, classifying, recording, and reproducing signals by digital and/or analog devices or techniques, while the scope of "signal" includes audio, video, speech, image, musical, multimedia, data/content, geophysical, sonar/radar, bio/medical, sensation, etc. Networking suggests transportation of such multimedia contents among nodes in communication and/or computer networks, to facilitate the ultimate Internet. Theory, technologies, protocols and standards, applications/ services, practice and implementation of wired/wireless networking are all within the scope of this series. We further extend the scope for 21st century life through the knowledge in robotics, machine learning, cognitive science, pattern recognition, quantum/biological/molecular computation and information processing, and applications to health and society advance.

- Communication/Computer Networking Technologies and Applications
- Queuing Theory, Optimization, Operation Research, Statistical Theory and Applications
- Multimedia/Speech/Video Processing, Theory and Applications of Signal Processing
- Computation and Information Processing, Machine Intelligence, Cognitive Science, and Decision

For a list of other books in this series, please visit www.riverpublishers.com

Digital Image and Signal Processing for Measurement Systems

Editors

Richard J. Duro
University of Coruña, Spain

and

Fernando López-Peña
University of Coruña, Spain

Aalborg

ISBN 978-87-92329-29-5 (hardback)

Published, sold and distributed by:
River Publishers
P.O. Box 1657
Algade 42
9000 Aalborg
Denmark

Tel.: +4536995 3197
www.riverpublishers.com

Table of Contents

Preface

This book provides an overview of advanced digital image and signal processing techniques that are currently being applied in the realm of measurement systems. It contains a selection of extended versions of the best papers presented at the Sixth IEEE International Workshop on Intelligent Data Acquisition and Advanced Computing Systems: Technology and Applications IDAACS 2011 related to this topic and encompass applications that go from multidimensional imaging to evoked potential detection in brain computer interfaces. The objective was to provide a broad spectrum of measurement applications so that the different techniques and approaches could be presented.

The Sixth IEEE International Workshop on Intelligent Data Acquisition and Advanced Computing Systems: Technology and Applications (IDAACS) was held in Prague, Czech Republic, September 15–17, 2011. The workshop was organized by the IEEE Ukraine I&M/CI Joint Societies Chapter, the Research Institute of Intelligent Computer Systems, the Ternopil National Economic University, the Glushkov Institute of Cybernetics, the National Academy of Science, Ukraine and the Faculty of Electrical Engineering at the Czech Technical University in Prague, Czech Republic. IDAACS 2011 was the sixth edition of a well-established bi-annual forum that has been increasing in participants since the first workshop conducted in Foros (Crimea), Ukraine in 2001.

The IDAACS workshop series is established as a forum for high quality reports on state-of-the-art theory, technology and applications of intelligent data acquisition and advanced computer systems. All of these techniques and applications have experienced a rapid expansion in the last few years that has resulted in more intelligent, sensitive, and accurate methods for the acquisition of data and its processing applied to manufacturing process control and inspection, environmental and medical monitoring and diagnostics, as well as intelligent information gathering and analyses for the purpose security and safety.

The success of IDAACS arises not only from the importance of the topics it focuses on, but also because of its nature as a unique forum for establishing scientific contacts between research teams and scientists from different countries. This purpose has become one of the main reasons for the rapid success of IDAACS, as it turns out to be one of the few events in this area of research where Western and former Eastern European scientists can discuss and exchange ideas and information, allowing them to characterize common and articulated research activities and creating the environment for establishing joint research collaborations. It provides an opportunity for all the participants to discuss topics with colleagues from different spheres such as academia, industry, and public and private research institutions.

Even though this book concentrates on providing insights into what is being done in the area of signal processing, the papers that were selected reflect the variety of research presented during the workshop as well as the very diverse fields that may benefit from these techniques.

The book concentrates on signal processing for measurement systems and its objective is to provide a general overview of the area and an appropriate introduction to the topics considered. This is achieved through 10 chapters devoted to current topics of research followed by different research groups within this area. These 10 chapters reflect advances corresponding to signals of different dimensionality. They go from mostly one dimensional signals in what would be the most traditional area of signal processing realm to three dimensional grey level images and the five dimensional RGB signals (two spatial and three color values) to signals of very high dimensionality such as hyperspectral signals that can go up to dimensionalities of more than one thousand.

The chapters have been thought out to provide an easy to follow introduction to the topics that are addressed, including the most relevant references, so that anyone interested in them can start their introduction to the topic through these references. At the same time, all of them correspond to different aspects of work in progress being carried out in various laboratories throughout the world and, therefore, provide information on the state of the art of some of these topics.

In terms of contents, the first chapter of the book, by Nikolay Chumerin, Nikolay Manyakov, Adrien Combaz, Arne Robben, Marijn van Vliet and Marc Van Hulle, discusses several decoding methods for the Steady State Visual Evoked Potential (SSVEP) paradigm as well as their use in Brain Computer Interfaces (BCIs). The chapter includes an introduction to the concept of BCI, its different categories and their relevance for speech and

motor disabled patients. The authors present processing and decoding methods that use either time-domain or spectral domain features and show their usability in a set of applications.
The second chapter, by Andrius Petrėnas, Vaidotas Marozas, Arūnas Lukoševičius, and Andrius Sakalauskas, contemplates the extraction of f-waves in electrocardiograms using echo state network based nonlinear adaptive filters. Echo state networks are an example of the new and very exciting dynamic recurrent neural network paradigm called reservoir computing that allows some very complex processing capabilities using a relatively simple training and adaptation procedure. The authors describe the technique in depth and present some examples of its application.
The third chapter deals with a higher level of what could be considered digital signal or message processing. That is, the adaptation of data streams to different platforms and formats. This problem has to do with appropriate and timely transcoding of data streams. In this case, Steven Pigeon and Stéphane Coulombe discuss the adaptation of multimedia messages in an optimal quality aware manner through the use of predictors. They provide a general state of the art of the area and specify the nature of the problem and some of its solutions.
A second block of the book addresses four applications that make use of images in order to provide measurements of different variables. This block contains four chapters, the first of which contemplates the problem of tracking the movements of IPMC actuators. This chapter, by Kyriakos Tsiakmakis and Theodore Laopoulos, describe the problem and present a comparison of a series of methods to solve it. The authors take into account the processing speed and use different architectures, such as GPU, to improve on this issue. Yuriy Kurylyak, Francesco Lamonaca and Domenico Grimaldi address in Chapter 5 the possibility of acquiring heart pulse signals using smartphone cameras. This is what is called the acquisition of the photoplethysmogram (PPG). The authors provide an overview of the state of the art in this type of signal acquisition process and study the problems for the applications of these techniques to measure pulse rate by using smartphone cameras and introducing some algorithms to make the process feasible. They provide the results of different tests using the technique.
In Chapter 6, Saowaluck Kaewkamnerd, Chairat Uthaipibull, Apichart Intarapanich, Montri Pannarat, Sastra Chaotheing and Sissades Tongsima describe an automated device for the detection and classification of Plasmodium species on thick blood films. These species cause Malaria and the authors discuss both, the biological aspects and the limitations of current diagnostic methods

before introducing an automated device for this purpose. The device consists in an image acquisition unit and its image analysis software module, which can be mounted on a conventional light microscope and operated by a health practitioner/technician with minimal training.
Finally, within this block, Ihor Paliy, Anatoly Sachenko and Ognian Boumbarov address the problem of detecting and tracking faces in images. In this chapter, the authors provide an extensive introduction to the problem and describe some techniques for the detection of faces in moving images as well as for tracking them for different applications. They propose combining two well-known techniques in order to improve the speed at which the images are processed, thus allowing time for other operations such as emotion detection.
A third block of the book contemplates the problem of processing signals of higher dimensionality. This block starts with a chapter by Valentin Ganchenko, Rauf Sadykhov, Alexander Doudkin, Albert Petrovsky and Tadeusz Pawlowski. It discusses the detection and recognition of special areas in agricultural fields from remote sensing images. In this case, the authors propose an effective method for processing vegetation cover color images produced with the help of high resolution digital cameras and address the problems related to these techniques.
The second chapter in this block, Chapter 9, is authored by Dora B. Heras, Francisco Argüello, J. López Gómez, Blanca Priego and Jose Antonio Becerra who describe a GPGPU implementation of artificial neural network based target detection and identification algorithms over hyperspectral images. The aim of their work is to try to achieve real time processing for this type of images. The authors present a very comprehensive description of the problem and related area as well as of GPU processing in this area. They show different examples and compare the results using different techniques.
Finally, the last chapter, by Blanca Priego, Daniel Souto, Francisco Bellas, Fernando López-Peña and Richard J. Duro, addresses the use of temporal processing artificial neural networks to produce classifications using the temporal characteristics that can be extracted from sequences of hyperspectral images. In other words, classification is achieved taking into account the temporal evolution of the process in the discrimination that must be made. To this end the authors describe a temporal delay based artificial neural architecture as well as its training algorithm and apply it to industrial processes.
The papers selected for this book are extended and improved versions of those presented at the workshop and as such are significantly expanded with respect to the original ones presented at the workshop. Obviously, this set of papers are just a sample of the dozens of presentations and results

that were seen at IDAACS 2011, but we do believe that they provide an overview of some of the problems in the area of signal and image processing for measurement systems and the approaches and techniques that relevant research groups within this area are employing to try to solve them. We hope the readers find them interesting, useful and that they can obtain a glimpse of this very interesting area.

The Editors
Richard J. Duro
Fernando López Peña

1

Processing and Decoding Steady-State Visual Evoked Potentials for Brain-Computer Interfaces

Nikolay Chumerin, Nikolay V. Manyakov, Marijn van Vliet, Arne Robben, Adrien Combaz and Marc M. Van Hulle

Laboratorium voor Neuro- en Psychofysiologie, KU Leuven, Campus Gasthuisberg, O&N 2, Herestraat 49, 3000 Leuven, Belgium; e-mail: {nikolay.chumerin, nikolayv.manyakov, marijn.vanvliet, arne.robben, adrien.combaz, marc.vanhulle}@med.kuleuven.be

Abstract

In this chapter, several decoding methods for the Steady State Visual Evoked Potential (SSVEP) paradigm are discussed, as well as their use in Brain Computer Interfaces (BCIs). The chapter starts with the concept of BCI, the different categories and their relevance for speech- and motor disabled patients. The SSVEP paradigm is explained in detail. The discussed processing and decoding methods employ either time-domain or spectral domain features. Finally, to show the usability of these methods and of SSVEP-based BCIs in general, three applications are described: a spelling system, the "Maze" game and the "Tower Defense" game. We conclude the chapter by addressing some challenges for future research.

Keywords: Brain-Computer Interfaces, electroencephalography, SSVEP, signal processing.

Richard J. Duro and Fernando López Peña (Eds.), Digital Image and Signal Processing for Measurement Systems, 1–33.

1.1 Brain-Computer Interface

While the idea of Brain Computer Interfaces (BCIs) appeared around the 1970s [42], BCI by itself received a lot of attention only in recent years, when technology made it possible to perform on-line computer-based monitoring and recordings of different aspects of brain activity. BCI can be defined as "*a communication system in which messages or commands that an individual sends to the external world do not pass through the brain's normal output pathways of peripheral nerves and muscles*" [48]. Thus, by measuring and interpreting brain activity directly, no muscular activity becomes necessary for communication. As a consequence, BCIs become especially useful for persons with severe motor- and speech disabilities such as Amyotrophic Lateral Sclerosis (ALS), Cerebrovascular Accident (CVA), etc. allowing them to communicate with external world overcoming there impairments [23, 25]. Such BCI ideas have already attracted attention not only in the scientific community, but also in the popular media and in different movies.[1]

Any BCI system consists of the following components: a brain activity recording device, a preprocessor, a decoder, and the external device, usually a robotic actuator or a display, where feedback is shown to the subject. Depending on the recorded brain activity and the used signals, BCI can be classified into *invasive* and *non-invasive*. Invasive BCIs are based on electrode arrays implanted in specific areas of the cortex [26, 27, 41] or just above the cortex (where electrocorticograms (ECoG) are recorded) [20], whereas non-invasive BCIs employ magnetoencephalography (MEG), functional magnetic resonance imaging (fMRI) and most often electroencephalography (EEG) [3, 4, 9].

1.1.1 Invasive BCI

The beginning of invasive BCIs can be traced back to 1999, when for the first time, it was shown that ensembles of cortical neurons could directly control a robotic manipulator [7]. Since then a steady increase in the number of publications can be observed. For a state-of-the-art of the invasive BCI, we refer to the review paper [19]. Invasive BCI can be divided into two categories, depending on the number of the recording sites. Some research groups constructed a BCI based on recording from a single cortical area (for example, the primary motor cortical area, M1), while others recorded from several areas, taking advantage from the distributed processing of information in the brain. On the other hand, invasive BCIs can also be divided based on the type of

[1] E.g., "Surrogates" movie (2009), series "House MD" season 5, episode 19 (2009).

signal used for decoding. It can be, for example, *action potentials* (*spikes*) or *local field potentials* (LFPs). In the first case, one records only from a few neurons, with most prominent tuning properties [37, 38], or from a large ensemble of neurons (hundreds of cells) [6,26,45]. The LFPs are more stable and can be recorded for longer period of time, which make them attractive for BCI applications [27,28,35]. Invasive BCIs can also be categorized according to their application. They are primary developed for the motor control of, for example, an arm actuator [19, 37, 38, 45]. This can be used for restoring the lost motoric abilities of patients. But, it should be mentioned that mostly all of these spike- or LFP-based BCI experiments have been performed only on monkeys, rather than on humans (for human invasive BCI, see [15, 16]). For such motoric BCIs, as a decoder, usually a linear regression of the spike firing rate into the position and velocity of the limb is considered. Another application of invasive BCIs is with cognitive neural prosthesis, which is aimed at relating the recording activity to the higher-level cognitive process that organize behavior. This can be used for decoding the mental state of the subject, its goals, and so on [30].

1.1.2 Non-invasive BCI

The non-invasive BCIs, which mostly exploit EEG recordings, in turn, can be categorized according to the brain signal evoking paradigm used. In one such category, which is also the topic of this book chapter (see Section 1.2 for more details), *visually evoked potentials* (VEPs) are explored, and its origins can be traced back to the beginning of BCI ideas (in 1970s) when Jacques Vidal constructed the first BCI [42]. As another category, we can mention the non-invasive BCIs that rely on the detection of imaginary movements of the right and the left hands. These methods exploit *slow cortical potentials* (SCP) [3, 18], *event-related desynchronization* (ERD) on the mu- and beta-rhythm [32, 47], and the *readiness potential* (bereitschaftspotential) [4]. The detection of other mental tasks (e.g., cube rotation, number subtraction, word association [29]) also belong to this category. Additionally to the mentioned paradigms, one can also distinguish BCIs that rely on the "oddbal" *event-related potential* (ERP) in the parietal cortex, where an ERP is a stereotyped electrophysiological response to an internal or external stimulus [21]. The most known and explored ERP is the P300. It can be detected while the subject is classifying two types of events with one of the events occurring much less frequently than the other ("rare event"). The rare events elicit ERPs consisting of an enhanced positive-going signal component with a latency of

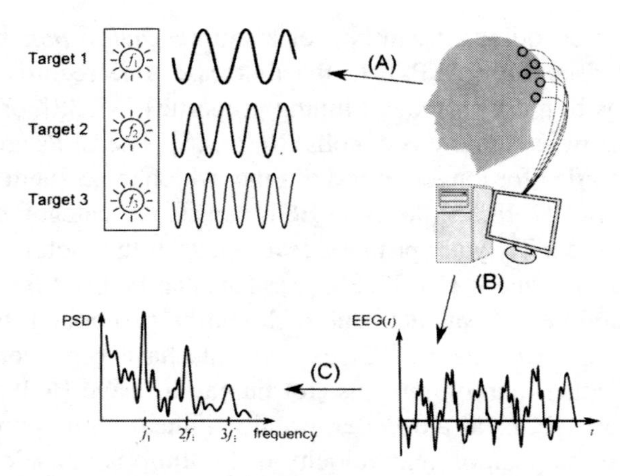

Figure 1.1 Schema of SSVEP decoding approach: (A) a subject looks at Target 1, flickering at frequency f_1, (B) noisy EEG-signals are recorded, (C) the power spectral density plot of the EEG signal (estimated over a sufficiently large window) shows dominant peaks at f_1, $2f_1$ and $3f_1$.

about 300 ms [33]. In order to detect the ERP in the signal, one trial is usually not enough and several trials must be averaged to reduce additive noise and other irrelevant activity in the recorded signals. The ability to detect ERPs can be used in a BCI paradigm such as the P300 mind-typer [9,13,39], where subject can spell words by looking at the randomly flashed symbols.

1.2 Steady-State Visual Evoked Potential

A BCI based on *Steady-State Visual Evoked Potential* (SSVEP) relies on the psychophysiological properties of EEG brain responses recorded from the occipital pole during the periodic presentation of identical visual stimuli (i.e., flickering stimuli). When the periodic presentation is at a sufficiently high rate (not less than 6 Hz), the individual transient visual responses overlap, leading to a *steady state* signal: the signal resonates at the stimulus rate and its multipliers [21]. This means that, when the subject is looking at stimuli flickering at the frequency f_1, the frequencies $f_1, 2f_1, 3f_1, \ldots$ can be detected in the Fourier transform of the EEG signal recorded from the occipital pole, as schematically illustrated in Figure 1.1.

Since the amplitude of a typical EEG signal decreases as $1/f$ in the spectral domain [1], the higher harmonics become less prominent. Furthermore, the SSVEP is embedded in other on-going brain activity and (recording) noise. Thus, when considering a too small recording interval, erroneous detections are quite likely to occur. To overcome this problem, averaging over several time intervals [8], recording over longer time intervals [44], and/or preliminary training [12, 22, 24] are often used for increasing the *signal-to-noise ratio* (SNR) and the detectability of the responses. Moreover, an efficient SSVEP-based BCI (or, shorter, SSVEP BCI) should be able to reliably detect SSVEP induced by several possible ($f_1, \ldots, f_n$) stimulation frequencies (see Figure 1.1), which makes the SSVEP detection problem even more complex, calling for an efficient signal processing and decoding algorithm.

SSVEP BCI can be considered as a dependent one according to the classification proposed in [48]. The dependent BCI does not use the brain's normal output pathways (for example, the brain's activation of muscles for typing a letter) to carry the message, but activity in these pathways (e.g., muscles) is needed to generate the brain activity (e.g., EEG) that does carry it. In the case of SSVEP BCI, the brain's output channel is EEG, but the generation of the EEG signal depends on the gaze direction, and therefore on extraocular muscles and the cranial nerves that activate them. A dependent BCI is essentially an alternative method for detecting messages carried in the brain's normal output pathways. According to this, for example, SSVEP BCI can be viewed as a way to detect the gaze direction by monitoring EEG rather than by monitoring eye position directly. Therefore, for the those patients that also lack extraocular muscle control, this BCI is inapplicable. However, for others, the SSVEP BCI is more feasible than other systems. It has the advantages of a high information transfer rate (the amount of information communicated per unit time) [2] and little (or no) user training [44].

As a stimulation device for SSVEP BCI, either light-emitting diodes (LEDs) or computer screen (LCD or CRT monitors) are used [50]. While the LEDs can evoke more prominent SSVEP responses [50] at any desirable frequency, they require additional equipment (considering that the feedback is presented on the monitor). Thus, SSVEP-based BCI systems mostly rely on computer screen for stimulation in order to combine stimulation and feedback presentation devices. And, as a consequence, they have some limitations: the stimulation frequencies become related to the refresh rate of the computer screen [43] (see the way for stimulation construction in Section 1.3.2), and restricted to specific (subject-dependent) frequency bands to obtain good responses [24]; the harmonics of some stimulation frequencies could interfere with one another (and their harmonics), leading to a deterioration of the decoding performance [43]. Thus, taking into account these restrictions, only a limited number of targets could be used in monitor-based SSVEP BCI.

An SSVEP BCI could be build as a system with *synchronous* and *asynchronous* modes. First one assumes that the subject observes the stimulus for a fixed predefined amount of time after which the classification is performed. This mode requires either putting some long timing of stimulation to satisfy all subjects' personal brain responses or to perform preliminary training/calibration for adjusting stimulation timing for each person. The asynchronous mode assumes that the stimulation and decoding go in parallel, thus allowing doing a proper classification, when the amount of data is sufficient for this. The comparison of those two modes are discussed in detail in Section 1.5.1 in the context of SSVEP BCI applications.

1.3 System Design

1.3.1 EEG Data Acquisition

We considered two EEG recording devices for the applications discussed in this chapter: an EEG device with a setup that is commonly considered in BCI research, thus, for in-lab environment, and a cheap, commercially-available device, specially developed for entertainment purposes.

The first one is a prototype of an ultra low-power eight channel *wireless* EEG system, which consists of two parts: an amplifier coupled with a wireless transmitter (see Figure 1.2a) and a USB stick receiver (Figure 1.2b). Denoting the number of the EEG channels by N_s (the subscript s stands for "source"), for the *imec* EEG device we have $N_s = 8$. This system was developed by *imec*,[2] and built around their ultra-low power 8-channel EEG amplifier chip [49]. The acquired EEG data is sampled using 12 bit/channel/sample and then transmitted at sample rate of $F_s = 1000$ Hz for each channel. We used an electrode cap with large filling holes, and sockets for mounting active Ag/AgCl electrodes (ActiCap, Brain Products) (Figure 1.2c). The recordings were made with electrodes located on the occipital pole (covering primary visual cortex), namely at positions P3, Pz, P4, PO9, O1, Oz, O2, PO10, according to the international 10–20 electrode placement system. The reference and the ground electrodes were placed on the left and right mastoids, respectively. The electrode positions are illustrated in Figure 1.2d.

[2] `http://www.imec.be`

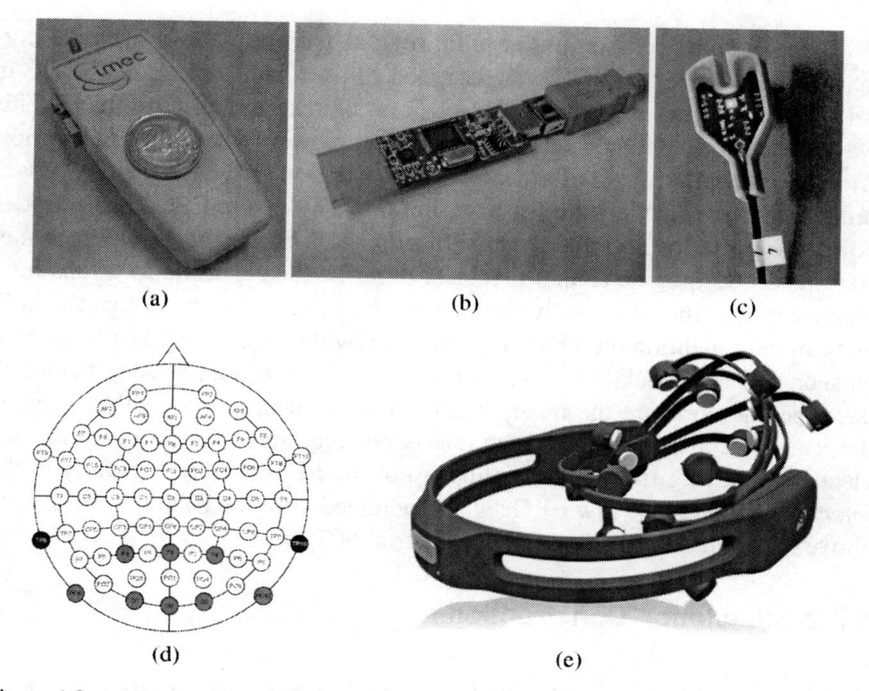

Figure 1.2 (a) Wireless 8 channels amplifier. (b) USB stick receiver. (c) Active electrode. (d) Locations of the electrodes on the scalp. (e) Emotiv EPOC headset.

The raw EEG signals are filtered above 3 Hz with a fourth order zero-phase digital Butterworth filter so as to remove the DC component and the low frequency drift. A notch filter is also applied to remove the 50 Hz powerline interference.

The second device is the EPOC (Figure 1.2e), developed by *Emotiv*.[3] This headset has $N_s = 14$ saline sensors placed for normal use approximately at positions AF3, AF4, F3, F4, F7, F8, FC5, FC6, P7, P8, T7, T8, O1, O2. The data is wirelessly transmitted to a computer with a sampling frequency of $F_s = 128$ Hz for each channel, at a resolution 14 bit/channel/sample. The choice of this device was mostly motivated by its low price (starting from $300) and wide availability (more than 30000 devices have already been sold). Thus, the implementation of a BCI with this device is potentialy aimed for a broad audience.

---

[3] `http://www.emotiv.com`

Since we are accessing other brain regions (primary above occipital cortex) that the ones the EPOC was designed for, we had to place the EPOC in a 180 °-rotated (in horizontal plain) position on the head of the subject. This way, the electrodes could reach the occipital region (where SSVEP is most strongly present), instead of the more anterior region for which the device was initially designed. After the rotation, the majority of the EPOC's electrodes cover the posterior regions of the subject's skull. Since the EPOC is a one-size-fits-all design, we cannot precisely describe the electrode locations for a given subject, since it strongly depends on the geometry of the subject's skull. We can only mention the brain area covered by the electrodes. While it could be seen as a drawback from a scientific point of view (not allowing to clearly describe and compare the results between the subjects), it actually increases the usability of the headset since one is not required to precisely place the electrodes, saving time in the setting-up of the EEG device. Similarly to the *imec* EEG device, the raw EEG signals obtained with the EPOC were filtered above 3 Hz with an additional notch filter at 50 Hz.

1.3.2 Stimulation Construction

In our applications we have used a laptop with a bright 15,4" LCD screen with refresh rate close to 60 Hz. In order to arrive at a visual stimulation with stable frequencies, we show an intense stimulus for k frames, and a less intense stimulus for the next l frames, hence, the flickering period of the stimulus is $k + l$ frames and the corresponding stimulus frequency is $r/(k + l)$, where r is the screen's refresh rate. Using this simple strategy, one can stimulate the subject with the frequencies that are dividers of the screen refresh rate: 30 Hz (60/2), 20 Hz (60/3), 15 Hz (60/4), 12 Hz (60/5), 10 Hz (60/6), 8.57 Hz (60/7), 7.5 Hz (60/8), 6.66 Hz (60/9), and 6 Hz (60/10).

1.4 Decoding Methods

In general, methods for SSVEP detection can be classified into frequency- and time-based ones. While former looks directly into power spectral density at frequencies used in a BCI system with the aim of monitoring the increase relative to some baseline (viewed in this chapter in terms of *signal-to-noise ratio* (SNR)), latter one directly exploits the fact, that SSVEP is a sort of ERP locked to the stimulation (with repeated pattern).

1.4.1 Classification in the Frequency Domain

As was already mentioned in Section 1.2, the recorded EEG data contain not only SSVEP-induced component, but also other brain activity and noise. Thus, it is useful not to directly perform decoding, by rather do some preprocessing in before to enhance the desired SSVEP components in the recorded EEG. For this reason, consideration of multiple EEG channels can be seen as beneficial for SSVEP analysis, since this allows to perform some spatial filtering (construction of weighted combination of the recorded N_s "source" signals). For example, in [44] it was shown that a suitable bipolar combination of EEG electrodes suppresses noise, resulting in increase in the SNR. Thus, here we start from the description of the spatial filtering approach (Section 1.4.1.1) followed by the decoding/classification strategy (Section 1.4.1.2).

1.4.1.1 Spatial Filtering: The Minimum Energy Combination

In [14], a spatial filtering technique is proposed called the *Minimum (Noise) Energy Combination* (MNEC) method. The idea of this technique is to find a linear combination of the channels that decreases the noise level of the resulting weighted signals at the specific frequencies we want to detect (namely, the frequencies of the oscillations evoked by the periodically flickering stimuli, and their harmonics). This can be done in two steps. Firstly, all information related to the frequencies of interest must be eliminated from the recorded signals. The resulting signals contain only information that is "uninteresting" in the context of SSVEP detection, and, therefore, could be considered as noise components of the original signals. Secondly, we look for a linear combination that minimizes the variance of the weighted sum of the "noisy" signals obtained in the first step. Eventually, we apply this linear combination to the original signals, resulting in signals with a lower level of noise.

The first step can be done by subtracting from the EEG signal all the components corresponding to the stimulation frequencies and their harmonics. Formally, this can be done in the following way. Let us consider the input signal, sampled over a time window of duration T with sampling frequency F_s, as a matrix $\mathbf{X}$ with (N_s) channels in columns and samples in rows. Then, one needs to construct a matrix $\mathbf{A}$, which should have the same number of rows as $\mathbf{X}$ and as the number of columns twice the number of all considered frequencies (including harmonics). For a given time instant t_i (corresponding to the i-th sample in $\mathbf{X}$) and frequency f_j (from the full list of stimulation frequencies including the harmonics), the corresponding elements $a_{i,2j-1}$ and $a_{i,2j}$ of the matrix **A** are computed as $a_{i,2j-1} = \sin(2\pi f_j t_i)$ and $a_{i,2j} = \cos(2\pi f_j t_i)$. For example, considering only $n_f = 2$ frequencies with their $N_h = 2$ harmonics and a time interval of $T = 2$ seconds, sampled at $F_s = 1000$ Hz, the matrix **A** would have $2n_f(1+N_h) = 2 \cdot 2 \cdot 3 = 12$ columns and $T \cdot F_s = 2000$ rows. The most "interesting" components of the signal **X** can be obtained from **A** by a projection determined by the matrix $\mathbf{P_A} = \mathbf{A}(\mathbf{A}^T\mathbf{A})^{-1}\mathbf{A}^T$. Using $\mathbf{P_A}$ the original signal *without* the "interesting" information is estimated as $\tilde{\mathbf{X}} = \mathbf{X} - \mathbf{P_A}\mathbf{X}$. Those remaining signals $\tilde{\mathbf{X}}$ can be considered as noise components of the original signals (i.e., brain activity not related to the visual stimulation).

In the second step, we use an approach based on *Principal Component Analysis* (PCA) to find a linear combination of the input data for which the noise variance is minimal. A PCA transforms a number of possibly correlated variables into uncorrelated ones, called principal components, defined as projections of the input data onto the corresponding principal vectors. By convention, the first principal component captures the largest variance, the second principal component the second largest variance, and so on. Given that the input data comes from the previous step, and contains mostly noise, the projection onto the last principal component direction is the desired linear combination of the channels, i.e., the one that reduces the noise in the best way (i.e., minimizing the noise variance).

The conventional PCA approach estimates the principal vectors as eigenvectors of the covariance matrix $\Sigma = E\{\tilde{\mathbf{X}}^T\tilde{\mathbf{X}}\}$, where $E\{\cdot\}$ denotes the statistical expectancy.[4] For N_s-dimensional EEG signal, matrix Σ has size $N_s \times N_s$ and is positive semidefinite. Therefore, it is possible to find a set of N_s orthonormal eigenvectors (represented as columns of a matrix V), such that $\Lambda = V\Sigma V^T$, where Λ is a diagonal matrix of the corresponding eigenvalues $\lambda_1 \geq \lambda_2 \geq \cdots \geq \lambda_{N_s} \geq 0$. Then, the K last (smallest) eigenvalues are selected such that K is maximal, and $\sum_{k=1}^{K} \lambda_{N_s-k+1} / \sum_{j=1}^{N_s} \lambda_j < 0.1$ is satisfied. The corresponding K eigenvectors, arranged as columns of a matrix V_K, specify a linear transformation that efficiently reduces the noise power in the signal $\tilde{\mathbf{X}}$. The same noise-reducing property of V_K is valid for the original signal **X**. Assuming that V_K would reduce the variance of the noise more than the variance of the signal of interest, the signal that is spatially filtered in this way, $\mathbf{S} = V_K\mathbf{X}$, would have greater (or, at least, not smaller) SNR than original recorded EEG signals [14].

---

[4] Since the original signal is high-pass filtered above 3 Hz, the DC component is removed and, therefore, the filtered data are centered (i.e., the mean is close to zero).

1.4.1.2 Classification

The straight-forward approach to select one frequency (among several possible candidates) present in the analyzed signal is based on a direct analysis of the signal power function $P(f)$ that is defined as follows:

$$P(f) = \left(\sum_t s(t)\sin(2\pi f t)\right)^2 + \left(\sum_t s(t)\cos(2\pi f t)\right)^2, \qquad (1.1)$$

where $s(t)$ is the signal after spatial filtering. Note that the right-hand part of this equation is the squared Discrete Fourier Transform magnitude at the frequency of interest [14]. The "winner" frequency f^* can then be selected as the frequency with maximal (among all considered frequencies $f_1, f_2, \ldots, f_{n_f}$) power amplitude:

$$f^* = \arg\max_{f_1,\ldots,f_{n_f}} P(f). \qquad (1.2)$$

Unfortunately, in the case of EEGs, this direct method is not applicable due to the nature of the EEG signal: the corresponding power function decreases (similarly to $1/f$) with increasing f [1]. In this case, the true dominant frequency could have an power amplitude less than the other considered lower frequencies. In [44] it was shown that the SNR does not decrease with increasing frequency, but remains nearly constant. Relying on this finding, one can select the "winner" frequency as the one which the maximal SNR $P(f)/\sigma(f)$, where $\sigma(f)$ is an estimation of the noise power for frequency f.

The noise power estimation is not a trivial task. One way to do this is to record extra EEG data from the subject, without visual stimulation. In this case, the power of the considered frequencies in the recorded signal should correspond to the noise level. Despite its apparent simplicity, this method has at least two drawbacks: (1) an extra (calibration) EEG recording session is needed, and (2) the noise level changes over time and the pre-estimated values could significantly deviate from the actual ones. To overcome these drawbacks, we need an efficient on-line method of noise power estimation. As a possible solution, one can try to approximate the desired noise power $\sigma(\tilde{f})$ for a frequency of interest $\tilde{f}$ using values of $P(f)$ from a close neighborhood $O(\tilde{f})$ of the considered frequency $\tilde{f}$. A simple averaging $\sigma(\tilde{f}) \approx \langle P(f)\rangle_{f\in O(\tilde{f})\setminus\tilde{f}}$ produces unstable (jittering) estimates if the size of the neighborhood $O(\tilde{f})$ is small. Additionally, a large neighborhood could contain several frequencies of interest that could bias the estimate of $\sigma(\tilde{f})$.

In our work, we have used an approximation of noise based on an autoregressive modeling of the data, after excluding all information about the flickering, i.e., of signals $\tilde{\mathbf{S}} = V_K\tilde{\mathbf{X}}$ (see Section 1.4.1.1). The rationale behind this approach is that the autoregressive model can be considered as a filter (working through convolution), in terms of ordinary products between the transformed signals and the filter coefficients in the frequency domain. Since we assume that the prediction error in the autoregressive model is uncorrelated white noise, we have a flat power spectral density for it with a magnitude that is a function of the variance of the noise. Thus, the Fourier transformations of the regression coefficients a_j (estimated, for example, with the use of the Yule-Walker equations) show us the influence of the frequency content of particular signals on the white noise variance ($\tilde{\sigma}$). By assessing such transforms, we can obtain an approximation of the power of the signal $\tilde{\mathbf{S}}$. More formally, we have:

$$\sigma(f) = \frac{\pi T}{4} \frac{\tilde{\sigma}^2}{|1 - \sum_{j=1}^{p} a_j \exp(-2\pi i j f / F_s)|}, \tag{1.3}$$

where T is the length of the signal, $i = \sqrt{-1}$, p is the order of the regression model and F_s is the sampling frequency. Since for the detection of each stimulation frequency, we use several channels and several harmonics, we could combine separate values of the SNR as:

$$T(f) = \sum_{i=1}^{N} \sum_{k=1}^{K} w_{ik} P_i(kf)/\sigma_i(kf), \tag{1.4}$$

where i is the channel index and k is the harmonic index. The "winner" frequency f^* was defined as the frequency having the largest index T among all frequencies of interest

$$f^* = \underset{f_1,\dots,f_n}{\arg\max}\ T(f). \tag{1.5}$$

Normally, equal weight values ($w_{ik} = 1/NK$) are used for estimation of $T(f)$ (considering that SNR at all harmonics are treated equally) [8, 14], leading to the *minimum noise energy combination* (MNEC) method. But this choice could not be always convenient. Thus, in [11] it was proposed to consider these weights as parameters, by adjusting which the system could be adapted for a particular subject and/or particular recording session of the subject. To train the weights one can re-use data from some calibration stage, where the desired outputs of the classifier are known *a priori* due to the calibration stage design. We will refer to this method the *weighted minimum noise energy combination* (wMNEC). Note, that the number of the combinations K (see Section 1.4.1.1) could be different for the data coming from the different recording sessions. This, in turn, can make impossible to apply pre-trained weights w_{ik} to the non-training data. In wMNEC we solve this problem by fixing the value of K to its maximal possible value N_s.

The above-mentioned weighting procedure can be represented by an artificial linear neural network. As input we use the SNR coefficients $P_i(kf)/\sigma_i(kf)$ for every channel and every harmonic. Thus, for an N_s electrode EEG system and by considering the fundamental stimulation frequency and its two harmonics, we have $3N_s$ elements in the input vector. As the output $\tilde{T}$, a fixed positive value (+1) for the case, when the input SNRs corresponds to a stimulation frequency, and zero otherwise are assigned. The training can be performed using least-square algorithm with additional restrictions (of non-negativity) on the weight values.

When training this network, one estimates values $\tilde{T}(f_i)$ for each stimulation frequency f_i, given considered EEG data. The "winner" frequency, again, is then selected as the frequency having largest index $\tilde{T}$ among all frequencies of interest f_i.

Comparison between those two classification approaches (MNEC and wMNEC) is presented further in this chapter, as a results of their validation for such SSVEP BCI application, as the "Maze" game (see Section 1.5.2).

1.4.2 Classification in the Time Domain

Other approaches to classify SSVEPs consists of looking at the average response expected for each of the flickering stimuli. For this, the recorded EEG signal of length t (ms) was divided into $n_i = [t/f_i]$ non-overlapping, consecutive intervals ($[\cdot]$ denotes the integer part of the division). For example, in 2000 ms long EEG recordings of assumed 10 Hz visual stimulation there are $2000/10 = 20$ such intervals of duration 100 ms[5] ([1, 100], [101, 200],...). This procedure is repeated for the recorded EEG assuming all stimulation frequencies used in the BCI setup. After that, the average response for all such intervals, for each frequency, is computed. Such averaging is necessary because the recorded signal is a superposition of *all* ongoing brain activities. By averaging the recordings, those that are time-locked to a known event

[5] The length of one period.

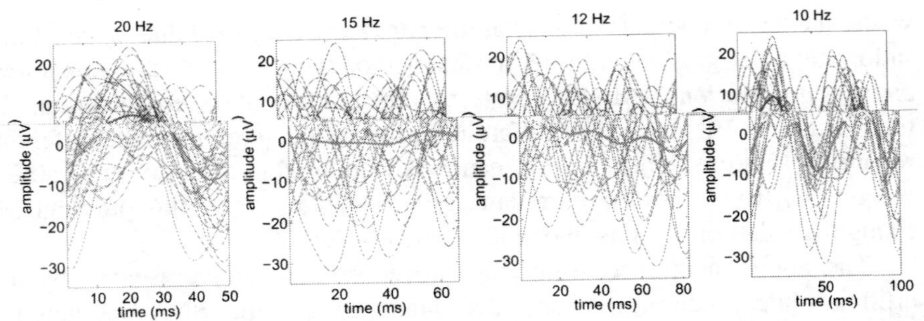

Figure 1.3 Individual traces of EEG activity (thin blue curves) and their averages (thick red curves) time locked to the stimuli onset. Each individual trace shows changes in electrode Oz. The lengths of the shown traces correspond to the durations of the flickering periods for 3, 4, 5 and 6 frames (from left to right panel), and with a screen refreshing rate close to 60 Hz (thus, 20, 15, 12, and 10 Hz visual stimulation). The subject was looking to the stimulus flickering at 20 Hz (the period is three video frames or 50 ms). One observes that, in the left panel, we obtain one complete period for the average trace, and in the right panel, two complete periods, while in the other panels, the average trace is almost flat.

are extracted as evoked potentials, whereas those that are not related to the stimulus presentation are averaged out. The stronger the evoked potentials, the fewer trials are needed, and *vice versa*. To illustrate this principle, Figure 1.3 shows the result of averaging, for a 2 s recording interval, while the subject was looking at a stimulus flickering at a frequency of 20 Hz. It can be observed that, for the intervals with assumptions of the stimulations at frequencies 12 and 15 Hz, the averaged signals are close to zero, while for those used for 10 and 20 Hz, a clear average response is visible. Note that the average response does not exactly look like period(s) of a sinusoid, because the 20 Hz stimulus was constructed using two consecutive frames of intensification and a next frame of no intensification. Additionally to this, not only principal frequency f_i of the stimulation can be presented in SSVEP responses, but also its harmonics $2f_i, 3f_i, \ldots$. There is also some latency present in the responses since the evoked potentials do not appear immediately after the stimuli onset. It could also be seen that, in the interval used for detecting the 10 Hz oscillation, the average curve consists of two periods. This is as expected, since a 20 Hz oscillation has exactly two whole periods in a 100 ms interval.

As the means for SSVEP decoding based on described time locked averages, we consider here the following two algorithms.

1.4.2.1 Stimulus-Locked Inter-Trace Correlation (SLIC)

This method is based on the fact, that constructed above individual period-length SSVEP responses (blue) exhibit good correlation between each other (and, as a consequence, with the their averaged curve (red)), while we assume correct stimulation frequency. This is visible, for example, in Figure 1.3 (left) for our 20 Hz oscillation. Simultaneously, previously constructed individual traces (blue) as for assumed other possible stimulation frequencies (for example, 15 and 12 Hz, which are represented in the two middle panels in Figure 1.3) have small level of correlation between each other (and their averaged curves). Thus, correlation coefficient can be taken as a measure for distinguishing the stimulation frequency subject is looking at. By estimating correlation coefficient between all possible pairs of individual responses (blue curves) within each cut and taking their median values, one constructs feature set for further classification [22]. The classification can be done by building all possible one-versus-all classifiers (f_i against all other stimulations used in the SSVEP BCI system) and searching for the highest outcome (the biggest distance to separating boundary in normalized feature space). If this outcome exceeds some predefined threshold, we can conclude about the stimulation frequency subject is looking at. As a classifier, simple *Linear Discriminant Analysis* (LDA) can be used, leading to the good results [22].

But it is worth mentioning that the previously described method has some limitations. As one can see from Figure 1.3, the correlation coefficients for cuts with assumptions of 10 Hz and 20 Hz oscillations should be close to each other. Thus, previously described SLIC strategy can potentially make a mistake, when there are visual stimulations with frequencies, that are divider of one another. To overcome this, we have to avoid the use of such frequencies in our stimulation, when we are stick to SLIC decoding method. While this can be easily done using external LED stimulations, this limits the number of possible encoded targets in the case of computer screen as a stimulation device (see Sections 1.2 and 1.3.2). As a some remedy for this problem, the method described further (see Section 1.4.2.2) can be used.

The SLIC method was also initially developed for only one EEG electrode. For its use in a case of multielectrode recordings, one can extend a feature subset by adding correspondent medians of correlation coefficients from other channels. In order to further improve the method, one can perform spatial filtering before SLIC in order to maximize separability between classes (SSVEP responses for repetitive stimulation with different frequencies). As an example of such a strategy, we present here an algorithm based on brain recordings from N_s channels for classification between events, when

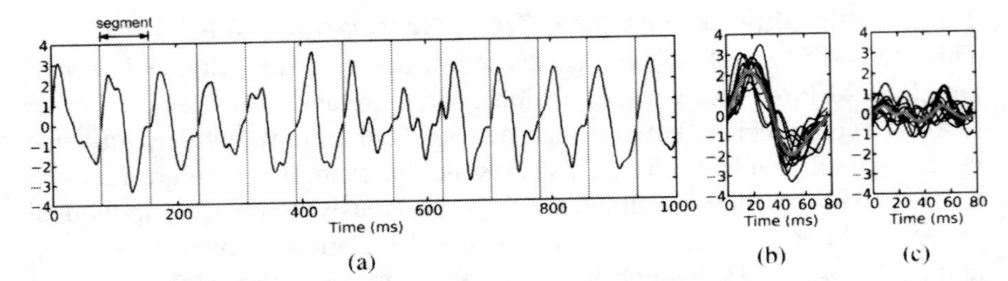

Figure 1.4 Detection of a 12 Hz SSVEP signal, recorded by the *imec* device. (a) A one second window, subdivided into 12 segments. The signal shown is not from a single electrode, but is one of the ICs resulting from the ICA step. (b) All extracted segments from the recording shown in the panel (a). The mean is plotted as a thick (red) curve. (c) Segments extracted from a window where no SSVEP stimulus was shown, with the mean plotted as a thick (red) curve. Note that the correlation between the trials and the mean is much lower than those shown in the center plot.

subject is either looking into flickering with frequency f Hz stimulation or not looking at stimulation at all. Such classifier was used in the SSVEP-based computer game “Tower Defense”, described in this chapter as an application of SSVEP BCI (see Section 1.5.3). Figure 1.4 presents a visualization of the process outlined below, which uses independent component analysis (ICA, by means of JADE algorithm [5]) as a spatial filtering for incorporation of information from several channels.

All of the resulting independent components (ICs) are divided in windows (thus, not the complete recorded EEG interval is considered as the whole entity, but rather its parts for accounting for SSVEP variability due, for example, subject’s lost of concentration on flickering stimulus) of a pre-defined length l_w seconds (which could be subject dependent) with a fixed overlap of 500 ms. Each such window is split into non-overlapping segments of length $l_s = F_s/f$ samples, where F_s is the sample rate of the signal and f is the frequency of the SSVEP stimulus.

The splitting operation as described above yields an array $\mathbf{W}$ with a dimensionality of #windows × #ICs × #segments × #samples, iterated by i, j, k and l respectively. From this array, matrix $\mathbf{R}$ is constructed, which, for each window, and each IC, contains the likelihood of an SSVEP signal being present. To determine $\mathbf{R}$, the correlation coefficients between each segment and the *average* of all segments is calculated (note, that this is slightly modified SLIC approach). The obtained correlation coefficients are themselves averaged to yield a single value between -1 and $+1$, which is normalized to

[0, 1]. From matrix **R**, vector **r**, containing a single value for each window, is calculated by taking the maximum of each row of **R**:

$$\mathbf{R}_{ij} = 0.5 + 0.5 \cdot \operatorname*{mean}_{k} \operatorname*{corr}_{l} \left(\mathbf{W}_{ijkl}, \operatorname*{mean}_{m} \mathbf{W}_{ijml} \right), \tag{1.6}$$

$$\mathbf{r}_i = \max_{j} \mathbf{R}_{ij}. \tag{1.7}$$

The final step is to threshold the vector **r** using two threshold values t_h and t_l. To determine these, the data collected during the calibration period were analyzed:

$$t_h = \min\left(\text{mean}\,\mathbf{s}, \frac{\text{mean}\,\mathbf{s} + \max\mathbf{f}}{2}\right), \tag{1.8}$$

$$t_l = \max\left(\text{mean}\,\mathbf{f}, \frac{\text{mean}\,\mathbf{f} + t_h}{2}\right), \tag{1.9}$$

where **s** denotes the values of **r** during which the SSVEP stimulus was shown and **f** denotes the values of **r** where the subject was looking at a fixation cross. The thresholded version of **r**, denoted $\mathbf{r}'$, then becomes:

$$\mathbf{r}'_i = \begin{cases} 0, & \text{if } i = 0, \\ 1, & \text{if } i > 0 \text{ and } \mathbf{r}_i > t_h \text{ and } \mathbf{r}'_{i-1} = 0, \\ 0, & \text{if } i > 0 \text{ and } \mathbf{r}_i < t_l \text{ and } \mathbf{r}'_{i-1} = 1, \\ \mathbf{r}'_{i-1}, & \text{otherwise}, \end{cases} \tag{1.10}$$

where i iterates over each value of **r**. So windows of data are continuously classified, indicating if an SSVEP response is present or not.

1.4.2.2 Classification Based on Time Value Features

In order to overcome some limitations of the SLIC methods and allow the use of time domain classifier for the case of stimuli with frequencies, which could be dividers of one another, one can directly use time amplitude features from *averaged* waveforms (see red curves in Figure 1.3). Thus, the essential difference with respect to the previous SLIC method is in a feature subset. As a classifiers, one can use simple *linear discriminant analysis* (LDA), since in BCI domain linear classifiers in general give better generalization performance than nonlinear ones [25]. These classifiers are constructed so as to discriminate the stimulus flickering frequency f_i from all other flickering frequencies, and for the case when the subject does not look at the flickering stimuli at all. As a result of such LDA classification, we have several posterior probabilities p_i, which characterize the likelihoods of a subject's gaze on the stimulus flickering at frequency f_i. If all probabilities p_i are smaller then 0.5, we conclude that the subject does not look at the flickering stimuli. In all other cases, we take as an indication on which stimulus the subject's gaze is the flickering frequency f_i with the largest posterior probability p_i.

Since we normally use visual stimulation with frequencies up to 20 Hz, and no more then two harmonics of SSVEP responses give real influence into decoding performance, we can downsample our data to a lower resolution, if it is possible (for example, for *imec* device with its $F_s = 1000$ Hz it is desirable to do this even for reducing the computational load). Additionally to this, we take only those time instants, for which the p-values were smaller than 0.05 (in training data), using a Student t-test between two conditions: averaged response in interval corresponding to the given stimulus with flickering frequency f_i versus the case when the subject is looking at other stimulus with another flickering frequency, or looking at no stimulus at all. This feature selection procedure, based on a filter approach, enables us to restrict ourselves to relevant time instants only.

All what was described above is valid only for the case when we have a single electrode. In the case of N_s electrodes, the same feature selection was performed for each electrode, but the LDA classifiers were built based on pooled features from all electrodes.

1.5 Applications

In order to validate SSVEP-based BCI we present here several applications, where users were able to type or play different games with use of their brain only. Those applications are also used for assessing previously described methods and algorithms.

1.5.1 SSVEP-Based Mind Spelling

As the first application, we present here a typing system based on the brain spelling device. The subject is presented with a screen with a set of characters arranged as an 8×8 matrix. The matrix is divided into four quadrants (sub-matrices of 4×4 characters) with different color background. The background of each quadrant is flickering with a particular and unique frequency, allowing the subject to select one group of characters through his/her SSVEP responses while (s)he gazes onto corresponding flickering quadrant. After the desired quadrant is selected, it is zoomed in to cover the entire screen and replace the initial 8×8 matrix. In the next stage the procedure is repeated: a 4×4 matrix is also split into four quadrants from which the subject can select only one. Eventually, after three selections, the system detects the desired quadrant by the subject character [36].

This application was used to compare synchronous and asynchronous modes during decoding based on MNEC strategy (see Section 1.4.1.2). In the synchronous mode the stimulation, signal processing and decoding are sequential: the stimulation lasts for a fixed time Δt, after which the acquired EEG-signals are processed to detect one out of four stimulation frequencies. This is different with respect to the asynchronous mode, where all system's components work in parallel: the signal processing and decoding are done during the stimulation phase and while the EEG signals are being recorded. Decoding starts after a short initial pause Δt_p after beginning of the visual stimulation. During this time the system keeps collecting EEG data. If after Δt_p seconds the collected data allows the classifier to make a "firm" decision (when $T(f)$ in MNEC method is greater than some quality threshold Q), this decision is considered as the "final" for this selection stage and the system goes to the next selection stage. Otherwise, the classifier tries to detect the winner frequency using more data, which have been acquired during a bit longer period $\Delta t_p + \Delta t_c$, where Δt_c is the time needed for the classifier to do the first classification attempt. The process repeats until the decision is made or the stimulation time exceeds the time thresholds $\Delta t_{\max}$ (five seconds in the described example). In the latter case, a most probable classification result is given.

Eight healthy subjects (aged 24–60 with an average age of 35, two female and six male) with no previous BCI experience participated in on-line experiment using *imec* EEG recording device (see Section 1.3.1), where they typed characters/words of their choice based on five seconds synchronous mode. Averaged among all subjects typing accuracy was 81%, with the chance level $100/64 = 1.5625\%$.

To make a qualitative comparison between synchronous and asynchronous modes, data recorded with previous on-line typing underwent classification also based on asynchronous decoding. But here we should mention that this mode also works on-line, and it was applied in a way that mimics on-line decoding. Table 1.1 shows the averaged detection percentages for different initial pauses Δt_p and quality thresholds Q. Additionally, Table 1.2 shows the corresponding averaged detection times. Note, that in some cells we have time bigger then $\Delta t_{\max} = 5$ second. This due to the fact, that table shows time required for stimulation *with* classification. Results indicate, that the higher Q

Table 1.1 Accuracy for different initial pauses Δt_p and quality thresholds Q.

		Quality threshold Q				
% detected		1.1	1.3	1.5	1.7	1.9
	0.5	15%	20%	36%	47%	57%
Δt_p [s]	1	37%	47%	58%	60%	65%
	1.5	44%	56%	62%	64%	66%

Table 1.2 Averaged detection time for different initial pause and threshold.

		Quality threshold Q				
Avg time [s]		1.1	1.3	1.5	1.7	1.9
	0.5	0.55	0.97	2.34	3.41	4.35
Δt_p [s]	1	1.12	2.25	3.56	4.41	5.12
	1.5	1.74	3.11	4.38	5.20	5.71

Table 1.3 Classification results and time per person for four command asynchronous typing together with general detection accuracy.

Subject ID	A	B	C	D	E	F	G	H
% correct	94	100	100	100	94	100	95	100
average time [s]	2.04	2.66	2.05	2.65	6.36	2.55	5.12	4.86

is, the better the classification results but the slower the detection time. This is as expected because the classified frequency needs to stand out more. This takes longer to achieve, but once this threshold is reached, it is more plausible that the classified SSVEP-frequency is the correct one. Higher initial pauses also yield better classification results and slower detection times. A possible explanation is that the SSVEP-response is not prominent enough if the initial pause is too short, because of the latency of responses or time required to set a steady mode.

Table 1.3 contains the typing accuracy per subject in asynchronous mode. The first row gives the detection percentages. All subjects manage to achieve near perfect classification results. The second row gives the average detection times. Here is quite a large inter-subject variability.

We also made a comparison between synchronous and asynchronous modes based on the theoretical *information transfer rate* (ITR) [46], which specifies how many bits per minute the system can theoretically communicate. It implies that we assume a zero time for changing from one selected target to the next. The ITR averaged over all our subjects was used for the assessment, since we wanted to compare the asynchronous with the synchronous mode, where the duration of the stimulation was fixed before the experiment, and does not depend on the subject. We can conclude from

Table 1.4 Averaged ITR [bits/min] for different modes and four targets.

Mode	Synchronous					Asynchronous
Initial pause (Δt_p) [s]	1 s	2 s	3 s	4 s	5 s	
Averaged ITR	35.7	33.4	28.8	22.9	19.0	38.2

Table 1.4, that, in general, the asynchronous mode ($Q = 1.5$ and $\Delta t_p = 1.5$) yields higher ITRs than the synchronous one. Examining the performance of each individual subject for asynchronous typing, we see that the theoretical ITRs are between 17.57 and 59.16 bit/min.

1.5.2 The Maze Game

As another application of SSVEP-based BCI, we developed so-called the "Maze" game [10]. The goal is to navigate a *player character* (avatar), depicted as Homer Simpson's head, to the target (i.e., a donut) through a maze (see Figure 1.5). The game has several pre-defined levels of increasing complexity. A random maze mode is also available. The player can control the avatar by looking at flickering arrows (showing the direction of the avatar's next move) placed in the periphery of the maze. Each arrow is flickering with its own unique frequency taken from the selected frequency band (see Section 1.5.2.1). The selection of the frequencies can be predefined or set according to the player's preferences.

The game is implemented in Matlab 2010b (`http://www.mathworks.com/products/matlab/`) with Psychotoolbox 3 [17] used for the accurate (in terms of timing) visualization of the flickering stimuli.

To reach a decision, the server needs to analyze the EEG data acquired over the last T seconds. In the game, T is one of the tuning parameters (must be set before the game starts), which controls the game latency. Decreasing T makes the game more responsive, but in the same time it makes the interaction less accurate, resulting in wrong navigation decisions. By default, a new portion of the EEG data is collected every 200 ms. The server analyzes the new (updated) data window and detects the dominant frequency using the (w)MNEC method (see Section 1.4.1). The command corresponding to the selected frequency is sent to the client also every 200 ms, thus, the server's update frequency is 5 Hz.

For the final selection of the command to be executed by the client we use the following approach based on weighting of the elements in the queue of the last m commands sent by the server. Each entry of the queue has a predefined weight ("age"), which linearly decreases from $w_{\max}$ (the most recent element)

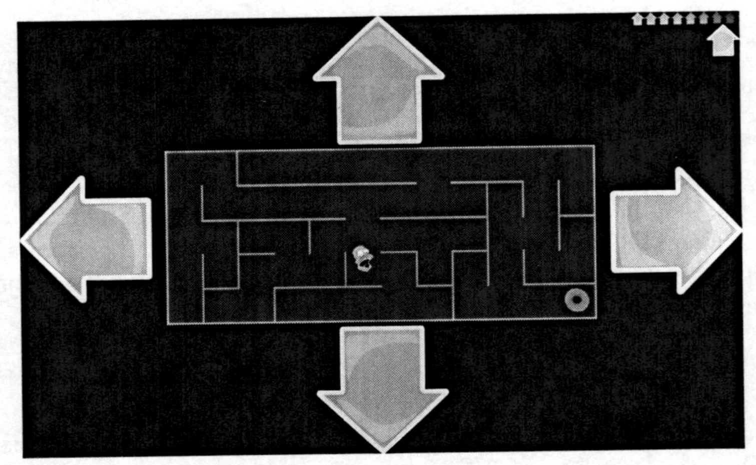

Figure 1.5 Snapshot of the "Maze" game. The decision queue is shown in the upper-right corner as a series of ($m = 8$) arrows, the intensities of which correspond to the weights ("ages") of the decisions (see text). The "final decision" (made on the basis of the decision queue) is depicted as the larger arrow just below the decision queue.

to $w_{\min}$ (the oldest element in the queue). The default values of the weights $w_{\max} = 1$ and $w_{\min} = 0.1$ can be changed in order to adapt the decision making mechanism. The "candidate" for the "final winner" is selected as a command with the maximal cumulative weight. The "candidate" becomes the "final winner" if its cumulative weight exceeds an empirically chosen threshold $\theta = \frac{m}{4}(w_{\max} + w_{\min})$, otherwise no decision is made.

Since command selection is made based on previously recorded EEG, the game control has an unavoidable time lag. In order to "hide" this latency, we let the avatar change its navigation direction only in so-called decision points: as the avatar starts to move, it will not stop until it reaches the next decision points on its way. This allows the player to use this period of "uncontrolled avatar movement" for planning (by looking on appropriate flickering arrow) the next navigation direction. By the time the avatar reaches the next decision point, the EEG data window, which is to be analyzed, would already contain the SSVEP response corresponding to the next navigation direction.

1.5.2.1 Calibration Stage

The "Maze" game uses only four commands for navigating the avatar through the maze: "left", "up", "right" and "down", hence, four stimulation frequencies are needed. During our preliminary experiments, we noticed that the optimal set of stimulation frequencies is very subject dependent. This motivated us to introduce a calibration stage, preceding the actual game play, for locating the frequency band, consisting of four frequencies, that evoke prominent SSVEP responses in the subject's EEG signal. To this end, we propose a "scanning" procedure, consisting of several blocks. In each block, the subject is visually stimulated for 15 s by a flickering screen ($\approx 28^\circ \times 20^\circ$), after which a black screen is presented for 2 s. The number of blocks in the calibration stage is defined by the number of available stimulation frequencies, introduced in Section 1.3.2.

We grouped these frequencies into overlapping bands, for which each band contains four consecutive stimulation frequencies (e.g., band 1: [6 Hz, 6.66 Hz, 7.5 Hz, 8.57 Hz], band 2: [6.66 Hz, 7.5 Hz, 8.57 Hz, 10 Hz], and so on). After stimulation, we analyze the spectrograms of the recorded EEG signals, and select the "best" band of frequencies to be used in the game.

1.5.2.2 Influence of Window Size and Decision Queue Length on Accuracy

To assess the best window size T (and the decision queue length m), we have studied their influence on the classification accuracy. Six healthy subjects (all male, aged 24–34 with an average age of 28.3, four right-handed, one left-handed and one both-handed) participated in the experiment with *imec* prototype as a recording EEG device (see Section 1.3.1). Only one subject had prior experience with SSVEP-based BCI. For each subject, several sessions with different stimulation frequency sets were recorded, but we present the results only for those sessions, for which the stimulation frequencies coincide with the ones that are determined with the calibration stage. Each subject was presented with a specially designed level of the game, and was asked to consequently look at each one of four flickering arrows for 20 s followed by 10 s of rest, so the full round of four stimuli (flickering arrows) was $4 \times (20 + 10) = 120$ s. The stimulus to attend to was marked with the words "look here". Each recording session consisted of two rounds and, thus, lasted four minutes. The recorded EEG data where then analyzed off-line using exactly the same mechanism as in the game. In the case of training mode (as for wMNEC method (see Section 1.4.1.2)), the first round was used for training. By design, the true winner frequency is known for each moment of time, which enables us to estimate the accuracy.

Table 1.5 Classification accuracy (in %) as a function of window size T (s) and classification method in frequency domain (see Section 1.4.1).

T	method	S1	S2	S3	S4	S5	S6	Aver. $\langle\cdot\rangle$	$\langle$wMNEC$\rangle$ $-\langle$MNEC$\rangle$
1	MNEC	54.17	41.15	35.42	78.65	69.27	55.73	55.73	3.99
	wMNEC	54.69	46.88	43.23	81.77	70.83	60.94	59.72	
2	MNEC	59.78	50.54	51.09	93.48	82.07	66.30	67.21	9.78
	wMNEC	79.35	58.70	63.04	92.93	86.96	80.98	76.99	
3	MNEC	69.19	62.79	54.07	94.77	88.95	69.19	73.16	9.30
	wMNEC	84.30	68.60	61.63	99.42	94.19	86.63	82.46	
4	MNEC	77.44	67.07	52.44	95.12	90.24	75.61	76.32	6.30
	wMNEC	86.59	73.17	51.83	100.00	95.73	88.41	82.62	
5	MNEC	82.89	69.74	51.97	99.34	96.71	71.71	78.72	5.60
	wMNEC	90.13	75.66	57.24	100.00	97.37	85.53	84.32	

1.5.2.3 Results and Discussion

The results of the experiment described in Section 1.5.2.2 are shown in Table 1.5, allowing us to compare MNEC and wMNEC methods (see Section 1.4.1.2 for their descriptions). With the accuracy of the frequency classification we mean the ratio of the correct decisions with respect to all decisions made by the classifier. Note, that the chance level of accuracy in this experiment is 25%. From the results one can see that the weighted version of the decoder (wMNEC) outperforms the standard (averaged) one by approximately 7% in terms of accuracy.

Experimental results also suggest that, in general, the longer queues m of the decision making mechanism lead to a better accuracy of the game control. The drawback of the longer queues is an additional latency. To reduce the later, the server's update frequency (the actual one is 5 Hz) can be increased. This, in turn, increases the computational load (mostly on the server part).

Based on our experience (also supported by the data from Table 1.5), we can recommend to use the window size $T = 3$ s and the queue length $m = 5$ (or more) as default values for an acceptable gameplay.

Unfortunately, the *information transfer rate* (ITR) commonly used as a performance measure for BCIs, is not relevant for the game, at least in its actual form. By design, the locations of the decision points depend on the (randomly generated) maze, and, therefore, the decisions themselves are made at an irregular rate, which, in turn, does not allow for a proper ITR estimation.

A few more issues concerning the visual stimulation and the game design need to be discussed. Even though the visual stimulation in the calibration stage (one full-screen stimulus, see Section 1.5.2.1) differs from the one used in the game (four simultaneously flickering arrows, see Figure 1.5), we strongly believe that the frequencies selected in such a way are also well suited for the game control. This belief has been indirectly supported during our experiments (see Section 1.5.2.2): the frequency sets, different from the ones selected during the calibration stage, in most cases yield less accurate detections.

One of the drawbacks of SSVEP-based BCIs with dynamic environment and fixed locations of stimuli is the frequent change of the subject's gaze during the gameplay, which leads to a discontinuous visual stimulation. To avoid this, we introduced an optional mode where the stimuli (arrows) are locked close to the avatar and move with it during the game, which might make the game more comfortable to play.

Several subjects have noticed that the textured stimuli are easier to concentrate on than the uniform ones. Some of our subjects preferred the yellow color of the stimuli to the white color, which partially might be explained by a characteristic feature of the yellow light stimulation: it elicits an SSVEP response of a strength that is less dependent on the stimulation frequency than other colors [34].

1.5.3 Tower Defense Game

As the last application, where we assess usability of time based decoding algorithm 1.4.2, the "Tower Defense" game was developed [40]. The goal of this game is to protect a tower against waves of enemies, who shall appear at one or more fixed points in the game world and walk towards the tower. When an enemy reaches the tower, the player loses the game. To prevent that, the user can build a limited amount of defensive structures. The user needs to decide on the optimal location of these defenses, based on information about the number of enemies that will appear at which positions. Because the game should be suitable for all ages, no violence is being shown: the enemies are giant red balls, which disappear upon being hit. A compilation from multiple screenshots is shown in Figure 1.6, explaining the various elements of the game.

To control the game, the user needs some method to make a selection on the screen based on his/her brain activity. At the beginning of the level, the user makes a selection from several predefined locations to build defensive structures. When the user is satisfied with the layout, he/she can select the 'done' button, which will unleash the enemies. From that point on, the user

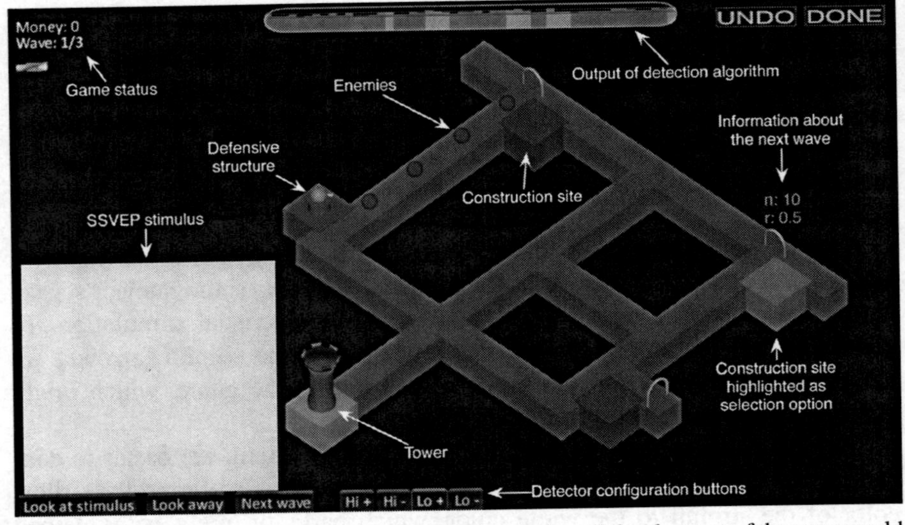

Figure 1.6 Compilation from multiple screenshots showing all the elements of the game world and the interface.

loses control until either all enemies have been defeated, or an enemy reaches the tower and the user loses the game. An undo option is also available, which will undo the last build command, enabling the user to correct mistakes made by either himself or the system.

Three levels were designed. The first level is used while explaining the game mechanics to the user and is simply a straight line with the tower at one end and the enemies appearing at the other. The user cannot make strategic mistakes in this level. The other two levels require the user to think about where to place the defensives, making them harder and more interesting at the same time.

1.5.3.1 SSVEP Stimulation

Only one stimulus is presented at the bottom-left corner of the screen, flickering at a fixed frequency. The system detects whether the user is looking at the stimulus or not. The selection options are highlighted one by one, for two seconds each. When the desired option is highlighted, the player looks at the flickering stimulus. When the system detects the presence of an SSVEP response, the currently highlighted option is selected. A small red dot is shown in the middle of the stimulus, which users indicated helps to keep the eyes focused.

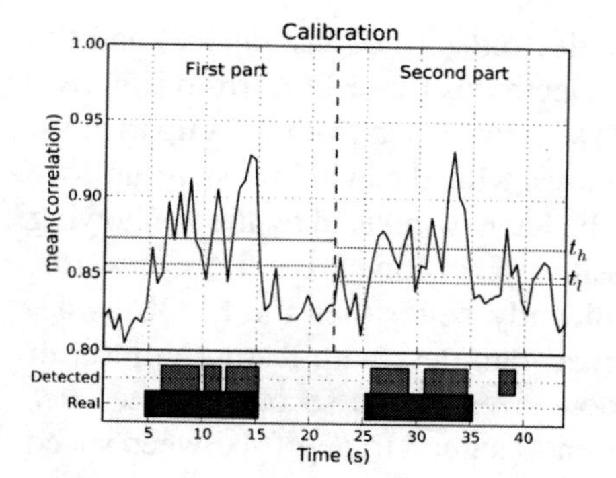

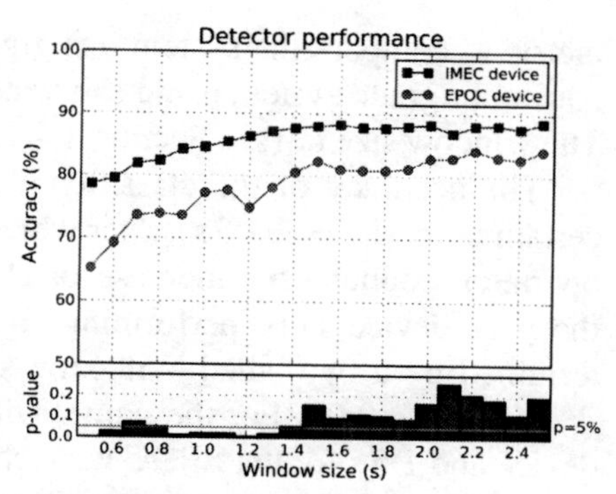

Figure 1.7 Left: detection during calibration period using the EPOC device on a subject with average performance. Shown are the vector **r** along with the threshold values t_h and t_l. Below are the detected periods of SSVEP activity along with the periods where the SSVEP stimulus was actually shown. The detector was trained on the first part and applied to the second part, and vise versa. In this configuration, a window size of 1.5 s was used and 85% of the windows were correctly classified. Right: performance of the detection algorithm for different window sizes during the calibration period (window step was fixed at 0.5 s). Shown is the accuracy (% windows correctly classified), averaged across 8 subjects, which all performed the calibration with both the imec and EPOC device. The p-values of a Wilcoxon signed rank test between the two devices is plotted at the bottom.

To obtain some data to determine optimal threshold for the SSVEP detection algorithm (Section 1.4.2.1) and to determine its performance, a short calibration is performed at the beginning of the game. The user looks at the center of the screen where a fixation cross is shown for five seconds, followed by ten seconds of an SSVEP stimulus (width and height 5 °), ten seconds fixation cross, ten seconds SSVEP stimulus and, finally, ten seconds fixation cross.

1.5.3.2 Results

To quantitatively determine the performance of the detection algorithm, it was run on the calibration data. Eight users (aged 23–34, mean 26.75, std. 4.26, two female and six male) completed the calibration period with both the *imec* and the EPOC devices (see Section 1.3.1), before playing the game. The detected SSVEP periods were compared with the actual periods during which the SSVEP stimulus was shown (Figure 1.7). For this offline analysis, the data were split into two parts, with the ICA and calculation of the threshold values being performed on the first part, and then applied to the second part, and vice versa. The percentage of correctly classified windows was used as a metric to compare the system using gel electrodes (the *imec* device) and the consumer grade system using salt water electrodes (the EPOC from Emotiv). The window size l_w (see Section 1.4.2.1) was increased from 0.5 s up to 2.5 s.

The accuracy of the classifier increases with the window size, up to a certain point ($l_w = \pm 1.5$ s), after which the latency induced by the windowing operation counters the increase of classifier precision. From 1.5 s onwards, the *imec* device stops performing significantly better than the EPOC as determined by a two-tailed Wilcoxon signed rank test with testing criteria of $w \leq 4, p \geq 0.05$. For the game, window sizes of one second for the *imec* device and 1.5 s for the EPOC were chosen as a good trade-off between speed and accuracy.

Note that, during the game, each option is highlighted for two seconds, a duration which corresponds to 10–15 windows, depending on the device used. Only one of them has to be classified as containing SSVEP in order to make the selection. The shown accuracy in the figure is therefore only useful to compare the performance of the two devices, but does not say much about the actual performance during the game. The in-game performance is considerably harder to quantify, as the user compensates for delays, and given that the thresholds can be tweaked. In this study, seven users achieved proper control over the selection process and were able to complete all three levels. One user did not achieve control with any of the devices.

1.6 Conclusion

We presented the Steady State Visual Evoked Potential (SSVEP) paradigm: a stimulation technique which can be used in the development of Brain Computer Interfaces (BCIs). The SSVEP can be decoded by any of the four algorithms we presented in Section 1.4, but the choice is often driven by the application: if a training-stage is no issue, the technique using wMNEC gives better results than the method using MNEC. If only the condition ‘gazing at one flickering target versus not gazing at this target’ needs to be decoded, the SLIC technique is preferable. The time domain technique can be seen as an alternative to the method using wMNEC, which might be easier to implement for the use in an on-line (asynchronous) BCI.

To show the feasibility of the SSVEP paradigm in terms of a BCI, three applications were presented: a speller system, allowing a subject to spell characters one by one, and two games: the “Maze” game and the “Tower Defense” game. All results show that SSVEP can reliably be used as a stimulation paradigm and little or no time for training the subject and machine is required. Because of this property and the fact that only eye-gazing is a requirement, we state that these kinds of BCIs are especially useful for motor disabled patients. Applications as the speller system can significantly improve the life quality of people with serious motor function problems (i.e., patients suffering from amyotrophic lateral sclerosis, stroke, brain/spinal cord injury, cerebral palsy, muscular dystrophy, etc.).

Future challenges in SSVEP-BCI design can be found on the hardware side: in the development of electrodes and amplifiers which record with a higher SNR, while still being able to reliably function when lab-conditions are not available: at public events or just at the home of any person. On the software side new signal processing and machine learning techniques can boost BCI performance, but clever design of the interface is also of utter importance: how can as much information as possible be encoded while bit rate is kept low? In the spelling system the characters are grouped so only one out of four targets needs to be detected, by iteratively regrouping selected characters (as in a tree search) a final character can be selected. In the maze game only the possible directions of movements are selectable, instead of all reachable states, etc.. Following this idea, i.e., to restrict (or guide) the search to future states which are possible (or highly probable), the inclusion of predictive action models (like a word prediction module for mind spelling) will boost the communication rate or the interface. An other solutions is provided by combining SSVEP with other paradigms such as P300, imaginary movement, slow cortical potentials (SCPs) etc (see for example [31]).

A final point of attention which is often neglected is the validation of the BCI to the target group, i.e., motor disabled patients, for systems as the mind spelling application or on healthy subjects for games as the "Maze" game and the "Tower Defense" game. A clinical and qualitative review of any BCI is therefore crucial.

Acknowledgments

NC is supported by IST-2007-217077, NVM is supported by the research grant GOA 10/019, MvV is supported by IUAP P6/29, AC and AR are supported by IWT doctoral grants, MMVH is supported by PFV 10/008, CREA 07/027, G.0588.09, IUAP P6/29, GOA 10/019, IST-2007-217077.

References

[1] P. Allegrini, D. Menicucci, R. Bedini, L. Fronzoni, A. Gemignani, P. Grigolini, B.J. West, and P. Paradisi. Spontaneous brain activity as a source of ideal 1/f noise. *Physical Review E*, 80(6):061914, 2009.

[2] B. Allison, T. Luth, D. Valbuena, A. Teymourian, I. Volosyak, and A. Gräser. BCI demographics: How many (and what kinds of) people can use an SSVEP BCI? *IEEE Transactions on Neural Systems and Rehabilitation Engineering*, 18(2):107–116, 2010.

[3] N. Birbaumer, A. Kübler, N. Ghanayim, T. Hinterberger, J. Perelmouter, J. Kaiser, I. Iversen, B. Kotchoubey, N. Neumann, and H. Flor. The thought translation device (TTD) for completely paralyzed patients. *IEEE Transactions on Rehabilitation Engineering*, 8(2):190–193, 2000.

[4] B. Blankertz, G. Dornhege, M. Krauledat, K.R. Müller, and G. Curio. The non-invasive Berlin Brain-Computer Interface: Fast acquisition of effective performance in untrained subjects. *NeuroImage*, 37(2):539–550, 2007.

[5] J.F. Cardoso and A. Souloumiac. Blind beamforming for non-Gaussian signals. *IEEE Proceedings for Radar and Signal Processing*, 140(6):362–370, 1993.

[6] J.M. Carmena, M.A. Lebedev, C.S. Henriquez, and M.A.L. Nicolelis. Stable ensemble performance with singleneuron variability during reaching movements in primates. *Journal of Neuroscience*, 25(46):10712–10716, 2005.

[7] J.K. Chapin, K.A. Moxon, R.S. Markowitz, and M.A.L. Nicolelis. Real-time control of a robot arm using simultaneously recorded neurons in the motor cortex. *Nature Neuroscience*, 2:664–670, 1999.

[8] M. Cheng, X. Gao, S. Gao, and D. Xu. Design and implementation of a brain-computer interface with high transfer rates. *IEEE Transactions on Biomedical Engineering*, 49(10):1181–1186, 2002.

[9] N. Chumerin, N. Manyakov, A. Combaz, J. Suykens, R. Yazicioglu, T. Torfs, P. Merken, H. Neves, C. Van Hoof, and M. Van Hulle. P300 detection based on feature extraction in on-line Brain-Computer Interface. In *Proceedings of 32nd Annual Conference on Artificial Intelligence*, Paderborn, Germany. Lecture Notes in Computer Science, Vol. 5803/2009, pages 339–346. Springer, 2009.

[10] N. Chumerin, N.V. Manyakov, A. Combaz, A. Robben, M. van Vliet, and M.M. Van Hulle. Steady state visual evoked potential based computer gaming – The Maze. In *Proceedings of the 4th International ICST Conference on Intelligent Technologies for Interactive Entertainment (INTETAIN 2011)*, Genoa, Italy, 2011.

[11] N. Chumerin, N.V. Manyakov, A. Combaz, A. Robben, M. van Vliet, and M.M. Van Hulle. Subject-adaptive steady-state visual evoked potential detection for Brain-Computer Interface. In *Proceedings of the 6th IEEE International Conference on Intelligent Data Acquisition and Advanced Computing Systems: Technology and Applications*, 2011.

[12] R.G. de Peralta Menendez, J.M.M. Dias, J.A. Soares, H.A. Prado, and S.G. Andino. Multiclass brain computer interface based on visual attention. In *ESANN2009 Proceedings, European Symposium on Artificial Neural Networks*, Bruges, Belgium, pages 437–442, 2009.

[13] L.A. Farwell and E. Donchin. Talking off the top of your head: Toward a mental prosthesis utilizing event-related brain potentials. *Electroencephalography and Clinical Neurophysiology*, 70(6):510–523, 1988.

[14] O. Friman, I. Volosyak, and A. Graser. Multiple channel detection of steady-state visual evoked potentials for brain-computer interfaces. *IEEE Transactions on Biomedical Engineering*, 54(4):742–750, 2007.

[15] L.R. Hochberg, M.D. Serruya, G.M. Friehs, J.A. Mukand, M. Saleh, A.H. Caplan, A. Branner, D. Chen, R.D. Penn, and J.P. Donoghue. Neural ensemble control of prosthetic devices by human with tetraplegia. *Nature*, 442:164–171, 2006.

[16] P.R. Kennedy and R.A.E. Bakay. Restoration of neural output from a paralyzed patient by a direct brain connection. *Neuroreport*, 9(8):1707, 1998.

[17] M. Kleiner, D. Brainard, D. Pelli, A. Ingling, R. Murray, and C. Broussard. What's new in Psychtoolbox-3. *Perception*, 36:14, 2007.

[18] A.K. Kübler, B. Kotchoubey, J. Kaiser, J.R. Wolpaw, and N. Birbaumer. Brain-computer communication: unlocking the locked. *Psychological Bulletin*, 127(3):358–375, 2001.

[19] M.A. Lebedev and M.A.L. Nicolelis. Brain-machine interface: Past, present and future. *Trends in Neuroscience*, 29(9):536–546, 2005.

[20] E.C. Leuthardt, G. Schalk, J.R. Wolpaw, J.G. Ojemann, and D.W. Moran. A brain–computer interface using electrocorticographic signals in humans. *Journal of Neural Engineering*, 1:63, 2004.

[21] S.J. Luck. *An Introduction to the Event-Related Potential Technique*. The MIT Press, Cambridge, MA, 2005.

[22] A. Luo and T.J. Sullivan. A user-friendly SSVEP-based brain–computer interface using a time-domain classifier. *Journal of Neural Engineering*, 7:026010, 2010.

[23] J.N. Mak and J.R. Wolpaw. Clinical applications of brain-computer interfaces: Current state and future prospects. *IEEE Reviews in Biomedical Engineering*, 2:187–199, 2009.

[24] N.V. Manyakov, N. Chumerin, A. Combaz, A. Robben, and M.M. Van Hulle. Decoding SSVEP responses using time domain classification. In *Proceedings of the International Conference on Fuzzy Computation and 2nd International Conference on Neural Computation*, pages 376–380, 2010.

[25] N.V. Manyakov, N. Chumerin, A. Combaz, and M. Van Hulle. Comparison of classification methods for P300 Brain-Computer Interface on disabled subjects. *Computational Intelligence and Neuroscience*, 2011(519868):1–12, 2011.

[26] N.V. Manyakov and M.M. Van Hulle. Decoding grating orientation from microelectrode array recordings in monkey cortical area V4. *International Journal of Neural Systems*, 20(2):95–108, 2010.

[27] N.V. Manyakov, R. Vogels, and M.M. Van Hulle. Decoding stimulus-reward pairing from local field potentials recorded from monkey visual cortex. *IEEE Transactions on Neural Networks*, 21(12):1892–1902, 2010.

[28] C. Mehring, J. Rickert, E. Vaadia, S.C. de Oliveira, A. Aertsen, and S. Rotter. Inference of hand movements from local field potentials in monkey motor cortex. *Nature Neuroscience*, 6(12):1253–1254, 2003.

[29] J. del R. Millán, F. Renkens, J. Mouriño, and W. Gerstner. Noninvasive brain-actuated control of a mobile robot by human EEG. *IEEE Transactions on Biomedical Engineering*, 51(6):1026–1033, 2004.

[30] B. Pesaran, S. Musallam, and R.A. Andersen. Cognitive neural prosthetics. *Current Biology*, 16(3):R77–R80, 2006.

[31] G. Pfurtscheller, B.Z. Allison, C. Brunner, G. Bauernfeind, T. Solis-Escalante, R. Scherer, T.O. Zander, G. Mueller-Putz, C. Neuper, and N. Birbaumer. The hybrid BCI. *Frontiers in Neuroscience*, 4, 2010.

[32] G. Pfurtscheller, C. Guger, G. Müller, G. Krausz, and C. Neuper. Brain oscillations control hand orthosis in a tetraplegic. *Neuroscience Letters*, 292(3):211–214, 2000.

[33] W.S. Pritchard. Psychophysiology of P300. *Psychological Bulletin*, 89(3):506–540, 1981.

[34] D. Regan. An effect of stimulus colour on average steady-state potentials evoked in man. *Nature*, 210(5040):1056–1057, 1966.

[35] J. Rickert, S.C. de Oliveira, E. Vaadia, A. Aertsen, S. Rotter, and C. Mehring. Encoding of movement direction in different frequency ranges of motor cortical local field potentials. *Journal of Neuroscience*, 25(39):8815–8824, 2005.

[36] H. Segers, A. Combaz, N.V. Manyakov, N. Chumerin, K. Vanderperren, S. Van Huffel, and M.M. Van Hulle. Steady State Visual Evoked Potential (SSVEP)-based brain spelling system with synchronous and asynchronous typing modes. In *Proceedings of 15th Nordic-Baltic Conference on Biomedical Engineering and Medical Physics (NBC15)*, June 2011.

[37] M. Serruya, N.G. Hatsopoulos, L. Paninski, M.R. Fellows, and J.P. Donoghue. Instant neural control of a movement signal. *Nature*, 416:141–142, 2002.

[38] D.M. Taylor, S.I.H. Tillery, and A.B. Schwartz. Direct cortical control of 3D neuroprosthetic devices. *Science*, 296(5574):1829–1832, 2002.

[39] M. Thulasidas, C. Guan, and J. Wu. Robust classification of EEG signal for brain-computer interface. *IEEE Transaction on Neural Systems and Rehabilitation Engineering*, 14(1):24–29, 2006.

[40] M. van Vliet, A. Robben, N. Chumerin, N.V. Manyakov, A. Combaz, and M.M. Van Hulle. Designing a Brain-Computer Interface controlled video-game using consumer grade EEG hardware. In *Proceedings ISSNIP Biosignals and Biorobotics Conference*, 2012.

[41] M. Velliste, S. Perel, M.C. Spalding, A.S. Whitford, and A.B. Schwartz. Cortical control of a prosthetic arm for self-feeding. *Nature*, 453(7198):1098–1101, 2008.

[42] J.J. Vidal. Toward direct brain-computer communication. *Annual Review of Biophysics and Bioengineering*, (2):157–180, 1973.

[43] I. Volosyak, H. Cecotti, and A. Gräser. Impact of frequency selection on LCD screens for SSVEP based brain-computer interface. In *Proceedings of IWANN*, Part I, Lecture Notes in Computer Science, Vol. 5517, pages 706–713, Springer, 2009.

[44] Y. Wang, R. Wang, X. Gao, B. Hong, and S. Gao. A practical VEP-based brain-computer interface. *IEEE Transactions on Neural Systems and Rehabilitation Engineering*, 14(2):234–240, 2006.

[45] J. Wessberg, C.R. Stambaugh, J.D. Kralik, P.D. Beck, M. Laubach, J.K. Chapin, J. Kim, S.J. Biggs, M.A. Srinivasan, and M.A.L. Nicolelis. Real-time prediction of hand trajectory by ensembles of cortical neurons in primates. *Nature*, 408:361–365, 2000.

[46] J.R. Wolpaw, N. Birbaumer, W.J. Heetderks, D.J. McFarland, P.H. Peckham, G. Schalk, E. Donchin, L.A. Quatrano, C.J. Robinson, and T.M. Vaughan. Brain-computer inter-

face technology: A review of the first international meeting. *IEEE Transactions on Rehabilitation Engineering*, 8(2):164–173, 2000.

[47] J.R. Wolpaw, D.J. McFarland, and T.M. Vaughan. Brain-computer interface research at the Wadsworth Center. *IEEE Transactions on Rehabilitation Engineering*, 8(2):222–226, 2000.

[48] J.R. Wolpaw, N. Birbaumer, D.J. McFarland, G. Pfurtscheller, and T.M. Vaughan. Brain-computer interfaces for communication and control. *Clinical Neurophysiology*, 113:767–791, 2002.

[49] R.F. Yazicioglu, T. Torfs, P. Merken, J. Penders, V. Leonov, R. Puers, B. Gyselinckx, and C. Van Hoof. Ultra-low-power biopotential interfaces and their applications in wearable and implantable systems. *Microelectronics Journal*, 40(9):1313–1321, 2009.

[50] D. Zhu, J. Bieger, G.G. Molina, and R.M. Aarts. A survey of stimulation methods used in SSVEP-based BCIs. *Computational Intelligence and Neuroscience*, 2010:1–12, 2010.

2

Extraction of f-Waves in Electrocardiograms Using Echo State Network Based Nonlinear Adaptive Filters

Andrius Petrėnas, Vaidotas Marozas, Arūnas Lukoševičius and Andrius Sakalauskas

Biomedical Engineering Institute, Kaunas University of Technology, Kaunas, Lithuania; e-mail: andrius.petrenas@ktu.lt

Abstract

Atrial fibrillation (AF) is the most common arrhythmia in clinical practice. The paper introduces a new method for ventricular activity cancellation in AF from surface electrocardiogram (ECG) signals. The proposed method is based on AF signal extraction using adaptive echo state neural network (ESN). Adaptive ESN estimates a time-varying, nonlinear transfer function between two ECG leads and separates ventricular activity from atrial activity. The method was compared with conventional pre-whitened recursive least squares (RLS) based linear adaptive filter. Both algorithms were applied to surrogate ECG data with known component of AF signal. The results revealed that the ESN based nonlinear filter extracts f-waves more accurately than the conventional pre-whitened RLS based linear algorithm, especially in lower amplitude (< 0.05 mV) AF signals.

Keywords: atrial fibrillation, ventricular activity cancellation, reservoir computing, artificial neural network.

Richard J. Duro and Fernando López Peña (Eds.), Digital Image and Signal Processing for Measurement Systems, 35–70.

2.1 Significance of Atrial Fibrillation in Cardiology

Cardiovascular diseases occupy the first place in the morbidity structure [1]. In Europe, every second human death is influenced by the heart and vascular diseases. Cardiac arrhythmias are a rapidly spreading pathology worldwide [2]. It is believed that the rapid growth of heart rhythm disorders is affected by the emotional stress caused by the intensive pace of life, sedentary lifestyle, unhealthy diet, smoking, excessive drinking of alcohol [3].

The success of arrhythmia treatment depends on the stage when heart rhythm disorders are detected. Unfortunately, the symptoms of disease are confusing and arrhythmia is frequently diagnosed too late. In some cases rhythm disorders cannot be observed by the physicians in surface electrocardiograms (ECG) although a patient may be complaining of frequent, intense or irregular heart rhythms. Holter monitors are commonly used for the detection of the arrhythmic episodes. The patient wears an ECG recorder for one or more days and then the recorded data is analyzed by specialized software when the monitoring is finished.

Various types of arrhythmias are known, however atrial fibrillation (AF) is the most common, complicated and the most difficult to treat. AF involves up to 1% of the general population [4]. The Euro Heart Survey on AF showed that the total annual costs of treatment of AF reached 6.2 billion Euros for five studied countries [5]. AF decreases the quality of life and increases the risk of stroke and death. In addition, it is predicted that the number of patients with AF may double in the near future [2].

AF is usually classified into four main categories: (1) the first detected (only one diagnosed episode); (2) paroxysmal; (3) persistent, and (4) permanent. Totally different treatment strategies are used for these groups. Paroxysmal AF is self-terminating and continues for less than 7 days. Permanent AF is long term as compared to paroxysmal and cannot terminate spontaneously [6]. Prolonged AF usually leads to electrophysiological and anatomical changes of the atrium therefore more aggressive treatment is needed. Thus, monitoring of AF progress is important for timely detection of AF.

In general, AF is determined from the surface ECG. ECG provides a lot of information about heart rhythm disorders including AF. The ECG analysis process is based on the detection of feature points and the assessment of time and amplitude parameters. The main discriminators of AF are irregularity of heart rhythm and absence of P-waves (Figure 2.1). Regular P-waves are replaced by relatively high-frequency (3–10 Hz) chaotic f-waves. Recently,

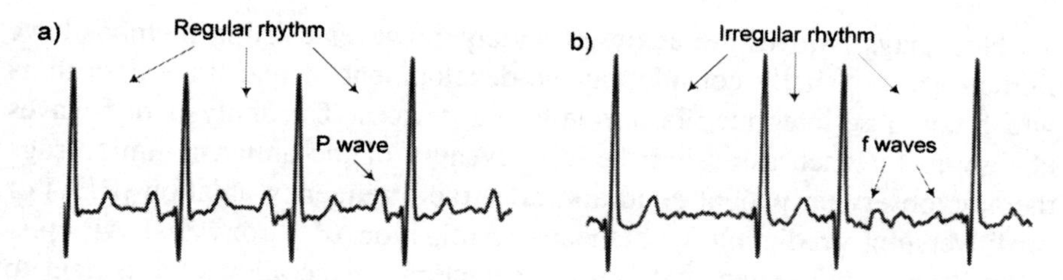

Figure 2.1 The ECG of normal heart rhythm (a) and atrial fibrillation (b).

the non-invasive evaluation of AF has attracted interest from many scientists and physicians [7]. However extraction of atrial activity from ECG is a much more challenging and complex task than it might appear at first sight. First of all, the electrical potential of atrial activity is mixed with the higher amplitude (2–50 times) electric potential of ventricular activity (QRST). In addition, the ECG signal is frequently distorted by various artifacts, e.g. patient movements, respiration, power line interference. For these reasons, delineation of f-waves from the ECG signals is still a complicated and not fully solved problem. It especially concerns real time AF monitoring systems. The complicated atrial activity extraction problem encourages the development of new digital signal processing algorithms [8].

There are many reasons to believe that objective monitoring of medical treatment may significantly facilitate the healing process. For example, f-waves morphology and spectral characteristics can provide important information about arrhythmia response to medical treatment [9]. Fibrillation rate can be obtained by using spectral analysis of f-wave sequences. The location of the largest spectral peak in the range 3–10 Hz is interpreted as the AF frequency. It was shown that there is a strong correlation ($r = 0.98$) between right atrium and surface ECG lead V_1 peak frequency [9]. However the correlation between right atrium and lead V_5 was considerably less ($r = 0.81$). This study revealed that it is possible to localize AF events using surface ECG. This assumption was reinforced by the work of Petrutiu et al. who have shown that lead V_1 reflects right atrium activity with strong correlation ($r = 0.98$) between surface and intracardiac "gold standard" ECG [10]. The opposite phenomenon was observed in lead V_9 which reflects left atrium activity while lead V_9 was recorded in an extended horizontal line from standard V_6 lead. These observations are very important to establish guidelines for developing AF monitoring devices.

Nowadays f-waves are analysed widely, however existing methods have limitations, especially considering the development of real time algorithms which could be integrated into monitoring devices. The analysis of f-waves is significant when assessing the effectiveness of the antiarrhythmic drugs therapy, observing patient condition after radiofrequency ablation (RFA) or cardioversion, predicting spontaneous termination of paroxysmal AF episodes. Providing a patient with an AF detecting device allows the patient to check his heart condition at the time he feels a possible arrhythmic episode. Recently portable AF detectors appeared on the market [11]. AF detector analyses heart rhythm and indicates whether there is a probability of AF or not. However, no announcements of devices capable of extracting and analyzing f-waves in real time were found in the literature. Efficient f-wave extraction methods allow assessing f-wave morphology more accurately. It can be anticipated, that real time f-wave processing in monitoring devices could facilitate the development of the new disease treatment strategies. It would be possible to provide the user an indication of the progress of the disease and the effectiveness of the medicines. For example, after the RFA procedure the AF is often asymptomatic despite the repeating AF attacks.

2.2 Atrial Electrical Activity Extraction Techniques

This section gives a brief overview of existing methods for atrial electrical activity extraction from surface ECG signals. Great attention was paid to the possible shortcomings of the methods also emphasizing the positive qualities. The second part of the section is designated to discuss possible future directions of method development considering implementations in continuous monitoring devices.

2.2.1 Methods for Atrial Electrical Activity Extraction from the Electrocardiogram

The simplest method for f-wave estimation is based on averaged QRST template subtraction from the ECG signals [12]. The average beat that represents ventricular electrical activity is obtained from an ensemble of time aligned QRST complexes. It is very important to ensure that the average beat and each QRST complex are aligned in time to each other, otherwise the resulting f-waves signal will contain QRST residuals. Then the time aligned QRST template is subtracted from each beat in the ECG signal. However this method suffers due to QRST morphology changes and may leave some ventricular activity as a residual after the subtraction operation. It was shown, that such cancellation is reliable for negligibly non-stationary ECG signals [13]. However the QRST complex morphology is very sensitive to the heart's electrical axis variations. The influence of QRST morphology alterations can be reduced by a spatiotemporal QRST elimination technique assuming that multi-lead ECG signals are available [12]. Lemay et al. [14] proposed method for the separation of QRS complexes and T waves based on the assumption that the repolarization wave (T) morphology varies depending on the heart rate (HR) while variability of a depolarization wave (QRS) shape is not related to HR. In this approach, the QRST complex is divided into two parts and separately averaged QRS and JQ interval templates are constructed and subtracted from the ECG signal in order to obtain f-waves. However, it is difficult to join interrupted intervals together because of mismatches between the levels of interval ends. The problem can be solved by filtering, for example using a zero phase low pass filter with cut off frequency of 50 Hz. Unfortunately, the zero-phase filters are based on forward and backward signal filtering and cannot be realized in real time. Lemay et al. [14] proposed another method which does not require beat averaging. The single beat cancellation method processes only one ECG cycle at a time and T wave morphology is estimated using all available lead information. Then the estimated T wave template is subtracted from the original ECG. However the atrial activity within the QRS complex is constructed from interpolation of the atrial activity signal in adjacent JQ intervals therefore the true f-waves information within the QRS interval is lost.

Castells et al. [15] proposed a methodology based on principal component analysis (PCA) which is well suited for Holter data analysis with a reduced number of ECG leads. Principal components are grouped to the three subspaces including ventricular activity, atrial activity and noise related components. The main advantage of PCA is robustness to QRST morphology variations however the performance is highly dependent on the algorithm used to identify an appropriate subspace. The PCA can be applied to multi lead ECG also [16]. Another group of statistical algorithms are based on the assumption that the atrial and ventricular electrical activities are generated by different electric sources and surface ECG is a linear sum of signals from these sources. Assumption based independent component analysis (ICA) methods have been proposed to distinguish between ECG signal generation sources. It was shown that ICA is suitable for the extraction of f-waves from multi lead ECG signals [17–20]. The ICA can be applied to a few lead ECG signals as well. For example considering the initial information of 64 lead

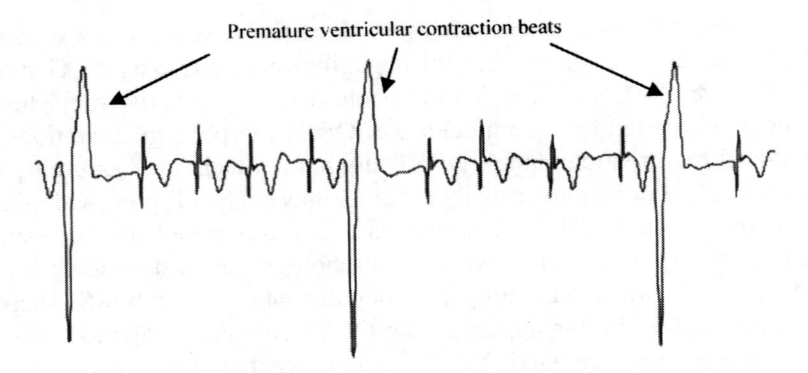

Figure 2.2 Premature ventricular contraction beats in ECG.

surface potential mapping the atrial activity was extracted using two leads only [21]. However the ICA is a multi lead technique in principle therefore the reduction of ECG leads decreases the quality of the extracted signal. The major drawback of ICA methods is the correct identification of atrial activity containing independent components thus the ICA methods need additional computations to separate the atrial activity source from the other extracted sources. Recently Llinares et al. [22] proposed a method for the extraction of atrial activity by using the ECG signal frequency domain information. The algorithm needs a priori knowledge about the dominant frequency of AF which is obtained from the TQ interval. TQ intervals become shorter when heart rate increases thus the reliability of the method decreases at higher heart rates. Besides the problems discussed, the AF signals may have premature ventricular contraction (PVC) beats (see Figure 2.2). The PVC beats differs from normal ECG beats in morphology and reduce the efficiency of QRST cancellation algorithms [23]. The PVC influence on the efficiency of the algorithms can be restrained by cancelling PVC beats in a pre-processing stage.

Previously considered methods are mostly used in multi-lead offline ECG processing systems, but these methods are less suitable for real time implementations. Moreover it is difficult to register an AF episode using diagnostic stationary equipment in the primary stage of paroxysmal AF because the first AF episodes are usually short (less than a minute) and therefore it is problematic to feel symptoms of the disease. Consequently it is important to develop a new algorithm which could be embedded in portable monitoring device and could be able to work with a reduced number of ECG leads.

2.2.2 Adaptive Filtering of Electrocardiogram Signals

The development of real time algorithm for f-waves extraction is a complicated task especially if it is restricted by a few lead applications. Digital filters are usually used for suppression of unwanted signal component from the analysed signal. Digital filters with fixed filter coefficients are suitable when signal and noise are stationary processes and their characteristics are known. The ECG of a healthy person can be considered as a quasi-stationary signal only if the signal is recorded in rest. However, ECG signals with AF are characterized by a high irregularity of ventricular activity. The PVC beats are also a considerable problem therefore the AF signals can be assigned to a highly non-stationary signal group. These non-stationary characteristics of ECG signals encourage scientists and engineers to search for more advanced methods such as nonlinear adaptive filters. Adaptive signal processing is widely used to improve ECG signal quality [24]. The researchers still deal with a classical adaptive filtering problem in order to remove power-line interference from the ECG signal minimally affecting the ECG spectrum [25]. Adaptive filters are used for ECG baseline wander removal also [26]. The ECG isoelectric line distorted by motion artifacts is restored using accelerometer data applied to a reference channel of the adaptive filter [27, 28]. Adaptive filters were successfully applied for the suppression of electromiographic (EMG) artifacts generated by muscle contraction [29]. Vice versa, the ECG signal is considered as an artifact and needs to be suppressed when EMG signals are studied [30].

Scientific literature analysis shows that there were attempts to design the algorithms for the real time ambulatory monitoring of AF signals. In the early 1990s Thakor and Zhu [24] proposed separating atrial activity from ventricular activity using an adaptive recurrent filter (ARF). The method requires a QRST detector in order to obtain the required signal as reference input of the ARF filter. The ARF based method was analyzed in [31] and the study revealed that the ARF is effective in the case of stable QRST morphology. However the rapid changes in QRST (e.g. PVCs) highly reduce the efficiency of the algorithm and the residuals of ventricular electrical activity remain unacceptably large.

The increasing computing power of microcontrollers and digital signal processors revives the interest in using the adaptive filters for the real time processing of AF signals. Adaptive filter based AF signal processing methods are single ECG lead methods and they also require an additional channel for the reference signal. Conventional adaptive filters are usually applied for solving linear problems. The growing interest in nonlinear adaptive filtering was shown by Camps-Valls et al. study [32]. They analyzed the performance of the linear least mean squares (LMS) algorithm, and the nonlinear FIR filter based artificial neural network (ANN) and the recurrent ANN (gamma network) methods for cancellation of the maternal signal in fetal ECG signals. The study results showed the nonlinear ANN based adaptive filter performing better than a conventional linear LMS algorithm. Regarding ventricular activity cancellation in ECG with AF signals, a similar approach was described by Vásquez et al. in [33]. This technique was realized using an Elman time delay artificial neural network (TDNN). The method estimates a time-varying, nonlinear transfer function between two ECG leads. The proposed approach is robust during noisy episodes and QRS morphology variations and does not require QRS detection. However the application of Elman TDNNs is limited by the slow and complicated network iterative training process, especially when the number of neurons is high (>20). The method's convergence is strongly related to the quality of the training data.

2.3 Artificial Neural Networks and Echo State Network Approach

The first part of this section introduces the basics of artificial neural networks. The rest of this section is devoted to concepts of reservoir computing and echo state networks in particular.

2.3.1 Concepts of Artificial Neural Networks

The information from the surrounding environment is processed with neural networks in humans and animals. Biological neural networks consist of billions of neurons exchanging electrical impulses between each other. Each neuron has approximately ten thousand connections to other neurons. Biological intelligence, including memory, is stored in the neurons and in the connections between them. The electrical impulses can be strengthened or attenuated in terms of relevance and thus the structure of neural network is changing all the time when new information reaches the brain. In the case of biological brains the strength of connections are adjusted by life experience.

Scientists and researchers have been trying to mimic these networks and develop computer algorithms for practical problem solving. Various artificial neural network structures have been proposed and many solutions for different problems were provided. ANN development is basically inspired by

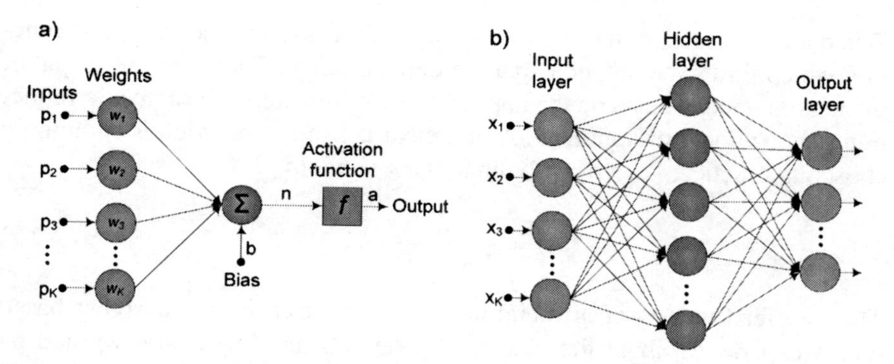

Figure 2.3 The basic structure of a multiple input artificial neuron (a) and feed-forward artificial neural network (b).

two main purposes: to understand the working principles of the biological brain and to apply ANNs for practical implementations. Artificial neurons are very simple abstractions of complex biological neurons and each artificial neuron performs simple mathematical operations consisting of multiplications and additions. However highly interconnected neurons create a powerful mathematical model that can solve poorly defined problems after training is completed. Artificial neurons usually have nonlinear activation function therefore ANNs are considered as nonlinear processing tools [34].

ANNs change their parameters (weights) during the learning process much like biological networks change the strength of connections. Then complex relationships between inputs and outputs are formed according to training data patterns. Figure 2.3 shows the structure of an elementary neuron (a) and the basic schematic of a feed-forward ANN (b). The most popular ANN structure consists of three layers, called input (passive) layer, hidden layer and output layer. Sometimes a neural networks can be constructed with several hidden layers, thus increasing the network complexity.

The artificial neuron usually consists of five main parts including one or more inputs pi, a set of weights for each input wi, bias b, activation function f and single neuron output a. The bias is also a weight except that it has a constant value which is summed along with weighted inputs. The bias can also be omitted and set to zero. More advanced neurons have a digital filter in series to the activation function (described in Section 2.4.2). The weighted sum of inputs is obtained by this equation:

$$n = \sum_{t} \mathbf{w}_i \mathbf{p}_i + b. \tag{2.1}$$

When the weighted sum of inputs is obtained, it is fed to the activation function which limits the output y to a pre-defined range. Therefore the output of the neuron depends also on the applied transfer function. The activation function may be linear of nonlinear. A nonlinear activation function is commonly chosen in practice, i.e. a hyperbolic tangent sigmoid:

$$a = \frac{2}{1 + e^{-2n}} - 1. \tag{2.2}$$

The transfer function of artificial neurons is selected by the designer based on his experience about the object. The weights and biases are adjusted by a learning algorithm that creates a relationship between inputs and outputs in order to achieve a desired goal. Thus the ANN is useless as long as the weighs connecting the neurons are not set properly. Commonly the weights are obtained by an iterative adaptation process. The adaptation (or learning) is accomplished using training patterns. The learning algorithm changes the network's behaviour depending on inputs and feedback received from the teacher. Regarding the feedback nature there are three main types of network training: supervised, unsupervised and reinforced. The learning is called supervised when the set of training input patterns and desired outputs is available. The teacher provides a target for each of input sets therefore the network has knowledge about the response it should provide. On the contrary, unsupervised learning is accomplished with self organizing networks using only input sets of data. The network receives no feedback from teacher thus unsupervised learning is based on similarities and differences between the input sets. Unsupervised learning is usually applied for data clustering. Reinforcement learning is a kind of supervised learning where the learning is executed trying to maximize a numerical reward signal. The network receives feedback from the teacher about the appropriateness of its response. If the response is correct, then the teacher gives information to the network that the response was appropriate. In the case of inappropriate response the teacher gives only information that the response was improper but does not provide information about what the response should be as in the case of supervised learning.

2.3.2 Reservoir Computing and Echo State Networks

Although conventional feed-forward ANNs are the most popular in practice, they are static and cannot deal with temporal patterns. Differently than ANNs, recurrent neural networks (RNNs) have feedback connections between neurons. The feedbacks are obtained by delaying the output value by one time step. The RNNs are quite different from feed-forward networks which have no backward connections and process information in only one direction. Such recurrent structure creates an internal dynamic memory. The internal memory allows RNNs to process dynamic temporal sequences and makes them more powerful than feed-forward networks. However, conventional RNNs have not met many applications in practice due to their long and complicated iterative training process and potential instability.

Recently a novel technique in RNN training called reservoir computing (RC) was introduced. The fundamentals of the reservoir computing idea were proposed independently by Herbert Jaeger under the name of Echo State Network (ESN) [35] and by Wolfgang Maass under the name of Liquid State Machines (LSM) [36]. The RC is a paradigm where a sparsely connected RNN (the reservoir) is generated randomly. The reservoir is a large hidden layer (50–1000 neurons) where the inputs are transformed by time depended nonlinear functions. The RC methods differ from other types of RNN particularly in the training nature. The most interesting thing is that the complexity of iterative RNN training is avoided by adapting only output connections. Surprisingly, the randomly generated RC networks outperform the more complicated fully trained RNNs in almost all cases [37].

Liquid state machines and echo state networks were developed for different purposes. LSM were designed from a view of computational neuroscience aiming to understand the principles of neural microcircuits. LSM use sophisticated and biologically more realistic spiking neurons than ESN. The ESN is developed for practical implementations. The various types of ESNs are under consistent development [38].

ESN is a recurrent neural network characterized by their specific architecture and flexible training method. The principle of ESN is to design a random and fixed RNN and drive it with an input signal. The generated RNN is not used for training and it is called a dynamic reservoir. Each excited neuron of the reservoir forms a nonlinear response signal. Then the output signal is obtained by a trainable linear combination of nonlinear responses. The ESN has several important adjustable parameters. The effectiveness of ESN is highly dependent on the appropriate selection of network size, spectral radius and input scaling. The network size corresponds to the number of neurons in the reservoir. The spectral radius (ρ) of the reservoir weight matrix is defined as the maximum absolute value of the eigenvalues of reservoir weights. Furthermore, it is related to the echo state property of ESN. The echo state property is ensured if $\rho(\mathbf{W}) < 1$. However it is not a sufficient condition in some cases [38]. The spectral radius should be selected between 0 and 1. A smaller value of spectral radius causes more rapid decay of reservoir dynamics. Therefore the spectral radius highly affects the length of memory and amount of nonlinearities in the reservoir.

Input scaling has a similar influence on the degree of reservoir dynamics as the spectral radius. Input scaling reduces (or increases) the values of input data. Therefore, for very small input scaling reservoir behavior is almost linear because the nonlinear parts of the sigmoid activation function are never excited. In the opposite case, the high values of input scaling drive the reservoir neurons to highly nonlinear regions of the activation functions. In some cases a feedback scaling can be set. The feedback scaling determines the feedback weights strength from the output layer to the reservoir and is usually useful for signal generation tasks. However the feedbacks increase the possibility of dynamical instability. Thus the feedback weights are set to zero in many approaches. The other interesting property of ESN is that it is not necessary to use a fully connected reservoir and it is sufficient to operate with sparsely connected (5–20%) reservoirs. Reduced computational complexity is an important advantage of reduced connectivity.

Main principles of a feed-forward, recurrent neural network and echo state neural network are depicted in Figure 2.4. Feed-forward ANNs (Figure 2.4a) have only direct connections between layers and thus no temporal memory. The fully connected RNNs have connections between all layers, but there are known simpler alternatives as those proposed by Elman [39]. Elman introduced a simple recurrent neural network which has recurrent connections only in the hidden layer. The Elman neural network is a kind of RNN that has the ability to maintain a copy of hidden layer neurons activations and pass back these activations to the inputs in the next state of the network (Figure 2.4b). All the weights in Elman RNNs are trained in an iterative manner. Echo state networks (Figure 2.4c) have a similar architecture to Elman neural networks when feedback weights are not used. The major difference between Elman neural networks and ESN is in the selection of the weights to be trained: all the weights must be trained in the Elman RNNs whereas only the output layer weights must be trained in ESNs.

ANNs are considered as black-box models and their structures are adjusted without any human knowledge. Therefore, in some cases, ANNs are criticized for being difficult to interpret. This problem becomes even more relevant in case of RNNs primarily due to the recursive connections. Besides this, RNNs present significant benefits over the other methods when mathematical approximation of dynamical systems is needed.

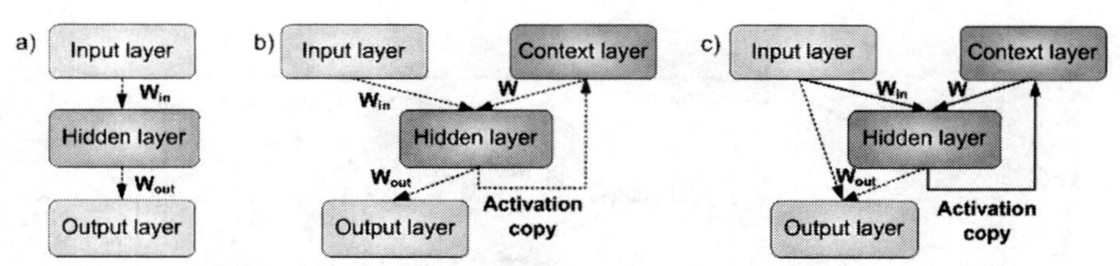

Figure 2.4 Schematic description of the main types of artificial neural network architectures: feed-forward (a), Elman simple recurrent (b), reservoir computing based RNN (c). The dashed lines indicate the adaptive weights. The connections depicted in solid lines are fixed and remain unchanged.

It is notable that information processing with large artificial neural networks is slow despite the fact that computational power of computers is increasing. This is because computers typically have a single processing unit that processes data in a serial manner while the human brain processes information in parallel. Thus even a considerably slower data processing in parallel may be more efficient than using a single powerful processor. Since ANNs require a considerable number of vector and matrix operations they are suitable for implementation on parallel processing cores, for instance, on Graphics Processing Units (GPUs). GPUs were developed exclusively for 3D graphics however these devices are becoming more popular for general purpose computations including ANNs [40].

2.4 Vetricular Electrical Activity Cancellation Using Adaptive Echo State Network

The present section describes a new method for f-waves signal extraction from ECG using an ESN based nonlinear adaptive filter. The aim of this chapter is to introduce a single lead method for f-waves estimation in the ECG signal which could be implemented in real time patient monitoring devices. In contrast to existing methods, the present approach could also be applied in situations where multi-lead methods are unsuitable, for example in the analysis of Holter monitoring data.

2.4.1 The Fundamental Principles of the Method

The proposed QRST cancellation method uses a classical adaptive filter approach: the needed signal is extracted from a mixture of signals using a reference signal which is modified by a filter with a time changing transfer

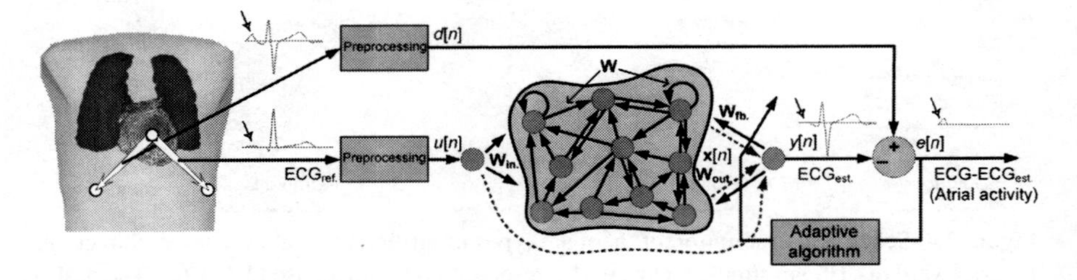

Figure 2.5 Atrial activity extraction scheme based on adaptive ESN. The dashed lines indicate adaptive weights; other weights are fixed and remain unchanged.

function controlled by an optimization algorithm. The underlying difference between ESN and conventional adaptive filters (LMS or RLS) lays in the fact that the amplitude transfer function of ESN is allowed to become nonlinear. This feature adds more degrees of freedom to the adaptation process and helps to increase the accuracy of the needed signal extraction.

The proposed method is explained in Figure 2.5. The method requires two ECG leads: an ECG signal with ventricular (QRST complex) and atrial activity (P-waves) expressed and a reference ECG signal with the prominent ventricular electrical activity but with less expressed atrial activity (low P wave amplitude). The existence of a reference point with low P-wave amplitude can be demonstrated using ECGSIM software [41], which allows studying the relationship between the electric activity of the heart and the resulting potentials on the thorax.

Three main blocks build the method for the extraction of f-waves: preprocessing, ESN and adaptive algorithm. The ESN is a large, fixed, recurrent neural network which serves as a random nonlinear excitable medium, whose high dimensional dynamical "echo" response to a driving input is used as non- orthogonal signal basis to reconstruct the desired output. Reservoir connecting weights $\mathbf{W}$ and input weights $\mathbf{W}_{\text{in}}$ are selected randomly during network initialization and remain fixed during the training phase. The only set of weights which is changed during training are the output weights $\mathbf{W}_{\text{out}}$. The inner neurons of the reservoir use nonlinear transformations (hyperbolic tangent) while the output neuron are a simple linear summation. The online adaptive algorithms that are used in linear adaptive filters (least mean squares or recursive least squares) can be employed to train the output weights W_{out} and to achieve the needed network dynamics, due to the linearity of output neurons.

The RLS algorithm is much quicker than the LMS algorithm in convergence due to recursively obtained filter coefficients. The filter coefficients calculated at time $n - 1$ are used to find the weights at time n. Fast convergence is an important criterion when dealing with the ECG signals, which are often distorted by motion artifacts. Rapid morphological changes in ECG can also occur because of heart rhythm disorder. The ventricular premature beats are a sufficiently common phenomenon in patients with AF and they are a good example of sudden change in ECG shape.

An adaptive RLS algorithm was used for the adaptation of the ESN output layer weights $\mathbf{W}_{\text{out}}[n](size 1 \times N)$ for each incoming sample. It was observed that after 7000–10000 iterations (processed samples) the conventional RLS algorithm becomes unstable. In order to increase the stability of the adaptive algorithm we applied a pre-whitened RLS algorithm as proposed in [42]. The ESN output weights $\mathbf{W}_{\text{out}}[n - 1]$ are adapted for each sample n in such a way that the difference between the target $d[n]$ and the filtered $y[n]$ signals lead to the error $e[n]$ (see Figure 2.5). The error signal $e[n]$ is the f-waves estimate.

The RLS algorithm requires a proper initialization of its parameters. During the initialization period of the adaptive algorithm the matrix $\mathbf{P}[0]$ is set equal to an identity matrix, which is divided by a very small scalar number:

$$\mathbf{P}[0] = \delta^{-1} \cdot \mathbf{I}, \tag{2.3}$$

where δ is a small positive constant (0.001 in this case), $\mathbf{I}$ an identity matrix of size $N \times N$, where N is the number of system parameters. The adaptive filter was initialized with the output layer coefficients set to zero:

$$\mathbf{W}_{\text{out}}[0] = 0. \tag{2.4}$$

The algorithm is repeated for each new sample n after initialization is completed. Let the input sample be $u[n]$, then a vector of reservoir activations $\mathbf{x}[n]$ at time step n is given by

$$\mathbf{x}[n] = f_{\text{res}}(\mathbf{W}\mathbf{x}[n - 1] + \mathbf{W}_{\text{in}}u[n]) + \boldsymbol{\eta}[n - 1] \tag{2.5}$$

where $\mathbf{W}$ is a weight matrix of size $N \times N$ corresponding to internal network connections, $\mathbf{W}_{\text{in}}$ is the input weight matrix of size $1 \times N$, f_{res} is a neuron activation function, $\boldsymbol{\eta}[n - 1]$ represents an artificial white noise component (leads to smaller output weights and better stability). The reservoir states $\mathbf{x}[n - 1]$ and input sample $u[n]$ are concatenated together with the previous output sample $y[n - 1]$ to a new column vector $\mathbf{c}[n]$:

$$\mathbf{c}[n] = [\mathbf{x}[n]; u[n]; y[n - 1]]. \tag{2.6}$$

Then the n-th sample of the neural network output will be

$$y[n] = f_{\text{out}}(\mathbf{W}_{\text{out}}[n-1] \cdot \mathbf{c}[n]). \tag{2.7}$$

The error between the target signal $d[n]$ and the estimated signal $y[n]$ is a desired f-waves signal:

$$e[n] = d[n] - y[n]. \tag{2.8}$$

The output weights of the neural network are updated as follows:

$$\mathbf{W}_{\text{out}}[n] = \mathbf{W}_{\text{out}}[n-1] + \frac{e[n] \cdot \mathbf{r}[n]}{\lambda + \|\mathbf{v}[n]\|^2}, \tag{2.9}$$

where the pre-whitened auxiliary vectors $\mathbf{v}[n]$ and $\mathbf{r}[n]$ are obtained as

$$\mathbf{v}[n] = \mathbf{P}[n-1] \cdot \mathbf{c}[n], \tag{2.10}$$

$$\mathbf{r}[n] = \mathbf{P}^T[n-1] \cdot \mathbf{v}[n]. \tag{2.11}$$

The matrix $\mathbf{P}[n]$ is updated as

$$\mathbf{P}[n] = \frac{\mathbf{P}[n-1] - k[n] \cdot \mathbf{v}[n] \cdot \mathbf{r}^T[n]}{\sqrt{\lambda}}. \tag{2.12}$$

The forgetting factor λ is an important parameter of the RLS algorithm and is commonly chosen from the interval $0.95 \leq \lambda \leq 1$. The expression for gain coefficient $k[n]$ is defined as:

$$k[n] = \frac{1}{\lambda + \|\mathbf{v}[n]\|^2 + \sqrt{\lambda}\sqrt{\lambda + \|\mathbf{v}[n]\|^2}}. \tag{2.13}$$

2.4.2 Echo State Network with Filter Neurons

Conventional ESN neurons with a sigmoid transfer function have no memory. Their output values depend only on current input values and do not depend on the past inputs. The neuron outputs of the dynamic reservoir are of an impulsive nature. The rapid impulses in neuron outputs can be considered as high frequency noise. An infinite impulse response (IIR) exponentially weighted moving average (EWMA) filter for signal smoothing is widely used and is expressed as [43]:

$$\omega[n] = (1-\alpha) \cdot \omega[n-1] + \alpha \cdot \psi[n], \tag{2.14}$$

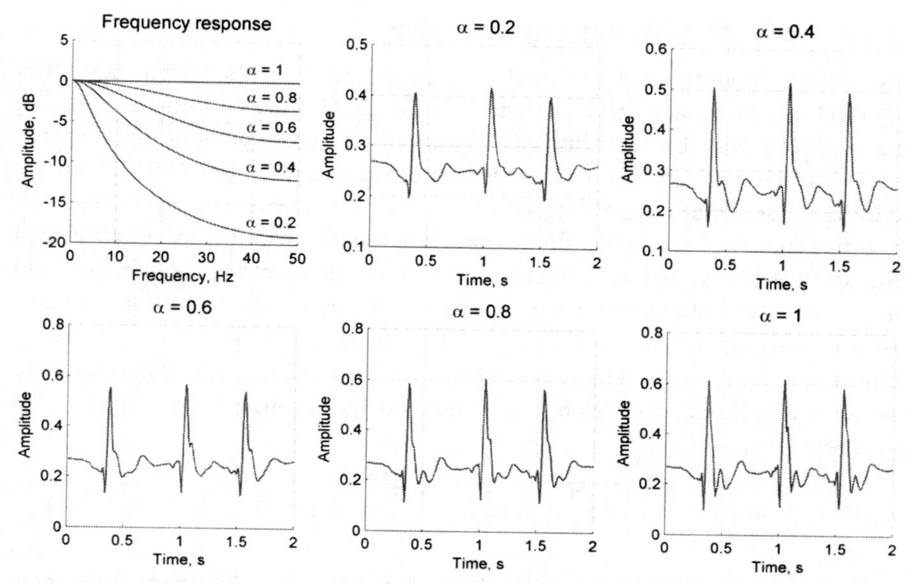

Figure 2.6 Influence of the exponentially weighted moving average filter's α coefficient on the neuron's output when the ECG signal is fed to the input of the ESN.

where $\psi[n]$ is the filter input and $\omega[n]$ the filter output. The coefficient α in Equation (2.14) determines the weighting factor, which decreases exponentially for each previous input. The filter acts as a smoothing filter and suppresses the higher frequency noise if coefficient α is selected from interval (0, 1). EWMA filters can be applied to the ESN reservoir neurons for output signal smoothing. Then the outputs of reservoir neurons are updated according to the previous neuron outputs and are found by replacing Equation (2.5) with (2.15):

$$\mathbf{x}[n] = (1-\alpha)\cdot\mathbf{x}[n-1]+\alpha(f_{\text{res}}(\mathbf{W}\mathbf{x}[n-1]-\mathbf{W}_{\text{in}}u[n])+\boldsymbol{\eta}[n-1]). \quad (2.15)$$

The influence of the value of coefficient α on the neuron's output is depicted in the illustrations of Figure 2.6. It can be observed that decreasing suppresses higher frequency components. Individual neurons of standard ESN have no memory and such neuron's response to the input ECG it is depicted when $\alpha = 1$.

Other types of digital filter based neurons were proposed in the scientific literature. In addition to the most widely used EWMA filter, the low pass [44] and band pass filters [45] could be used for solving specific problems.

2.4.3 Fibrillatory Waves Signal Model

In order to quantitatively evaluate the proposed f-waves signal extraction method, the atrial activity component in the ECG must be known. This can be achieved only by working with synthetic or semi-synthetic signals, i.e. surrogate signals. In this work, the surrogate signals were constructed by adding an artificial f-waves signal to the real ventricular activity signal. The additive model of surrogate signals was constructed with the assumption that the surface AF signal is a linear composition of electrical ventricular and atrial activity. Subsequent f-waves extraction operations allow the comparison of several different methods. Our main goal was to design an easily controlled f-waves model with simply changing parameters. The changing f-waves amplitude over time was simulated by amplitude modulation of a sine waveform:

$$\varsigma[n] = A \sin\left(2\pi \frac{f_c}{f_s} n\right) + \frac{M}{2}\left(\sin\left(2\pi \frac{f_c + f_m}{f_s} n\right) + \sin\left(2\pi \frac{f_c - f_m}{f_s} n\right)\right). \tag{2.16}$$

where M is the modulation coefficient, f_c the f-waves dominant frequency, f_m the modulation frequency, f_s the sampling frequency, A the amplitude. The simulated f-waves signal $\varsigma[n]$ was distorted by adding white noise $v[n]$ and making the f-waves more unpredictable. The resulting signal is given by the additive model (Figure 2.7b):

$$s'[n] = \varsigma[n] + k \cdot v[n]. \tag{2.17}$$

Here k determines the amount of white noise in the f-waves signal. In order to make the synthetic AF signal closer to the real signal it was processed to suppress the noise $s'[n]$ components, which are outside of the $s'[n]$ signal's bandwidths using 40 tap FIR type Wiener filter. If $s'[n]$ and $v[n]$ processes are uncorrelated the Wiener–Hopf equation in matrix form can be written as

$$(\mathbf{R}_d + \mathbf{R}_v) \cdot \mathbf{w}^o = \mathbf{p}_{dv}, \tag{2.18}$$

where $\mathbf{R}_d$ is the autocorrelation matrix for $s'[n]$, $\mathbf{R}_v$ the autocorrelation matrix for $v[n]$, $\mathbf{w}^o$ are the optimal filter coefficients, $\mathbf{p}_{dv}$ the cross correlation function between $s'[n]$ and $v[n]$ signals. The needed optimal coefficients are expressed as

$$\mathbf{w}^o = (\mathbf{R}_d + \mathbf{R}_v)^{-1} \cdot \mathbf{p}_{dv}. \tag{2.19}$$

The white noise corrupted signal $s'[n]$ is filtered using the Wiener coefficients (2.19) and the f-waves model $s[n]$ with known central frequency is

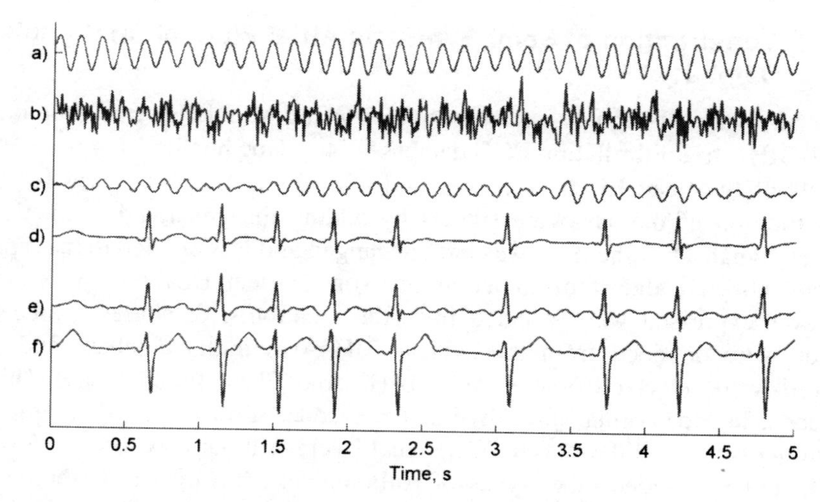

Figure 2.7 Construction of surrogate AF signals: (a) amplitude modulated 7 Hz signal, (b) amplitude modulated signal corrupted by additive white noise, (c) f-waves signal after filtering with optimal Wiener coefficients (amplified three times in the picture), (d) real AF signal episode without atrial activity expressed, (e) surrogate signal model, (f) reference real signal without atrial activity expressed.

constructed (Figure 2.7c). The dominant frequency of the f-waves was fixed at 7 Hz.

It is easy to change the dominant frequency of f-waves with the proposed model. The Wiener filter automatically finds the appropriate coefficients for signal smoothing. For this reason, there is no need to design a new filter when the dominant frequency of the f-waves is changed. Furthermore the amplitude frequency response of the Wiener filter is fundamentally different from the amplitude frequency response of classical filters and it is adapted to suppress the signal corrupting noise optimally with respect to the filter order. The simulated f-waves signal is added to the ventricular activity component of real ECGs with AF episodes but without visible f-waves (Figures 2.7d and f).

It should be noted that the f-waves morphology is individual for each patient thus it is difficult to create the adequate f-waves model which will be suitable for all cases.

2.4.4 Construction of Semi-Synthetic Atrial Fibrillation Signals Database

The surrogate AF signals were constructed using MIT-BIH Arrhythmia and MIT-BIH Atrial Fibrillation ECG databases [46]. One hundred 1-minute segments of two lead ECGs were selected. One ECG lead was used for the construction of the surrogate signals by adding the generated synthetic f-waves signal. The other lead was left unchanged and was applied to the inputs of both adaptive algorithms under comparison. Thirteen ECG records without f-waves expressed were selected from the mentioned databases. Three records were obtained from the MIT – BIH Arrhythmia database and ten records were selected from the MIT – BIH Atrial Fibrillation database. Thirty episodes include normal sinus rhythm, 66 episodes show AF and 4 other heart rhythm disorders. Fifty-seven ECG signals were without PVC beats, 35 with PVC and 8 included other types of pathological heart cycles. It should be mentioned that the surrogate signal data consists of 13 record groups which are different from each other in ECG morphology. The ECG morphology of the 1 minute length segments is similar in each signal group, only the extraction time is different. The requirement to have just ventricular activity is accomplished only partially. Several records had increasing amplitude f-waves in short segments. In part of the surrogate signals, normal sinus rhythm ECGs, had low amplitude visible P waves in both data and reference channels however this was not a problem as the adaptive algorithms eliminated P waves and minimized their impact on the final results.

The segments of real signals are corrupted with broadband noise. ECG signals were de-noised using a discrete wavelet transform based de-noising procedure with soft wavelet coefficient thresholding [47]. De-noised ventricular activity was added with an f-waves synthetic model. Considering that the f-waves amplitude is often varying in time it is reasonable to define the f-waves using RMS. We defined the f-waves as small (RMS $= 30\ \mu$V), medium (RMS $= 50\ \mu$V) and large (RMS $= 70\ \mu$V). The f-waves extraction algorithms should be able to work even with very small f-waves amplitude signals especially if the implementation in AF monitoring devices is considered. The database of surrogate signals consists of 10 ensembles whose distinguishing feature is the RMS value of the f-waves. A total of 10 sets of 100 surrogate signals with different RMS values were generated. The RMS value of the f-waves was changed from 10 to 100 μV in 10 μV steps. The final surrogate signals database consists of 1000 two channel semi-synthetic signals of 1 minute in length.

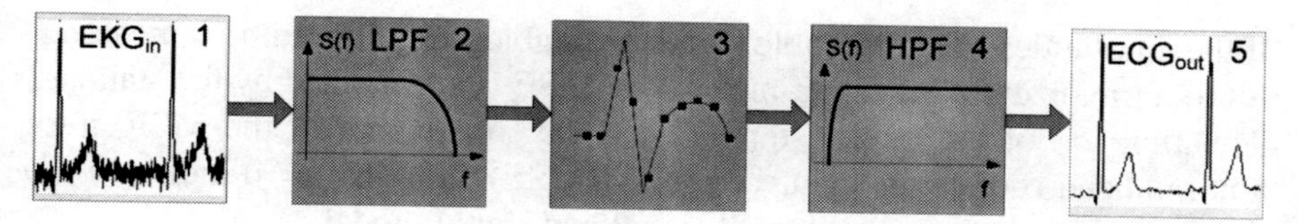

Figure 2.8 ECG signal pre-processing scheme.

2.4.5 Electrocardiogram Data Pre-processing

The neural network efficiency is highly dependent on the quality of the ECG signal, since pre-processing is required to remove the noise and interference.

We have used the following pre-processing stages for the ECG signals (see Figure 2.8):

1. The baseline wander was removed with an efficient interpolated high pass FIR filter (cut-off frequency 0.67 Hz).
2. The higher frequency noise was reduced with an equiripple low pass FIR filter (cut-off frequency 28 Hz).
3. In order to equalize the sampling frequency of the ECG signals obtained from different databases, the signals were re-sampled to 100 Hz.

2.5 Results of Atrial Activity Extraction

In this section we will discuss the main differences of the linear recursive least squares and nonlinear echo state network based adaptive filters. Much attention is paid to showing the importance of the parameters and their influence on the algorithms' performance.

2.5.1 Performance Measures

In order to evaluate the efficiency of the f-waves extraction algorithms the similarity of the estimated f-waves and the f-waves' model must be found. The mean squared error (MSE) is usually used as a performance measure. The MSE evaluates the error between the known signal model $s[n]$ and the estimated signal $e[n]$.

$$\mathrm{MSE} = \frac{1}{N}\sum_{n=0}^{N-1}(s[n] - e[n])^2. \tag{2.20}$$

If the amplitudes of comparable signal ensembles are different it is meaningful to normalize the MSE parameter. The MSE is normalized by the standard deviation (σ) of the model signal. Then the square root of the MSE gives a normalized root mean square error (NRMSE). NRMSE = 0 indicates the ideal matching between estimated signal and signal model.

$$\text{NRMSE} = \sqrt{\frac{1}{N\sigma}\sum_{n=0}^{N-1}(s[n]-e[n])^2}. \tag{2.21}$$

The cross-correlation coefficient is the second useful parameter for the evaluation of the algorithms. The cross-correlation coefficient evaluates the similarity of morphology between two signals and is easily interpreted. The Pearson cross-correlation coefficient between the estimated signal and its model is obtained through

$$r = \frac{\sum_{n=1}^{N}(s[n]-\bar{s}[n])\cdot(e[n]-\bar{e}[n])}{\sqrt{\sum_{n=1}^{N}(s[n]-\bar{s}[n])^2\cdot\sum_{n=1}^{N}(e[n]-\bar{e}[n])^2}}. \tag{2.22}$$

The cross-correlation coefficient is automatically normalized to within predefined limits $[-1, 1]$ and this is an important advantage with respect to the MSE. However, the cross-correlation coefficient evaluates just the morphological similarity of signals and ignores the bias. Less effective algorithms can suppress the f-waves amplitude although partial morphological similarity is maintained. For this reason, the cross-correlation coefficient and the NRMSE parameters are both important when evaluating the efficiency of f-waves extraction algorithms.

2.5.2 Initialization of RLS Based Linear Adaptive Filter Parameters

The efficiency of the classical RLS algorithm highly depends on its forgetting factor λ parameter. In practice, λ values are chosen from the interval of $0.95 \leq \lambda \leq 1$. If the smaller λ values are chosen, then the adaptive filter is more sensitive to recent samples than with larger ones. The number of adaptive coefficients is another important parameter of the RLS based adaptive filter. In order to obtain an appropriate forgetting factor value and the number of adaptive coefficients, we evaluated the influence of these parameters with regards to the NRMSE error. The calculations were performed with an ensemble of average amplitude f-waves (RMS = 50 μV) signals.

The influence of the forgetting factor λ and the number of adaptive coefficients N was evaluated in two stages: firstly, the parameters were changed in a wide range of values and afterwards the range was narrowed taking into account the results of the first stage. In the first stage the value of the forgetting factor was changed from 0.95 to 1 with a step size of 0.01. The number of filter coefficients was changed from 5 to 200 in a logarithmic scale. For each of the 100 surrogate signals the NRMSE was obtained and the average value was calculated. The calculations were repeated for each parameter change. The lowest NRMSE value corresponds to the optimal set of parameters.

The results of these calculations showed that the lowest NRMSE values were obtained when the parameters were changed within the intervals $0.99 \leq \lambda \leq 1$ and $50 \leq N \leq 100$. In the second stage of parameters optimization, the forgetting factor λ was changed within the interval 0.99 to 1 with a step size 0.002. The adaptive coefficients N were changed from 30 to 80 with a step size 10. It was observed (see top illustration of Figure 2.9a) that the NRMSE decreases with increasing number of filter coefficients only when $\lambda \rightarrow 1$. The NRMSE decreases until the number of adaptive coefficients reaches 100 and significantly increases when the number of coefficients approaches 200. It was observed that for a small number of adaptive coefficients (up to 20) the influence of λ decreases but becomes more important when the number of filter coefficients is increased. A more detailed analysis showed that the linear RLS based adaptive filter is most effective when $\lambda = 0.998$ and $N = 50$ ($NRMSE_{RLS} = 2.77$) when computation time is considered. It is necessary to emphasize that the computational cost of the RLS algorithm is the square of the filter length.

2.5.3 Initialization of ESN Based Nonlinear Adaptive Filter Parameters

The echo state network based nonlinear adaptive filter (NAF) has more adjustable parameters when compared to the RLS based linear filter (LAF). The feedback weights, output scaling and activation functions of the output neuron have not been changed in the optimization process. The feedback weights from the output layer to the reservoir were set to zero. The output scaling proportionately impacts the amplitude of the output signal; therefore it was fixed to 1. The other parameters of the adaptive ESN were changed systematically in each step of the optimization process. The first optimization step was carried out in an analogous manner to that described for the RLS based linear filter. The aim was to obtain the forgetting factor and the number

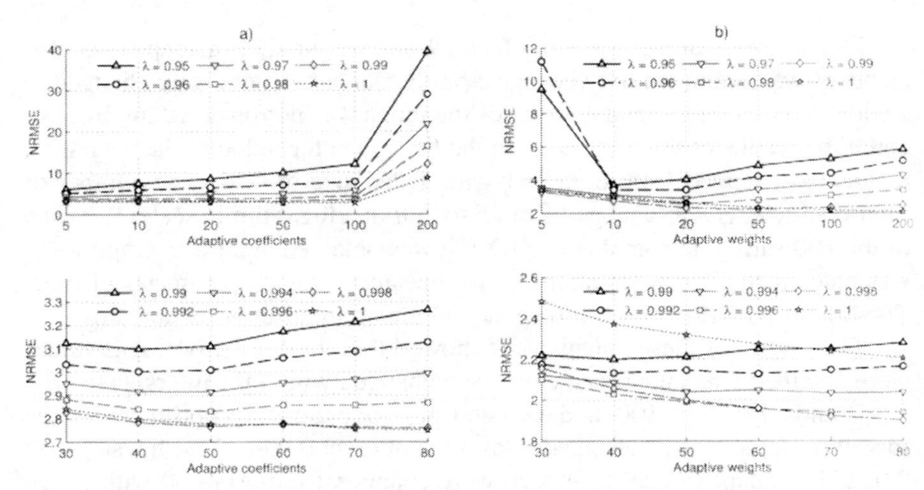

Figure 2.9 The efficiency of adaptive filters as a function of the number of coefficients: (a) RLS based linear filter, (b) ESN based nonlinear filter.

of filter coefficients, or in this case, the number of output neurons with which the algorithm is most effective. The ECG signal before entering the adaptive ESN was normalized to the interval [0, 1]. The ESN output signal (f-waves) was restored to its initial levels after the extraction of f-waves was completed.

Optimization was started with the following ESN parameters: reservoir connectivity $\delta = 20\%$, EWMA filter coefficient $\alpha = 0.25$, spectral radius $\rho = 1$, input scaling $\gamma = 1$, reservoir noise level $\nu = 0.00001$, reservoir neurons' activation function – hyperbolic tangent sigmoid. The experiment revealed (see top illustration of Figure 2.9b) that the NRMSE decreases if the number of reservoir neurons increases even when $\lambda = 1$. By comparison, the RLS based linear filter performance deteriorated when the number of adaptive coefficients reached 100. However, the adaptive ESN efficiency decreases significantly for an increasing number of neurons when $\lambda < 1$. The repeated experiment in a narrowed range showed that the adaptive ESN is most effective when $\lambda = 0.998$ and $N = 70$ ($\text{NRMSE}_{\text{ESN}} = 1.92$) considering that increasing the number of system parameters leads to more computations. The results obtained from the parameters optimization of both adaptive filters are sufficient to prove that the nonlinear adaptive filter is more effective than the linear filter for solving the f-waves extraction problem. However, it is important to emphasize that the ESN based nonlinear adaptive filter needs additional calculations for updating the reservoir states.

The second optimization step was carried out for the evaluation of the influence of the spectral radius and input scaling. The reservoir spectral radius normalizes the randomly generated weight matrix within the spectral radius range. A correctly selected spectral radius keeps the reservoir stable. However, smaller weights reduce the length of memory. Reservoir spectral radius (ρ) is associated to input scaling (γ). Input scaling changes the amplitude of the input signal proportionally and it is selected according to the problem. If the input scaling value is chosen to be close to zero then the hyperbolic tangent activation function is excited with values close to zero and the neuron works in an almost linear mode. On the contrary, the highest nonlinearity is obtained when γ approaches to 1. Furthermore higher values of ρ and γ increase the temporal memory length of the reservoir. However, the upper values for ρ and γ are limited by possible instability of the reservoir, thus these parameters are usually chosen as less than 1. The spectral radius and input scaling were changed within $0.2 \leq \rho, \gamma \leq 1$ with a step of 0.2. As it was expected, the highest efficiency for the adaptive ESN was achieved with the largest values of ρ and γ (Figure 2.10a). The values ρ and γ were equated to 1 for the following experiments thus extending reservoir memory to the maximum.

The third optimization step was based on the assumption that the efficiency of the reservoir does not deteriorate when reservoir neuron connectivity is highly reduced. The reservoir connectivity shows the percentage of connections between neurons which are different from 0. The aim of this experiment was to obtain the reservoir connectivity dependence on reservoir size (N). The reservoir size was changed from 5 to 200 neurons using a logarithmic scale. Reservoir connectivity was changed from 1 to 100% in the same way. The connectivity of 1% indicates that only 1/100 of connections between reservoir neurons are not equal to zero. A reservoir connectivity of 100% represents that all connections are active. The experiment confirmed that it is not necessary to have all the connections active trying to ensure the same effectiveness. The results show that if the number of neurons is $N \geq 50$ the algorithm is equally effective even at 10% of all connections (Figure 2.10b). Summarizing, it is sufficient to have only one tenth of non-zero connections in order to achieve a similar efficiency for the ESN.

The purpose of the fourth experiment was to determine the optimal coefficient α for the EWMA filter in the reservoir neurons. The EWMA filter works in low-pass filter mode when $\alpha < 1$. Coefficient α was changed from 0 to 1 in 0.1 steps. A ESN with 70 hidden neurons and a reservoir connectivity of 10% was generated. Other parameters were set according to the values

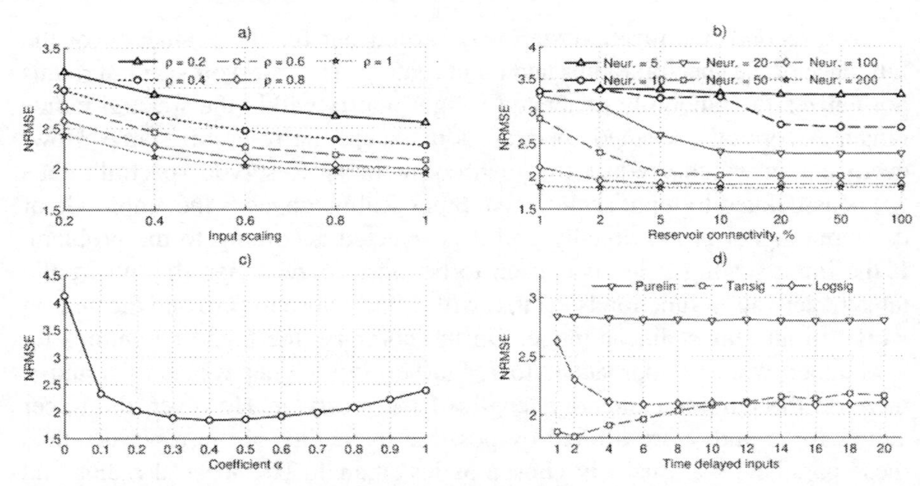

Figure 2.10 ESN parameter initialization: (a) the selection of spectral radius and input scaling, (b) reservoir connectivity influence on ESN performance, (c) EWMA filter coefficient selection, (d) influence of the reservoir activation function on ESN efficiency.

obtained in previous experiments. It was found that the ESN is most effective when $0.3 \leq \alpha \leq 0.5$. The efficiency of the algorithm decreases rapidly when $\alpha \rightarrow 1$. In the case of a standard ESN, when $\alpha = 1$, the efficiency obtained for the algorithm was similar to the conventional RLS algorithm ($NRMSE_{ESN} = 2.39$). This effect is caused by the rapidly changing internal dynamics. Therefore it is important to choose a correct EWMA filter for reservoir state filtering during the processing of ECG signals. The lowest value of the NRMSE was obtained when $\alpha = 0.4$ (Figure 2.10c).

The fifth optimization step addressed the analysis of the influence of different activation functions on the efficiency of the algorithms. Linear, hyperbolic tangent and logarithmic activation functions were tried. The experiment was extended by constructing a tapped delay line for the input signal. The tapped delay line helps to increase the memory length of neural networks. The results show (Figure 2.10d) that the adaptive ESN extracts f-waves most effectively when the hyperbolic tangent activation function and a tapped delay line with two inputs are applied. However, for simplification of the algorithm only one input delay line was used in subsequent experiments.

Summing up, the ESN based nonlinear adaptive filter works most efficiently when the following parameters are chosen: forgetting factor $\lambda = 0.998$, number of neurons in the reservoir $N = 70$, reservoir connectivity $\delta = 10\%$, EWMA filter coefficient $\alpha = 0.4$, input scaling $\gamma = 1$ and spectral radius $\rho = 1$. It must be emphasized that the selected parameters are the most efficient ESN parameters for the whole ensemble of surrogate signals and are not optimal for any given individual signal.

2.5.4 The Stability Issues of Recursive Least Squares

The RLS algorithm has well known attractive advantages, for instance, fast adaptation. However, there are major shortcomings which limit its practical applications. The main shortcomings are: computational cost and possible instability. A large amount of computations is becoming a less important problem due to the rapid development of digital electronics but the instability issue still remains. Rounding errors are the main cause of the instability. The errors are summed over time and cause a rapid growth of the error. The optimal parameters which were investigated earlier were used in the proposed filters for stability testing. The sum of adaptive weights was used as the algorithm stability measure. The stability was tested for both (linear and nonlinear) filters using two test signals with a duration of 30 min. The test signals are depicted in the illustrations of Figure 2.11. Two premature ventricular contraction beats can be observed in the upper signal and a motion artifact in the lower one. The location of the first PVC beat is at 7 min and 34 s, the peak of the second PVC is at 13 min 17 s and the artifact in the reference signal appears at 18 min 30 s.

We observed that the sum of adaptive weights of the classical RLS grows uncontrollably after ~1–2 minutes and thus the error becomes extremely large (Figures 2.11c and d). Fortunately, using the pre-whitened RLS algorithm [42] the stability problems were solved and the modified algorithm maintained stability for the whole length of the 30 minute signal (Figures 2.11e and f). A slight increase of the weights was observed in the time instant when the first PVC beat occurred. It is worth noting that the sum of adaptive weights is less than 5 for the RLS based linear filter. It means that the algorithm operates with small numbers (about one hundredth part of a unit). However in the case of the nonlinear adaptive ESN filter, the sum of the weights is in range of 4000–12000 (see Figure 2.11f). Thus each weight of the ESN is in the order of tens and hundreds when the number of adaptive weights of the ESN is 70.

The sum of weights of ESN based nonlinear filter can be reduced by applying white noise to the reservoir states, however the quality of ESN output signal would be decreased due to increased noise component.

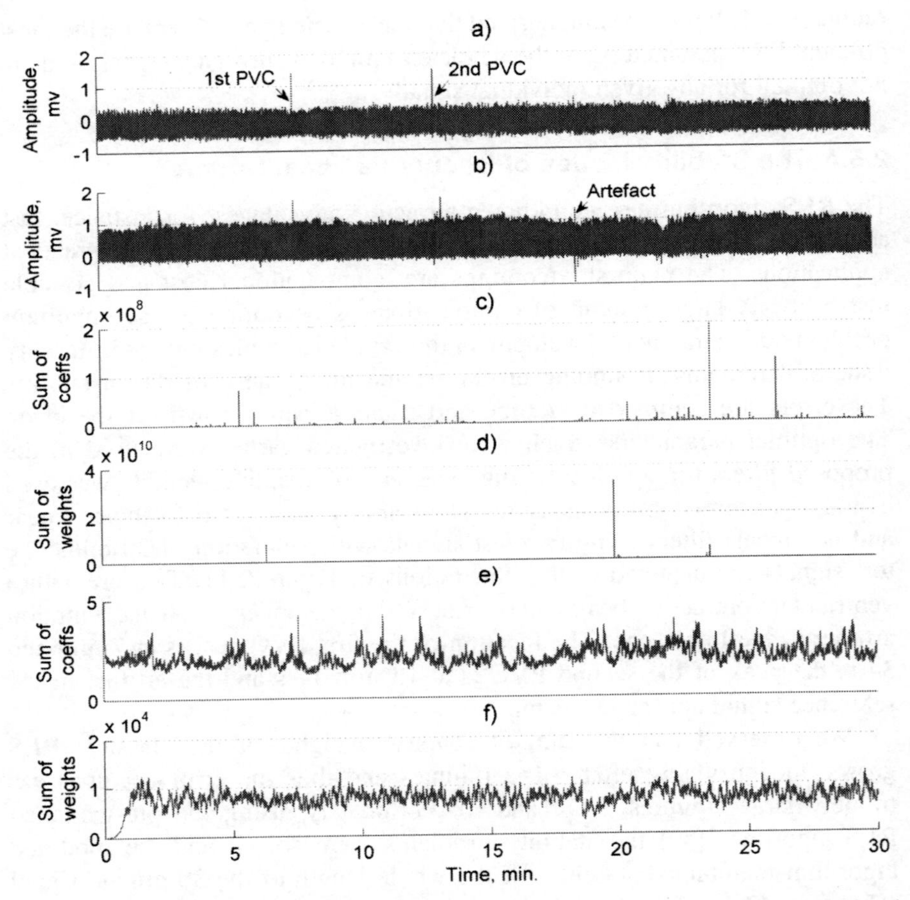

Figure 2.11 RLS stability testing: (a) target signal (with f-waves), (b) reference signal (without f-waves), (c) sum of weights using classical RLS in the linear filter, (d) sum of weights using classical RLS in the nonlinear filter, (e) sum of weights using pre-whitened RLS in the linear filter, (f) sum of weights using pre-whitened RLS in the nonlinear filter.

2.5.5 The Results of f-Waves Extraction

As was expected, the NRMSE of the proposed algorithms decreases for growing RMS values of the f-waves (Figure 2.12a). Interestingly, the growth of f-waves amplitude reduces the difference between the f-waves extraction algorithms. We have found that the ESN based adaptive nonlinear algorithm is more precise than the RLS based linear filter in estimating the morphology of the target signal. Therefore we suppose that the reduced differences

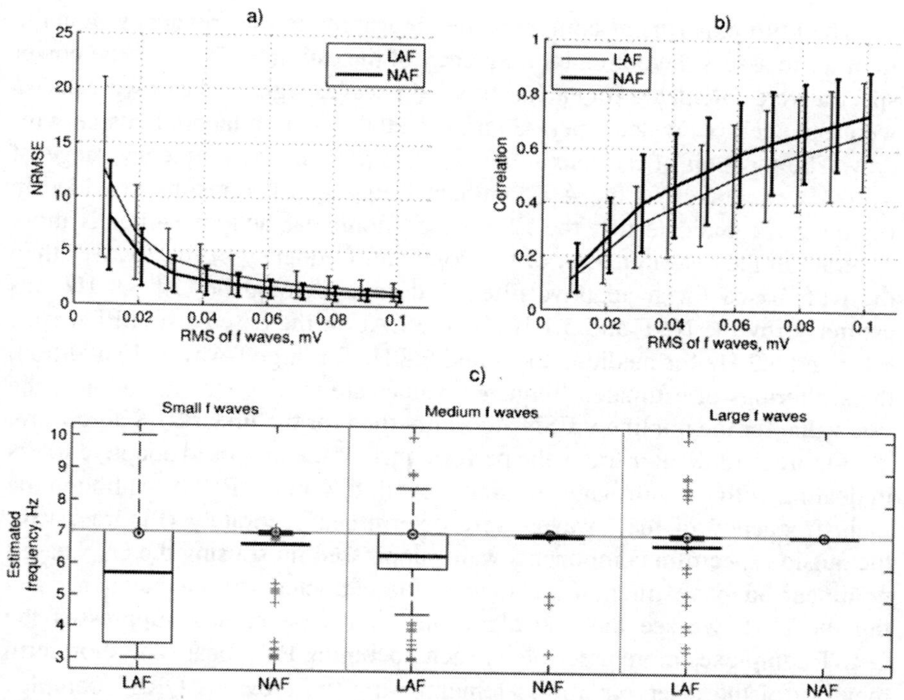

Figure 2.12 Summary of simulation results: NRMSE (a) and correlation coefficient (b) as a function of the f-waves signal strength (RMS value) and algorithms under comparison (variables are expressed as the mean +/- one standard deviation); the accuracy of the estimated dominant frequency of the f-waves (c) (the dominant frequency is depicted by a line, the median by a point, the edges of the box are determined by the 25th and 75th percentiles, the separate crosses are considered as outliers).

between the investigated algorithms were caused by the decreased influence of the proposed algorithms' accuracy. Figure 2.12b depicts the variation of averaged correlation coefficient versus the RMS of synthetic f-waves. The dependency of the correlation coefficient on f-waves signal strengths (RMS) reveals a systematic superiority of the ESN based nonlinear filter versus the linear one. Only at very low amplitudes of f-waves both methods are similar in the sense of achieved correlation coefficients. Thus it can be concluded that the ESN based nonlinear filter is more accurate than the linear filter in extracting signal amplitudes (lower bias or NRMSE) and morphology (higher correlation coefficients) in a wide range of signal strengths.

The third experiment compares the algorithms in the frequency domain. In order to assess the dominant frequency of the estimated f-waves the power spectra were calculated for each estimated f-waves signal. The power spectra were obtained by Welch's periodogram method using a tapered cosine window. The location of the maximum spectral peak in the frequency range of 3–10 Hz was taken as the AF dominant frequency. The results provided in Figure 2.12c indicate that the ESN based nonlinear adaptive filter is more accurate in the determination of the dominant frequency of the f-waves than the RLS based linear adaptive filter. A dominant frequency of 6.9 Hz was estimated by the NAF and 5.1 Hz by the LAF in the case of small f-waves, 7 Hz and 6.2 Hz for medium and 7 and 6.8 Hz for large f-waves. In addition, the scatterings of estimated frequency values are more concentrated near the true value for the nonlinear ESN algorithm than for the linear RLS algorithm.

Figure 2.13 demonstrates the performance of the proposed adaptive filters in dealing with the surrogate AF signal exhibiting many PVCs. Although the main frequency of the f-waves were determined accurately (Figure 2.13b) the outside spectrum components were suppressed most using the ESN based nonlinear adaptive filter. If we look at the extracted f-waves sequence using the NAF we see that the algorithm adapts better and suppresses the QRST complexes more precisely at each upcoming PVC beat. The short term memory of the reservoir allows remembering the previous QRST complex morphology and the algorithm reacts to the upcoming beat. Such effect was not observed when studying the LAF algorithm. Bearing in mind that PVC beats is a common phenomenon in elder patients this adaptation property of the NAF is very important.

The quantitative evaluation of the algorithms' efficiency is impossible using real ECG signals because of unknown atrial activity components and thus only qualitative comparisons are possible. Figure 2.14 demonstrates the performance of adaptive ESN in dealing with the real AF signals with large (upper illustrations) and small (lower illustrations) amplitude f-waves. The *learning set/n02* and *learning set/n03* records were selected from the AF Termination Challenge Database [46]. A spectral maximum in the range of 3–10 Hz was chosen as the fibrillation frequency of the estimated f-waves signal. A dominant frequency of 5.65 Hz was obtained using both linear and nonlinear filters for an AF signal with large f-waves (Figure 2.14 top illustrations). However, the dominant frequency estimate was different (6.16 Hz for the nonlinear and 6.92 Hz for the linear filter) in the case of smaller amplitude f-waves (Figure 2.14 bottom illustrations). The higher estimated frequency

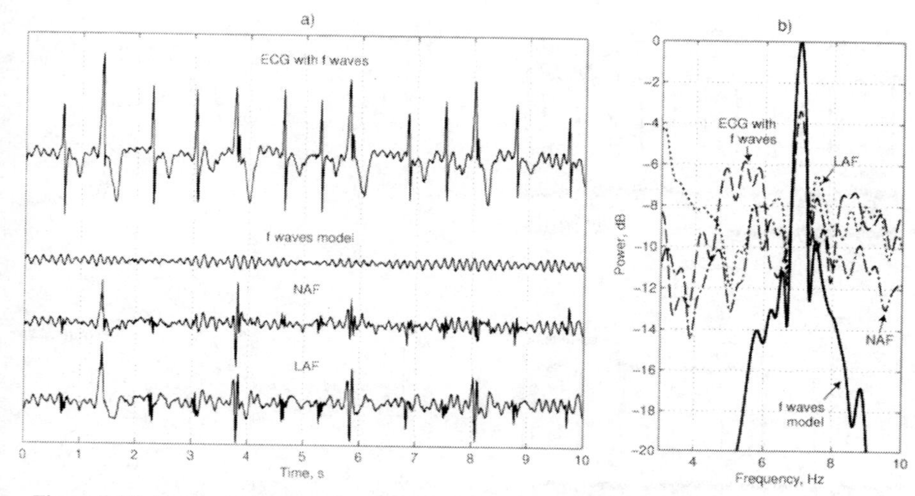

Figure 2.13 Extracted f-waves from ECG with PVC beats in time and frequency domains.

using the LAF can be influenced by QRST complex residuals in the f-waves signal.

If we look at the extracted f-waves sequences we see that the QRST complexes are mostly suppressed using the adaptive ESN based nonlinear filter, especially in the case of small amplitude f-waves.

2.6 Discussion and Conclusions

Various approaches for atrial activity extraction from ECG were recently proposed. It was mentioned that the development of such methods is very important for the identification of potential candidates for cardioversion or interventional procedures [48]. We think that real time AF monitoring is another important direction where information extracted from f-waves could be used for e.g. drug selection and dosage estimation. In this study we developed and evaluated a new algorithm for f-waves extraction.

The proposed algorithm is more complex and requires more computations than other algorithm under comparison. The pre-whitened RLS algorithm requires $4N^2 + O(N)$ multiplication operations per iteration for its implementation [42]. The proposed ESN based adaptive algorithm needs additional mathematical operations to update the reservoir states. However, despite this a 2.6 GHz dual core processor takes only about 200 μs to process a sample. Nowadays computational complexity is becoming a less important problem

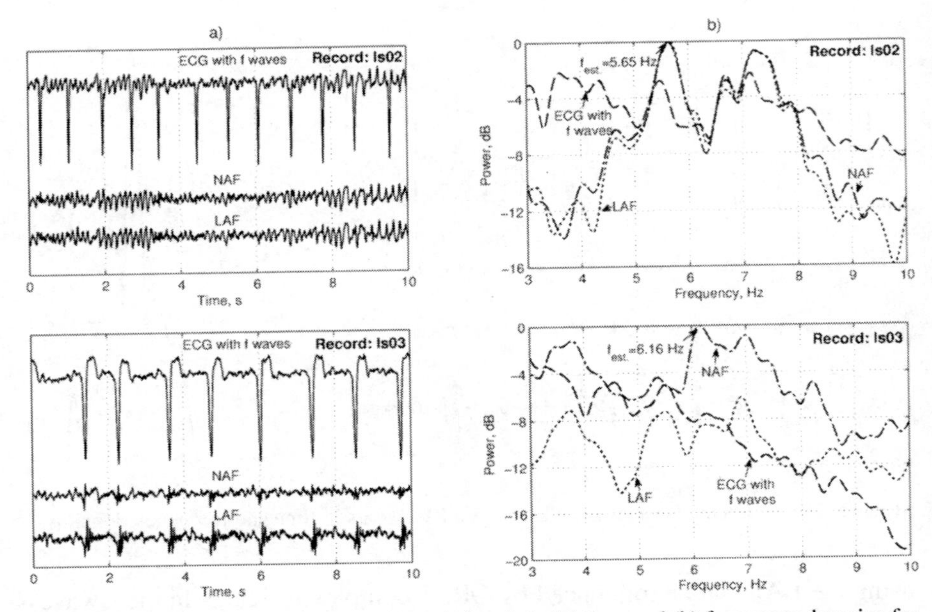

Figure 2.14 Extracted f-waves from the real signals in (a) time and (b) frequency domains for the corresponding ECG records.

because of the rapidly developing electronics. Furthermore the problem could be shaded by optimizing the reservoir via the elimination of inefficient neurons [49]. Although the algorithm performs better when larger reservoirs are used we have used a relatively small reservoir consisting of 70 neurons after considering the drastically increasing computational demands with increasing size of the reservoir.

The experiments with surrogate signals revealed that the nonlinear filter extracts f-waves more accurately than the conventional pre-whitened RLS based linear algorithm. The maximum benefit of an adaptive ESN was obtained in the case of small f-waves amplitudes. Similar performance for the compared algorithms was observed in the case of larger f-wave amplitudes. The dominating frequency of f-waves was identified almost perfectly using a nonlinear filter, even for small AF signal amplitudes.

The proposed method requires a reference signal; however this is becoming a less important problem because of the rapidly developing intelligent textiles with integrated conductive electrodes. The reference signal could be obtained by integrating an electrode in a selected place on the textile fabric where the atrial activity component is restrained. Furthermore we have demonstrated the superiority of the proposed method for AF episodes when fast changes occur (PVCs) in the ECG morphology.

Summing up, the results of the present study demonstrate that the adaptive ESN based nonlinear filter has the potential to be implemented in digital signal processors for long term AF monitoring.

References

[1] D. Lloyd-Jones, R. Adams, M. Carnethon et al. Heart disease and stroke statistics 2009 update. A report from the American Heart Association statistics committee and stroke statistics subcommittee. *Circulation*, 119(3):e2, 2009.

[2] Y. Miyasaka, M.E. Barnes, B.J. Gersh et al. Secular trends in incidence of atrial fibrillation in Olmsted County, Minnesota, 1980 to 2000, and implications on the projections for future prevalence. *Circulation*, 114:119–125, 2006.

[3] P. Koskinen and M. Kupari. Alcohol and cardiac arrhythmias. *British Medical Journal*, 304:1394–1395, 1992.

[4] V. Fuster, et al. ACC/AHA/ESC 2006 guidelines for the management of patients with atrial fibrillation – Executive summary. *Journal of the American College of Cardiology*, 48(4):854–906, 2006.

[5] A. Ringborg, et al. Costs of atrial fibrillation in five European countries: results from the Euro Heart Survey on atrial fibrillation. *Europace*, 10:403–411, 2008.

[6] M.A. Allessie, P.A. Boyden, A.J. Camm, et al. Pathophysiology and prevention of atrial fibrillation. *Circulation*, 103:769–777, 2001.

[7] L. Sornmo, M. Stridh, D. Husser, et al. Analysis of atrial fibrillation: From electrocardiogram signal processing to clinical management. *Philosophical Transactions of the Royal Society – Series A*, 367(1887):235–253, 2009.

[8] L. Mainardi, L. Sornmo, and S. Cerutti. *Understanding Atrial Fibrillation: The Signal Processing Contribution*. Morgan and Claypool Publishers, 242 pp., 2008.

[9] A. Bollmann, N. Kanuru, K. McTeague, et al. Frequency analysis of human atrial fibrillation using the surface electrocardiogram and its response to ibutilide. *American Journal of Cardiology*, 81:1439–1445, 1998.

[10] S. Petrutiu, A.V. Sahakian, W.G. Fisher, and S. Swiryn. Manifestation of left atrial events in the surface electrocardiogram during atrial fibrillation. *Computers in Cardiology*, pp. 1–4, 2006.

[11] Atrial fibrillation monitor, http://www.afibalert.com/, accessed on 1 February 2012.

[12] M. Stridh and L. Sornmo. Spatiotemporal QRST cancellation techniques for analysis of atrial fibrillation. *IEEE Transactions on Biomedical Engineering*, 48(1):105–111, 2001.

[13] Q. Xi, A.V. Sahakian, and S. Swiryn. The effect of QRS cancellation on atrial fibrillatory wave signal characteristics in the surface electrocardiogram. *Journal of Electrocardiology*, 36(3):243–249, 2003.

[14] M. Lemay, J.M. Vesin, A. van Oosterom et al. Cancellation of ventricular activity in the ECG: evaluation of novel and existing methods. *IEEE Transactions on Biomedical Engineering*, 54(3):542–546, 2007.

[15] F. Castells, C. Mora, J.J. Rieta et al. Estimation of atrial fibrillatory wave from single-lead atrial fibrillation electrocardiograms using principal component analysis concepts. *Medical and Biological Engineering and Computing*, 43(5):557–560, 2005.

[16] P. Langley, J.P. Bourke, and A. Murray. Frequency analysis of atrial fibrillation. *Computers in Cardiology* , 27:65–68, 2000.

[17] J.J. Rieta, F. Castells, C. Sanchez, et al. Atrial activity extraction for atrial fibrillation analysis using blind source separation. *IEEE Transactions on Biomedical Engineering*, 51(7):1176–1186, 2004.

[18] F. Castells, J.J. Rieta, J. Millet, and V. Zarzoso. Spatiotemporal blind source separation approach to atrial activity estimation in atrial tachyarrhythmias. *IEEE Transactions on Biomedical Engineering*, 52(2):258–267, 2005.

[19] R. Llinares and J. Igual. Application of constrained independent component analysis algorithms in electrocardiogram arrhythmias. *Artificial Intelligence in Medicine*, 47(2):121–133, 2009.

[20] V. Zarzoso and P. Comon. Automated extraction of atrial fibrillation activity from the surface ECG using independent component analysis in the frequency domain. In *IFMBE Proceedings 25/IV*, pp. 395–398, 2009.

[21] J. Igual, R. Llinares, M.S. Guillem, and J. Millet. Optimal localization of leads in atrial fibrillation episodes. In *Proceedings of IEEE International Conference on Acoustics, Speech and Signal Processing*, Vol. 2, pp. II, 2006.

[22] R. Llinares, J. Igual, and J. Miro-Borras. A fixed point algorithm for extracting the atrial activity in the frequency domain. *Computers in Biology and Medicine*, 40:943–949, 2010.

[23] A. Martiñez, R. Alcaraz, and J.J. Rieta. Ectopic beats canceler for improved atrial activity extraction from holter recordings of Atrial Fibrillation. *Computers in Cardiology*, 1015–1018, 2010.

[24] N.V. Thakor and Y.-S Zhu. Applications of adaptive filtering to ECG analysis: noise cancellation and arrhythmia detection. *IEEE Transactions on Biomedical Engineering*, 38(8):785–794, 1991.

[25] S.M.M. Martens, M. Mischi, S.G. Oei, and J.W.M. Bergmans. An improved adaptive power line interference canceller for electrocardiography. *IEEE Transactions on Biomedical Engineering*, 53(11):2220–2231, 2006.

[26] V.K. Pandey. Adaptive filtering for baseline wander removal in ECG. In *Information Technology and Applications in Biomedicine*, pp. 1–4, 2010.

[27] M.A.D. Raya and L.G. Sison. Adaptive noise cancelling of motion artifact in stress ECG signals using accelerometer. *Engineering in Medicine and Biology*, 2:1756–1757, 2002.

[28] N. Martini, M. Milanesi, N. Vanello, et al. A real-time adaptive filtering approach to motion artefacts removal from ECG signals. *International Journal of Biomedical Engineering and Technology*, 3(3/4):233–245, 2010.

[29] Y. Bensadoun, E. Novakov, and K. Raoof. Multidimensional adaptive method for cancelling EMG signal from the ECG signal. *Engineering in Medicine and Biology Society*, 1:173–174, 1995.

[30] G. Lu, J.-S. Brittain, P. Holland, et al. Removing ECG noise from surface EMG signals using adaptive filtering. *Neuroscience Letters*, 462(1):14–19, 2009.

[31] M. Cesarelli, P. Bifulco, and M. Bracale. An algorithm for the detection of the atrial fibrillation from the surface ECG for an of home-care evaluation of the implanted at-

rial defibrillators. In *Proceedings of the Mediterranean Conference on Medical and Biological Engineering and Computing*, p. 5, 1998.

[32] G. Camps-Valls, M. Martnez-Sober, E. Soria-Olivas, et al. Foetal ECG recovery using dynamic neural networks. *Artificial Intelligence in Medicine*, 31(3):197–209, 2004.

[33] C. Vasquez, A. Hernandez, F. Mora, et al. Atrial activity enhancement by Wiener filtering using an artificial neural network. *IEEE Transactions on Biomedical Engineering*, 48(8):940–944, 2001.

[34] K. Gurney. Neural networks for perceptual processing: from simulation tools to theories. *Philosophical Transactions of the Royal Society B*, 362(1479):339–353, 2007.

[35] H. Jaeger. The "echo state" approach to analysing and training recurrent neural networks. GMD Report 148, pp. 43, 2001.

[36] W. Maass, T. Natschlager, and H. Markram. Real-time computing without stable states: a new framework for neural computation based on perturbations . *Neural Computation*, 14(11):2531–2560, 2002.

[37] Luko?evi?ius, M. and H. Jaeger. Reservoir computing approaches to recurrent neural network training. *Computer Science Review*, 3 Iss. 3, 2009, pp. 127-149.

[38] M. Lukoševičius. Reservoir computing and self-organized neural hierarchies. Ph.D. Dissertation, Jacobs University Bremen, Bremen, Germany, 2011.

[39] J. Elman. Distributed representations, simple recurrent networks, and grammatical structure. *Machine Learning*, 7(2/3):195–224, 1991.

[40] D.B. Heras, F. Arguello, J.L. Gomez, et al. Towards real-time hyperspectral image processing, a GP-GPU implementation of target identification. In *Proceedings of IEEE 6th International Conference on Intelligent Data Acquisition and Advanced Computing Systems (IDAACS)*, Vol. 1, pp. 316–321, 2011.

[41] T.F. Oostendorp and A. van Oosterom. ECGSIM: An interactive tool for the study of the relation between the electric activity of the heart and the QRST waveforms at the body surface. In *Proceedings of the International Conference of IEEE Engineering in Medicine and Biology Society*, Vol. 5, pp. 3559–3562, 2004.

[42] S.C. Douglas. Numerically-robust O(N2) RLS algorithms using least- squares prewhitening. *Proceedings of IEEE International Conference on Acoustics, Speech, and Signal Processing*, Vol. 1, pp. 412–415, 2000.

[43] L. Sornmo and P. Laguna. *Bioelectrical Signal Processing in Cardiac and Neurological Applications*. Elsevier (Academic Press), Amsterdam, 2005.

[44] G. Holzmann and H. Hauser. Echo state networks with filter neuron and delay&sum readout. *Neural Networks*, 23(2):244–256, 2010.

[45] U. Siewart and W. Wustlich. Echo-state networks with bandpass neurons: towards generic time-scale-independent reservoir structures. Internal Status Report, PLANET Intelligent Systems GmbH, pp. 24, 2007.

[46] A.L. Goldberger, et al. PhysioBank, PhysioToolkit, and PhysioNet: Components of a new research resource for complex physiologic signals. *Circulation*, 101:e215–e220, 2000.

[47] D.L. Donoho. De-noising by soft-thresholding. *IEEE Transactions on Information Theory*, 41(3):613–627, 1995.

[48] S.A. Chen and C.T. Tai. Is analysis of fibrillatory waves useful for treatment of atrial fibrillation? *Journal of Cardiovascular Electrophysiology*, 15:918–919, 2004.

[49] X. Dutoit, V. Brussel, and M. Nuttin. A first attempt of reservoir pruning for classification problems. In *Proceedings of European Symposium on Artificial Neural Networks*, pp. 507-512, 2007.

3

Optimal Quality-Aware Predictor-Based Adaptation of Multimedia Messages

Steven Pigeon[1] and Stéphane Coulombe[2]

[1]*Department of Computer Science and Operation Research, University of Montréal, Canada; e-mail: pigeon@iro.umontreal.ca*
[2]*Department of Software and Information Technologies Engineering, École de Technologie Supérieure, Université du Québec, Canada*

Abstract

The Multimedia Messaging Service (MMS) platform allows messages composed of various multimedia attachments to be exchanged between users of mobile devices. However, these heterogeneous devices exhibit different capabilities regarding what media types, resolution, and maximal message size they support making the adaptation of messages mandatory to ensure compatibility between sending and receiving devices. The challenge is therefore to adapt messages so that they satisfy the receiving device's constraints in a way that both maximizes the user experience and minimizes the computational cost of adaptation. Minimizing computational cost will help cope with high-volume traffic while maximizing user experience, as estimated by the perceived quality of adapted messages, will secure the service provider's user base. In this work, we propose a generic adaptation scheme based on predictors for file size and image quality resulting from transcoding parameters applied to a given image that will explicitly maximize perceived quality as estimated by the structural similarity image quality index. We will further show that our proposed method is resilient to the imprecision of the predictors and that it yields significantly better quality at a greatly reduced computational cost compared to other methods proposed in prior art.

Richard J. Duro and Fernando López Peña (Eds.), Digital Image and Signal Processing for Measurement Systems, 71–99.

Keywords: MMS, multimedia adaptation, dynamic programming, structural similarity, user experience, multimedia optimization.

3.1 Introduction

From the users' point of view, Multimedia Messaging Service, or MMS, merely provide a convenient mean of exchanging messages composed of various multimedia attachments such as audio, still image, and video, between mobile terminal users. If, for a user, sending a message is a simple matter, as it suffices him to assemble attachments and select recipients, the service provider's task is significantly more complex as he must ensure not only fast delivery of messages but also interoperability between his users' heterogeneous mobile devices [19].

Ensuring interoperability means that each attachment (and the message as a whole) is potentially adapted so that it satisfies all of the receiving terminal's constraints. Such constraints specify the maximum message size the device can receive and decode and other limitations such as the maximal image resolution and the types of media the device can interpret correctly. For image attachments (for the most part JPEG images taken from the mobile device's camera), adaptation will consist in adjusting resolution, compression parameters, and sometimes even file format, so that the resulting image is compatible with the receiving terminal's capabilities. Unfortunately, adaptation cannot be devolved to the receiving terminal because the standard precludes terminals from receiving larger messages than what they can handle as specified by their limitations [8, 9, 35]. Having the sending terminal adapt the message is also quite impractical as it supposes it has a description of the receiving terminal, which in practice it does not have, and in the case of multiple recipients, adaptation would be too intensive for a such device where battery life and computing power are limited. Therefore, adaptation *must* be performed server-side.

The task of adapting a single message with a few image attachments does not seem daunting given the relatively small message sizes allowed by the MMS standard (which we discuss in more details in the next section), especially when one considers a server-class computer to perform the adaptation. Even adapting a JPEG image to fit given constraints such as resolution and file size seems trivial. However, adapting an image to specific constraints is an intensive process, as estimating efficiently and accurately the file size resulting from a given transformation remains a challenge [20, 21, 39, 40]. Adapting a whole message to fit the receiving device's constraints so that it maximizes overall perceived quality remains a complex operation [41] that we will discuss at length in this chapter. Finally, one must realize that a service provider does not have to adapt the occasional message, but to adapt potentially very large numbers of messages in a timely fashion. One therefore concludes that adaptation must be performed rapidly in order to cope with high-volume MMS traffic, and that it must be done in a way that yields a superior user experience, in particular yielding images with a high perceived quality.

Of course, one may be tempted to resort to trivial adaptation strategies, for example indiscriminately shrinking images to thumbnails and discarding troublesome media such as video, but this is clearly unacceptable as users expect not only the messages to be delivered in a timely fashion, but also to be adapted in a way that does not incur objectionable degradation. Other simple strategies, such as forwarding the messages to an e-mail account [10, 30] or sending an URL by SMS where the users can fetch the message via the device's browser [18], will also lessen the user experience by failing to provide the integration of services one expects from MMS capable devices. Therefore, the provider must perform the best possible adaptation of messages in order to maximize the user experience of his customers, preferably doing so at the lowest possible computational cost.

Minimizing computational cost is a necessity beyond merely ensuring diligent delivery of messages between users. In 2009, MMS traffic grew 48% in the United States alone, where 34.5 billion multimedia messages were exchanged [27]. With a shift towards green technologies and energy-efficient data centers, it seems unrealistic to hope coping with such a traffic with an annual growth of nearly 50% using naïve adaptation methods and merely adding new servers (with all the complications it implies) to meet demand. The solution to mitigate this problem is to use better algorithms that perform message adaptation in a computationally efficient way without sacrificing user experience.

It is in this context that we propose a novel and computationally efficient method of multimedia message adaptation that explicitly maximizes user experience under the constraints of the receiving terminal. In this chapter, we will expose our solution where we propose to not only to maximize explicitly resulting quality of adapted messages using an efficient algorithm, but to speed-up optimization significantly by the use of *predictors*. Predictors are fast algorithms that predict the resulting file size and perceived quality of an image on which were applied *transcoding parameters*, parameters that describes how to transform the image – for example by specifying a scaling factor to change resolution and other parameters to affect the level of compression.

This chapter is structured as follows. In the next section, Section 3.2, we introduce the reader to the environment of multimedia messaging service and its constraints. In Section 3.3, we discuss the various approaches proposed in literature to address the problem of MMS adaptation as well as adaptation of media in other contexts. We also discuss the nature of quality measure and their essential role in adaptation. We expose the details of our proposed solution in Section 3.4. Section 3.5 describes our simulations and presents the results obtained. We discuss and interpret the results in Section 3.6. Finally, the chapter closes with Section 3.7, where we summarize our contributions and present perspectives for future work.

3.2 The MMS Environment

In this section, we will cursorily introduce MMS messages and the MMS operating environment. We will discuss the challenges it poses for multimedia messages adaptation.

An MMS message is essentially a MIME e-mail message [13, 17, 23, 42] as defined by the MMS encapsulation specification [34] where one can attach a number of media files (with acceptable formats defined in [8, 9]). The presentation of the media elements can be simple attachments, but they can be structured using HTML or XHTML [6, 15], or dynamically arranged in a slide-show-type presentation using Synchronized Multimedia Integration Language, or SMIL [4, 14, 28]. In other words, the MMS standard provides a number of means to ensure not only delivery of media attachment, but also that they are presented adequately.

Not only MMS-capable devices are heterogeneous (devices can be any one of the tens of thousands of different models of terminals or even a desktop computer), the MMS network itself is composed of a great number of different technologies. The mobile devices may use any of the many wireless technologies (GSM [2], GPRS, TDMA, CDMA, or 3G [1]) but fortunately, network specificities are abstracted through the Wireless Application Protocol (WAP) [33] that provides transport and through the Wireless Session Protocol that provides session control (for an introduction to the transaction-based protocol ensuring the delivery of the messages from the mobile device to the service provider's network servers, the reader is directed to [19]).

When an MMS message is sent, it is first processed by the service provider's server that determines how to route the message to its recipient. The

Table 3.1 Simplified MMS profiles.

Profile Name	Maximum Resolution	Maximum Message Size	Image Media
Image Basic	160 × 120	30 KB	Baseline JPEG, GIF87a/89a, WBMP
Image Rich	640 × 480	100 KB	Baseline JPEG, GIF87a/89a, WBMP
Megapixel	1600 × 1200	300 KB	Baseline JPEG, GIF87a/89a, WBMP
Content Rich	1600 × 1200	300 KB	Baseline JPEG, GIF87a/89a, WBMP, SVG

server first determines if the recipient is local or if it belongs to another service provider. In the latter case, the message is forwarded to the recipient's service provider using an MM4 network, the inter-working between MMS service centers. If the recipient is local, the service provider finds himself with basically two cases to consider. The first, and usual, case is when an MMS is sent as an MMS to a recipient on a mobile device. The second case to consider is when an e-mail is sent to an MMS-only address. The third and fourth cases, (sending an MMS to an e-mail address and sending MMS-like e-mails), are not very interesting as, unlike the two first cases, they do not need significant adaptation: as an MMS is essentially a MIME e-mail, it can be sent as-is and left for the e-mail client to display. The two first cases, however, may require adaptation as sending and receiving device may not have the same capabilities.

Device capabilities are standardized via *capability profiles*, or profiles for short. Table 3.1 present a simplified view of profiles, where constraints such as maximum image resolution, maximum message size, and supported image media format are listed. The profiles also specify the type of character encodings, presentations, and encapsulation each profile is to support, as well audio, speech, and video formats for attachments [30], but we will ignore these aspects in this work.

Therefore, when an MMS server receives an MMS message, it first *characterizes* it, that is, assesses its contents by examining presentation and individual attachments, and then determines if adaptation is needed given the receiving terminal's capabilities. To do so, the service provider may maintain a database associating recipients addresses to their corresponding devices. The device characterizations are maintained in a database that describes the devices, beyond MMS capabilities, containing information about screen resolution and other features, information that could be used, in principle, for finer adaptation of the messages. One such database, the Wireless Universal Resource File (WURFL) maintained by Luca Passani, contains approximately 11300 devices [37].

Using the database, the provider is capable of devising the best strategy for message adaptation, whether adaptation or mere pass-through. In some cases, it may be possible to simply relay the message as-is because its contents are already compatible (as determined by characterization) with the receiving terminal (for example, a message using Image Basic sent to a Megapixel-capable device). In any other case, adaptation must be performed.

3.3 MMS Adaptation

The goal of MMS message adaptation is to make a message compatible to the recipient's receiving device. To do so, we may need to *transcode* images, that is to change their resolutions, compression parameters or even compression format altogether. However, changing compression format may not be wanted in general. First because most of the images are "natural images" in JPEG format (as taken from camera phones and JPEG is generally efficient in this case), and second because the remaining images types will likely need to remain in their original formats to retain their specificities – animations in GIF89a format, for example, will need to be preserved. For this reason, and ultimately without loss of generality as we will discuss in section 3.6, we will consider here only the case of all-JPEG message adaptation.

In Section 3.3.1, we discuss prior art and other techniques relevant for the problem of MMS adaptation. In Section 3.3.2 we discuss the problem of visual quality assessment and motivate our choice of quality metric, the structural similarity of Wang et al. In Section 3.3.3 we establish the notation we will be using in the remainder of this chapter, as well as present a formalization of the problem we address.

3.3.1 Prior Art

A priori, one would think that adapting a Baseline JPEG image [7,38] against a given maximal file size is an easy task. One could think it suffices to adjust compression parameters until the desired file size is obtained, but for JPEG, especially at very high compression, that means introducing a host of conspicuous artifacts, such as color bleeding and blocking. Blocking occur when the compression is too aggressive and that DCT patches boundaries cease to match their neighbors' boundaries, and the discontinuities are visually displeasing [29, 52]. Color bleeding is a side effect of quantization in the usually sub-sampled chroma planes [56]. To avoid these artifacts and yet meet the target file size, one must rather consider a mixed strategy where resolution and compression are adjusted jointly, that is, also consider solutions where the resolution of the image is lowered but the image is compressed less aggressively, thus avoiding compression-related artifacts while reducing the overall file size significantly. Therefore, the best trade-off between resolution change and compression settings is to be found (while meeting the constraint of the desired file size), where "best" is measured by an objective quality measure highly correlated with the quality perceived by a human observer – we discuss the problem of the quality measure in Section refsubsection-image-quality-assessment.

Some solutions were proposed to recompress and/or adjust resolution in the (partially) compressed domain, where operations are performed on the DCT coefficients resulting from a partial decompression. In Ridge's method [44], the file size resulting from a new compression parameter is estimated by computing the code-length of the DCT coefficients after re-quantization. The compression parameter is adjusted (using a binary search) until the parameter yielding the file size closest to, but not exceeding, the target file size is found. This method is more efficient than completely decompressing and recompressing the image six or seven times (as the main parameter controlling compression, known as the *quality factor*, can vary from 1 to 100, according to the Independent JPEG Group semantics [5], and $\log_2 100 \approx 7$) but it is still very intensive; it is unclear what are the speed-ups one can expect from such a method. Furthermore, Ridge's method of bit-rate adaptation does not consider the case where resolution must be jointly optimized – although he does propose a (partially) compressed-domain resizing method.

Fast algorithms for resolution changes in the (partially) compressed domain were proposed [22, 32, 44]. These methods manipulate the DCT coefficients in order to merge four adjacent 8×8 tiles into a single 8×8 tile, effectively reducing the resolution by a factor of two in each direction. The operations necessary for the proposed power-of-two reduction algorithms are rather complex, and it is unclear how they compare, speed-wise, with a well-implemented classical pixel-domain filter. Additionally, these methods do not generalize well to arbitrary resolution changes, that is, scaling an image by a factor of, say, 0.9 (thus yielding an image with a resolution 90% of the original in both directions). To summarize, these algorithms are complex from a conceptual point of view, have unclear speed-up benefits (as speed-ups are reported only relative to the DCT operations, neglecting the remaining operations such as entropy (de)coding, which are not a negligible part of the operation), and are limited to reductions by a power of two.

Simultaneously adjusting scaling and quality factor in order to maximize perceived quality under the constraint of a maximal file size is therefore an expensive process if one proceeds by successive transcodings, even when using an efficient parameter-search method. As we will discuss in Section 3.4, the parameter search is further complicated by the fact that the file sizes of each attachment in a message are added together for the total message size; and therefore parameters must be optimized jointly across all images, leading to a potentially very large number of combination examined. In this case, it is impractical to proceed to optimization by performing a very great number of actual transcodings. To speed up optimization, we propose to avoid tentative transcodings by the use of predictors [20, 21, 39, 40] for the file size and quality resulting from applying a scaling factor and a quality factor to an image. We discuss predictors in Section 3.4.2.

Image and media adaptation was, of course, considered before in other contexts. For example, the problem was considered for mobile browsing where clients with varying reception bandwidths access media on the web, and where media is adapted not to satisfy the devices' display constraints, but to maximize quality of service as defined by download time [24, 49]. In these settings, resolution and image quality are sacrificed in order to provide a fast, responsive browsing experience. In a sense, these solutions address only a part of the problem we consider, that is, the part where we determine the largest possible file size to meet a constraint of download time given a device's bandwidth, but without consideration for the visual quality of the images.

Adapting media with explicit maximization of quality was proposed in Mohan et al. [31, 48]. They propose to maximize "content value" (a measure inversely proportional to distortion) rather directly visual quality, and the problem is formulated as a rate-distortion optimization problem. After observing that this particular optimization problem is difficult to solve directly if all possible transcoding parameters are considered, they adapt Shoham's et al. [47] quantization method to their framework, basically using a quantization of the parameter combination space they examine; that is, they will consider only a rather small number of precomputed profiles for optimization. In their paper, they present six profiles (from full resolution 24 bits image to "alt-text" description of the image) each corresponding to a generic device class such as a desktop computer or a PDA.

Others proposed techniques based on message understanding or region of interest extraction. In [55], Yan and Kankanhalli propose a model of MMS adaptation based on "multimedia simplification," where images are cropped around their most likely region of interest, and video clipped around their most salient moment as estimated using hints such as the loudness of its sound track. Chen et al. [16] propose an approach based on probable element relevance to simplify and adapt content of Web pages, discarding elements that are not immediately relevant to the requested topic. However, these techniques are deemed too computationally expensive to face the demands of high-volume transcoding, furthermore we may question their applicability for MMS. Indeed, it is unlikely that users would want their images cropped or video truncated, much less so if the system fails to provide optimal decisions every single time – from the *users*' point of view. Also, message simplification techniques based on relevance are probably inapplicable to an image-only MMS message containing possibly unrelated images chosen by the user; in a sense, for the user sending the message, the images are *all* related.

3.3.2 Image Quality Assessment

Accurately measuring perceived image quality – a crucial operation in media adaptation – remains problematic. The peak signal-to-noise ratio (PSNR) has been used extensively in literature, whether for sound or for image processing, but the objective measure of PSNR is not a good estimator of *perceived* quality of a transformed signal, and will leave us in the need of a better measure.

The PSNR is defined from the mean squared error (MSE) between an original signal $X=\{x_i\}_{i=1}^n$ and a reconstructed or transformed signal $\hat{X}=\{\hat{x}_i\}_{i=1}^n$. Between an original image X and a transformed image $\hat{X}$ (of width w and height h) the MSE is given by

$$\mathrm{MSE}(X, \hat{X}) = \frac{1}{wh} \sum_{i=1}^{w} \sum_{j=1}^{h} (x_{ij} - \hat{x}_{ij})^2 \tag{3.1}$$

and the PSNR by

$$\mathrm{PSNR}(X, \hat{X}) = 10 \log_{10} \frac{M^2}{\mathrm{MSE}(X, \hat{X})} = 20 \log_{10} \frac{M}{\sqrt{\mathrm{MSE}(X, \hat{X})}}, \tag{3.2}$$

where M is defined as the maximal value one x_j can take, for example 255 if $0 \leqslant x_j < 256$. While other variants define M as the maximum value found in either X or $\hat{X}$, using a fixed maximum allows to compare against a class of signals rather against a single pair. However, from the very formulation of eq. (3.1), it should be clear that the PSNR is not invariant to simple transformations that may not affect perceived image quality such as translations, biases (adding or subtracting a small constant from X), and in the specific case of images, modifying the colors by either reducing or enhancing color saturation. In fact, while PSNR will measure an image degradation, enhancing saturation and boosting contrast may result in an image *more* pleasing than the original image.

If PSNR is a poor indicator of image quality because it basically ignores the human psychovisual model, which we aught to use explicitly as it is user experience one wants to maximize in our context, we must therefore look at other image quality models. There are many models that try to model the human psychovisual response to images (see [46] for a survey and analysis), but one has to find the adequate trade-off between accuracy (for example, as measured by a ranking test with the mean opinion score [43]) and the computational cost (and other logistic considerations) of the measure. One such measure is the structural similarity (SSIM) of images [53, 54].

The structural similarity is based on the premise that images that are non-objectionable transformations of an original image lie nearby the original image on an image local luminance-contrast manifold. Taking into account the properties of the proposed structural similarity image space, the distance between an original luminance image X and a transformed luminance image $\hat{X}$ is given by

$$\mathrm{SSIM}(X, \hat{X}) = \frac{(2\mu_X \mu_{\hat{X}} + c_1)(2\sigma_{X\hat{X}} + c_2)}{(\mu_X^2 + \mu_{\hat{X}}^2 + c_1)(\sigma_X^2 + \sigma_{\hat{X}}^2 + c_2)},$$

where μ_X and $\mu_{\hat{X}}$ are the means of X and $\hat{X}$, respectively, and where σ_X^2, $\sigma_{\hat{X}}^2$ and $\sigma_{X\hat{X}}$, are the variance of X, $\hat{X}$ and the covariance of X and $\hat{X}$, respectively. The constants c_1 and c_2 are added for numerical stability. Wang et al. further propose the use of a version of SSIM defined at a point, where averages and variances are computed using a normalized 2D Gaussian weighting function on a $w \times w$ window (with w odd, and $w = 11$ in [53]). The SSIM at each of the image locations (i, j), denoted here $\mathrm{SSIM}_{ij}(X, \hat{X})$, are pooled to yield the MSSIM index:

$$\mathrm{MSSIM}(X, \hat{X}) = \frac{1}{wh} \sum_{i=1}^{w} \sum_{j=1}^{h} \mathrm{SSIM}_{ij}(X, \hat{X}) . \tag{3.3}$$

In this work, we will use the MSSIM as the estimator of perceived image quality.

The definition of MSSIM lets us measure the impact of changing compression parameters on image quality, but does not directly allow for the changes in resolution. The operation of scaling will affect the pixel-size of an image and therefore makes the direct application of measures like PSNR or SSIM impossible.

To deal with this situation, one has essentially three choices. Let X be the original image, and $\hat{X}$ the transformed image on which the scaling factor $0 < z \leqslant 1$ was applied (for example, a scaling factor of $z = 0.3$ will yield an image $\hat{X}$ with resolution of 30% of that of X). One can therefore compare X with $\hat{X}$ by scaling back $\hat{X}$ at X's original resolution; one can compare with X scaled down (but not compressed) at $\hat{X}$'s resolution; or one can scale both to an intermediary resolution, say the device's maximal image resolution. The first option assesses quality against the original, and in a sense measures the best possible reconstruction. The second option measures the best possible reconstruction at the size of $\hat{X}$, which favors thumbnailing. The last option measures the best possible reconstruction for a specific receiving device. In this work, we have opted for the first option. The last option implies that one has to train a large number of predictors (which we discuss in Section 3.4.2) to accommodate the various device screen resolutions.

3.3.3 Formalizing the Problem

In the previous sections, we discussed the general setting of the problem we consider in this work, from the structure of an MMS message to broad MMS environment, passing by prior art and a discussion on image quality assessment. We have given the reader a high-level view of the problem we address, but in this section, we will formalize the notation, the problem elements, and the problem itself.

A message $M = \{m_1, m_2, \ldots, m_n\}$ is composed of n images m_i, each with resolution $R(m_i) = (w_i, h_i)$, file size $S(m_i)$, and original quality factor $QF(m_i)$. A receiving device D, for our needs, is characterized by the maximum message size $S(D)$ and a maximum resolution $R(D) = (w_D, h_D)$, both dictated by the device profile. For example, a device D capable of Image Rich messages (see Table 3.1) would report $S(D) = 300$ KB and $R(D) = (640, 480)$.

The transcoding parameters series $T = \{t_1, t_2, \ldots, t_n\}$, with $t_i = (q_i, z_i)$, where q_i is the new quality factor and $0 < z_i \leqslant 1$ is the resolution scaling factor, describes the transformations to apply to a message M. The transcoding parameters t_i are applied to image m_i using function $\mathcal{T}(m_i, t_i)$, which yields a new image with resolution $z_i R(m_i) = (z_i w_i, z_i h_i)$ that is compressed with quality factor q_i with file size $S(\mathcal{T}(m_i, t_i))$.

Let $Q(m_i, \mathcal{T}(m_i, t_i))$ measure the quality of transcoded image $\mathcal{T}(m_i, t_i)$ relative to original image m_i. As we discussed in Section 3.3.2, whenever the resolution of the transcoded image differs from the original (that is, whenever $z \neq 1$), the transcoded image is scaled back to the original resolution for comparison.

We are therefore interested in finding the optimal transcoding parameters series T^* that maximizes an objective function $\mathcal{Q}(M, T)$ (which we define and discuss in Section 3.4.1), that is, to solve

$$T^* = \arg \max_{T \in \boldsymbol{T}(M,D)} \mathcal{Q}(M, T) \tag{3.4}$$

where $\boldsymbol{T}(M, D)$ is the set of all transcoding parameter series T that satisfies the constraints

$$\sum_{i=1}^{n} S(\mathcal{T}(m_i, t_i)) \leqslant S(D) \tag{3.5}$$

and

$$\begin{aligned} z_i \max(w_i, h_i) &\leqslant W_D\,, \\ z_i \min(w_i, h_i) &\leqslant H_D\,, \end{aligned} \tag{3.6}$$

for $i = 1, 2, \ldots, n$.

Eq. (3.4) expresses a generic objective function which we want to be indicative to perceived quality. Eq. (3.5) states that the sum of the transcoded images file sizes must be smaller or equal to the maximum message size for device D, while eqs. (3.6) express orientation-independent resolution constraints where the image can fit the device maximum resolution D in either portrait or landscape orientation. Let us note that eq. (3.5) ignores the cost of the presentation of the message which should be normally included in the optimization. To reflect the cost of the presentation layer (and of the rest of the message itself including headers), one could rewrite eq. (3.5) by replacing $S(D)$ with $S(D) - P(M, D)$, where $P(M, D)$ is the cost of the presentation of message M on device D, but we do not address the problem of adapting the presentation here.

3.4 Proposed Solution

In this section, we present the details of our proposed solution. In Section 3.4.1, we discuss the choice of the objective function for the problem of multipart messages. In Section 3.4.2, we discuss the need for predictors to speed-up optimization and we describe the predictors used in this work, the JQSP predictor. We also introduce the concept of *oracular predictors* and discuss their expected properties in relation to our proposed solution. In Section 3.4.3, we revisit constraints and objective function, proposing modification to accommodate predictors. We discuss the optimization algorithm used to maximize the objective function in Section 3.4.4. Lastly, we present and discuss comparative adaptation algorithms in Section 3.4.5.

3.4.1 Objective Function

In Section 3.3.3, we have proposed to use an objective function $\mathcal{Q}(M, T)$ that measures the quality of the message M transcoded using the transcoding parameters series T, without defining it. In Section 3.3.2 we discussed image quality metrics and we chose SSIM (or more exactly, MSSIM, as given by eq. (3.3)), as a quality metric. However, as SSIM is essentially a local correlation factor between original and distorted images, it yields values on $[-1, 1]$, but the useful range will be $[0, 1]$, where 0 already corresponds to a perfectly uncorrelated image (and -1 would correspond to an *anti-correlated* image). Constraining the quality measure $Q(m_i, \mathcal{T}(m_i, t_i))$ on $[0, 1]$ will allow us use the objective function

$$\mathcal{Q}(M, T) = \prod_{i=1}^{n} Q(m_i, \mathcal{T}(m_i, t_i)) . \tag{3.7}$$

The proposed objective function in eq. (3.7) presents only one of an infinite number of possible objective functions, but several aspects makes it especially suitable for the task considered. If maximizing eq. (3.7) is not the same as maximizing average quality of the transcoded images, the expected average of the $Q(m_i, \mathcal{T}(m_i, t_i))$ increases as eq. (3.7) increases, and the expected variance necessarily decreases [26,50,51]. While maximizing eq. (3.7) is not the same as maximizing the average quality of the transcoded image, it will still prevent choosing solutions where there is a significant difference between the best image and the worst, thus forcing balanced solutions.

3.4.2 On Predictors

However, as we mentioned in Section 3.3.1, actually performing a transcoding with given transcoding parameters to examine resulting file size and image quality is an extremely expensive process, and therefore it is impractical to maximize eq. (3.7) by performing an exponentially large number of transcodings. Rather than computing $S(\mathcal{T}(m_i, t_i))$ and $Q(m_i, \mathcal{T}(m_i, t_i))$ exactly by performing a transcoding, we will use predictors, $\widehat{S}(m_i, t_i)$ and $\widehat{Q}(m_i, t_i)$, both formulating their prediction from the characterization of the image m_i (such as its original file size $S(m_i)$, quality factor $QF(m_i)$, and resolution $R(m_i)$) and the transcoding parameters t_i.

In previous works we have proposed such predictors [20, 21, 39, 40] but in this work, we will use the predictors presented in [21], to which we will refer here as the JQSP, the JPEG Quality and Size Predictor. The salient point of the proposed predictors is that predictions are learned, rather than engineered, from a large image corpus on which was applied a large number of transcodings. The specific experimental conditions and the constitution of the corpus are detailed in section 3.5 and in [21].

However, if the file size and quality predictors from [21] are well behaved, only using these predictors would validate the predictors themselves more than our proposed solution for MMS adaptation, of which we want to show the efficiency and stability. To establish the upper-bound of obtainable quality for the proposed predictor-based algorithm, in addition to the JQSP, we will use *oracular* predictors that return the exact file size and quality resulting from applying transcoding parameters to a given image. Of course, the oracular predictors perform their pythian prognostication by actually transcoding the image and observing the exact resulting file size and quality.

The oracular predictors are especially well suited to characterize the graceful degradation – or lack thereof – of the proposed algorithm to predictor error. Since they are exact, oracular predictors can be used to simulate predictors with any error characteristics. In the experiment we performed, described further in section 3.5, we used, in addition to the JQSP and the exact oracular predictor, predictors with Gaussian relative errors of 1%, 2%, 5%, and 10%, 95% of the time. Relative error is computed as $|\hat{x} - x|/x$, for exact value x and predicted value $\hat{x}$.

3.4.3 Objective Function and Constraints, Revisited

Using the predictors, we re-write the objective function eq. (3.7) for message M and transcoding parameters series T as

$$\widehat{Q}(M, T) = \prod_{i=1}^{n} \widehat{Q}(m_i, t_i) \tag{3.8}$$

where $\widehat{Q}(m_i, t_i)$ formulates a prediction on the resulting quality of image m_i on which were applied the transcoding parameters t_i. Constraints will be modified to accommodate predictors. The size constraint, eq. (3.5), will be re-written as

$$\sum_{i=1}^{n} \widehat{S}(m_i, t_i) \leqslant S(D), \tag{3.9}$$

where $\widehat{S}(m_i, t_i)$ formulates a prediction on the resulting file size of image m_i on which were applied the transcoding parameters t_i. However, the constraints of eqs. (3.6) are left unchanged, as there is no uncertainty in the resulting image resolutions.

Then the problem becomes to find the optimal predicted transcoding parameter series

$$\widehat{T}^* = \underset{\widehat{T} \in \widehat{\boldsymbol{T}}(M,D)}{\arg\max} \; \widehat{Q}(M, \widehat{T}), \tag{3.10}$$

where $\widehat{\boldsymbol{T}}(M, D)$ is the set of all possible transcoding that (probably) satisfy the constraints of eqs. (3.9) and (3.6). The formation of $\widehat{\boldsymbol{T}}(M, D)$ is discussed in Section 3.4.4 and section 3.5.

3.4.4 Optimization Algorithm

To solve eq. (3.10) (or eq. (3.7)) efficiently, we will need an efficient algorithm, as it is impractical to test a combinatorial number of parameters to find the optimal transcoding parameter series. Fortunately, the optimization problem can be formulated as a *distribution of effort* problem [25], a classical problem in operations research with known efficient algorithms, where a limited number of resources must be distributed at a number of points in order to maximize a gain function under given constraints. Here the resources spent correspond to the file size of images, the total budget of which is determined by the maximum message size for the target device, and the gain function is the overall message quality as estimated by the objective function eq. (3.8), under the additional constraints of satisfying the receiving device resolution.

The particular form of eq. (3.7) (and eq. (3.8)) makes the problem amenable to efficient optimization algorithms [36], and in particular to dynamic programming or A^* search. In the A^* formulation, the problem of distribution of effort is usually modeled as a graph [25] where, after having reached an intermediate state s_i (where attachments up to attachment m_i were solved), the possible successor states of s_i, the $s_{i+1,j}$, if reachable (transiting from state s_i to state $s_{i+1,j}$ does not violate the constraints), are connected by edges corresponding to the possible transcoding operations of attachment m_{i+1}. Therefore, the edge between the state s_i and state $s_{i+1,j}$ is labeled by a transcoding $(q_{i+1,j}, z_{i+1,j})$. One solves for the best path in this graph, the path that maximizes the objective function while being admissible, that is, satisfying all constraints.

To minimize optimization time, one must therefore limit the number of successor states to be examined, that is, minimize the size of $\widehat{T}(M, D)$. In particular, $\widehat{T}(M, D)$ cannot contain an infinite number of elements. This implies that the set of all tuples (q_{ij}, z_{ij}) for attachment m_i considered must be quantized to a limited, ideally rather small, number of possibilities. However, it is unclear how one would prune combinations of q and z when considered jointly; and in our experiments, after the initial quantization discussed in Section 3.5, we limited ourselves to pruning the set of possible transcoding parameters for each attachment m_i by excluding all transcoding (probably) exceeding the device constraints, that is, we eliminated all transcoded with a predicted file size exceeding the maximum message size and all scalings exceeding the resolution of the device – all z such that $zR(m_i) > R(D)$.

After optimization, the algorithm yields $\widehat{T}^*$, the (probably) best transcoding to apply to message M to satisfy the constraints of device D. However, it may be that the predicted transcoding parameter series produce a message that exceeds the constraints of the receiving device. If the message exceeds the constraints of the receiving device, a scaling factor $0 \ll \alpha < 1$ is applied to the maximum message size and optimization is performed again. We rewrite

Table 3.2 Combination of resolution and quality factors forming the profiles used for algorithm "successive profiles".

Resolution	Quality Factors
640×480	90, 80, 70, 60
320×240	90, 80, 70, 60, 50
160×120	90, 80, 70, 60, 50, 40

the size constraint of eq. (3.9) at the r-th retry as

$$\sum_{i}^{n} S(T(m_i, t_i)) \leqslant \alpha^r\ S(D)\,. \tag{3.11}$$

The initial optimization, with $r = 0$, yields eq. (3.9). In our experiments, we set $\alpha = 0.9$.

3.4.5 Comparative Algorithms

To compare results from our proposed algorithm, we will use algorithms inspired from the previous literature (which is rather thin for this particular topic of MMS adaptation). The first algorithm, "successive profiles", inspired by the fixed adaptation strategy of Mohan et al. [31], will apply successively more restrictive profiles to all images until the transcoded message satisfies the receiving device constraints. For this algorithm, a profile defines both the maximum resolution of images and the quality factor with which they will be compressed. For example, a profile could limit the resolution to 640×480 and impose a quality factor of 90. The next profile, more restrictive, could impose the same resolution but a quality factor of 80, and so on. The profile considered for this algorithm are shown in Table 3.2. We will see in Section 3.5 that it is not helpful to have a great number of profiles.

The second comparative algorithm, "successive scaling", will only reduce the images' resolution while using a fixed, but otherwise reasonable, quality factor of 85, until it yields a message that satisfies the device constraints. The algorithm proceeds as follows. First, for each image m_i, the largest allowable scaling factor $0 < z_i \leqslant 1$ such that $z_i R(m_i) \leqslant R(D)$ is found: in order words, the image is initially scaled to the maximum resolution acceptable for the device. Adaptation proceeds by adjusting, at iteration $r = 1, 2, \ldots$, a global parameter β_r (initially $\beta_1 = 1$) that is applied to every image so that the scaling factor, at step r, for image m_i, is $\beta_r z_i$, yielding an image with resolution $\beta_r z_i R(m_i)$. Assuming the scaling factor controls quadratically the file size (as a scaling of z does not yield a file $O(z)$ times smaller, but $O(z^2)$ times smaller), a reasonable adjustment β_{r+1} (for $r > 1$) is given by

$$\beta_{r+1} = \alpha_2 \sqrt{\frac{S(D)}{S_r}}, \tag{3.12}$$

where α_2 is an dampening factor (set to 0.95 for our experiments) to ensure that the algorithm stops rapidly after only a few iterations, and S_r is the size of the message obtained at step r. The adaptation terminates when a message satisfying the device constraints is produced.

Both comparative algorithms heuristically maintain a balance between image scaling and quality factors to produce satisfactory message quality – one by using predefined profiles, possibly tried in a decreasing expected message size order, the other by adjusting only resolution but keeping a good quality using a fixed quality factor. One could think of other heuristic adaptation strategies, for example, fixing the images' resolution to at most $R(D)$ (leaving smaller images' resolution unchanged) and increase the compression aggressiveness by progressively lowering the quality factor until a message meeting the maximum message side is created. This would lead to messages with images with very conspicuous blocking artifacts (resulting from an aggressive JPEG compression), even more so as the number of attachments grows and that the limited message budget is split across the many images.

3.5 Simulations & Results

In this section, we detail our experimental setup. In Section 3.5.1, we discuss the image corpora used for the training of the predictors as well as for the formation of the test MMS messages. In Section 3.5.2, we describe the adaptation experiment itself, the various operating conditions including a description of the test machine, and we present the results thus obtained.

3.5.1 Image Corpora, Predictors, and Test Messages

The image corpus used to train the JQSP is formed of 70 000 JPEG images obtained from the web in 2008, using a crawler using high-profile web sites as origination points [39]. Mainly because of confidentiality, it is impossible to sample messages from an MMS provider's traffic, and we deemed that good surrogate for MMS traffic would be a web crawler sampling web sites with user-submitted content. Ideally, a service provider would replicate the experiments in [20,21,39,40] with a much larger number of images sampled from their actual traffic, possibly retaining information about the distribution of the number of attachments, average size and quality factor of original images, etc., none of which, most unfortunately, we had access to at the time of writing.

Each of the 70 000 images was subjected to 100 transcoding (corresponding to all combinations of all quality factors $q \in \{10, 20, \ldots, 100\}$ and scalings $z \in \{0.1, 0.2, \ldots, 1.0\}$) yielding 7 000 000 examples. Scaling was performed using a Blackman filter for its desirable properties [12], and both resizing and compression were performed using the Magick++ library [3]. The training procedure for JQSP is described in [21]. The JQSP predictor is designed to predict the optimal transcoding parameters (q, z) given a target file size, and not directly predict file size from transcoding parameters, contrary to other predictors presented in [39,40].

The target file sizes chosen to query the JQSP were selected so that they were spread 5% apart in relative size for a given attachment; thus greatly limiting the number of parameters to examine without jeopardizing quality of adaptation. The oracular predictors where constrained similarly to the training conditions of the JQSP predictor, that is, limited to quality factors $q \in \{10, 20, \ldots, 100\}$ and scalings $z \in \{0.1, 0.2, \ldots, 1.0\}$); infeasible transcodings (file sizes or resolution exceeding the device constraints) were pruned from the optimization.

The proposed algorithm used $\alpha=0.9$ in eq. (3.11), that allows retries if optimization fails to produce a message that satisfies the device constraints. The comparative successive profile algorithm used the profiles listed in Table 3.2. The successive scaling method, described in Section 3.4.5, used $\alpha_2=0.95$. Both constants α and α_2 were set arbitrarily to reasonable, yet likely non-optimal, values.

Test examples were also obtained from a web crawler but at a later time, in the fall of 2010 [41]. The resulting corpus contains 370 000 JPEG image images. For each of the 220 MMS test messages, five images were drawn at random from this second corpus, yielding messages with an average size of 1.4MB and average image resolution of 1140×838.

3.5.2 Adaptation Experiment

For the experiments, the target profile chosen was Image Rich (images limited to 640×480 and message size to 100KB, see Table 3.1), thus requiring an average 15 : 1 reduction ratio for the simulated messages. This difference in original message size and target was deliberate as it seemed that considering cases where only a moderate adaptation of, say 2 : or 3 : 1 would not prove our point as strongly.

The predictors used for the experiments were the JQSP predictors, an (exact) oracular predictor, and four additional oracular predictors with 1%, 2%, 5%, and 10% relative error 95% of the times, respectively.

On the messages formed, we proceeded to their adaptation by our proposed algorithm using the various predictors and by the two comparative algorithms, and measured key indicators of performance, namely capacity, the propension of the algorithm to use all of the available message budget; objective function, to measure how the algorithm maximizes overall quality as measured by eq. (3.7) (as it was observed *after* transcoding and not predicted, therefore eq. (3.8) was not applicable); average image quality as a measure of balance in the solutions; and lastly the number of actual transcoding performed by the various methods and the wall-time needed for adaptation. The test machine was an Intel T9600 64bits CPU at 2.8 GHz running Ubuntu Linux 10.04 LTS, the current version at the time of writing [41]. Scaling was performed using a Blackman filter [12] and transcoding performed using Magic++ [3] in single-threaded mode.

The capacity, the portion of the maximum message size used by the transcoded message, is shown for the different predictors and algorithms in Figure 3.1. The objective function score for the different algorithms are shown in Figure 3.2 and the average SSIM for each attachment is shown in Figure 3.3.

Figure 3.4 shows the objective function scores resulting from individual messages, each curve sorted in ascending order separately. The curves do no allow to compare the relative performance of adaptation algorithms on a same message, but do render the general behavior of the different algorithms and predictors. Figure 3.5 presents a similar graph for the average message SSIM.

Figure 3.6 shows the times in seconds for the dynamic programming algorithm using the JQSP predictor, the successive scalings and successive profiles algorithms for our single threaded implementation on our test machine described previously. Oracular times are not shown, as not applicable to an actual implementation (and since a great number of actual transcoding are performed, we have times two orders of magnitude greater). Lastly, Table 3.3 shows the average number of transcodings and retries (the number of times the algorithm must restart with new constraints) for different algorithms. Let us note that in our experiment, the fail rate is zero, as all transcodings are eventually successful. For a transcoding to fail, we would need to have an

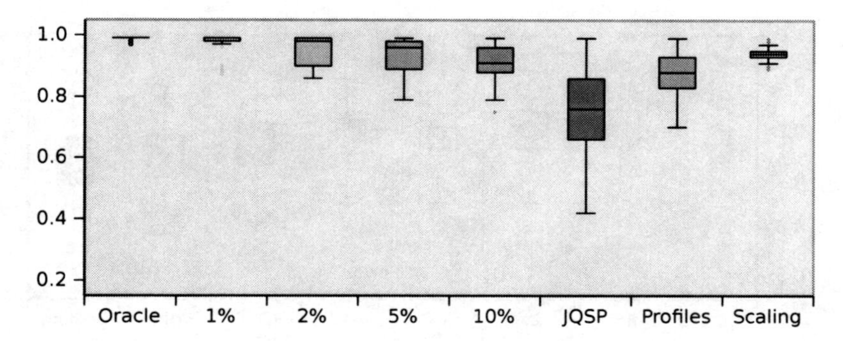

Figure 3.1 Message capacity distributions, by algorithms and predictors.

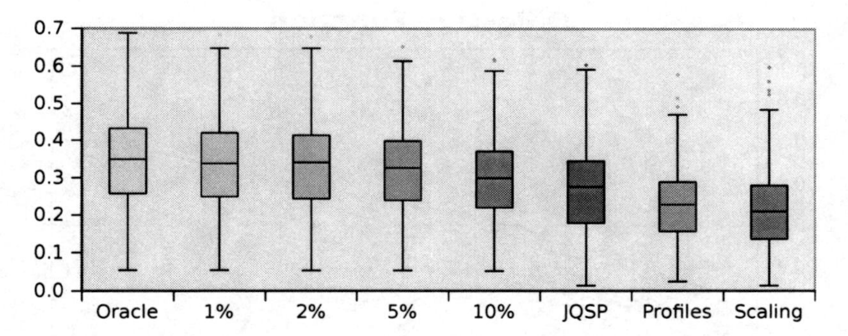

Figure 3.2 Objective function score distributions, by algorithms and predictors.

image large enough so that with a quality factor of 10 and a scaling of 10% it still exceeds the device constraints, and there were no such images in the test messages. If such a case would arise, one must think of a contingency method, possibly dropping attachments or splitting the message across many messages – issues we do not address in this work.

3.6 Discussion

One surprising thing, shown in Figures 3.1 and 3.2 is that is not sufficient to merely maximize capacity to achieve high adaptation quality; but that capacity must be maximized as a side-effect of quality-aware optimization. The great number of profiles used in the successive profiles algorithm and the efficient parameter search in the successive scaling algorithm allows them to get close to the capacity, but as neither explicitly maximize message qual-

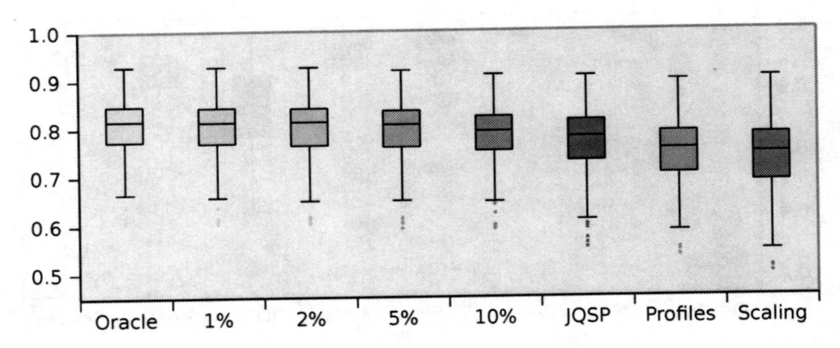

Figure 3.3 Average SSIM distributions, by algorithms and predictors.

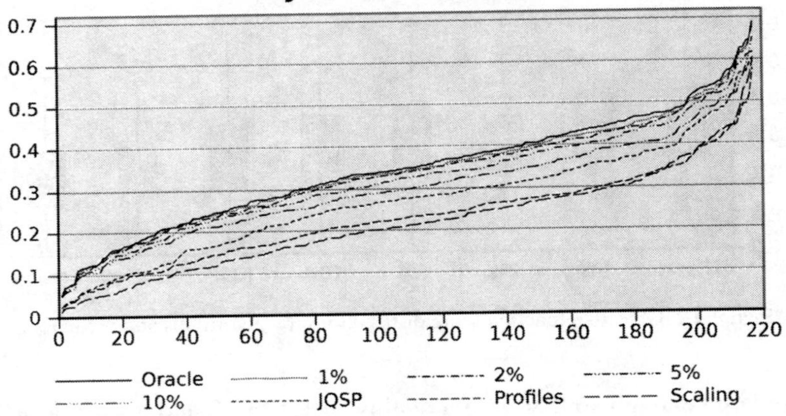

Figure 3.4 Objective function scores, by algorithms and predictors.

Table 3.3 Compared averages of algorithms for a message with five attachments.

Optimization Algorithm	Transcodings	Retries	Objective Function
Oracle	5.00	0.00	0.35
1%	6.03	0.21	0.34
2%	6.54	0.31	0.33
5%	7.19	0.43	0.32
10%	8.25	0.65	0.30
JQSP	5.55	0.13	0.27
Profiles	33.36	5.67	0.22
Scalings	15.02	2.00	0.23

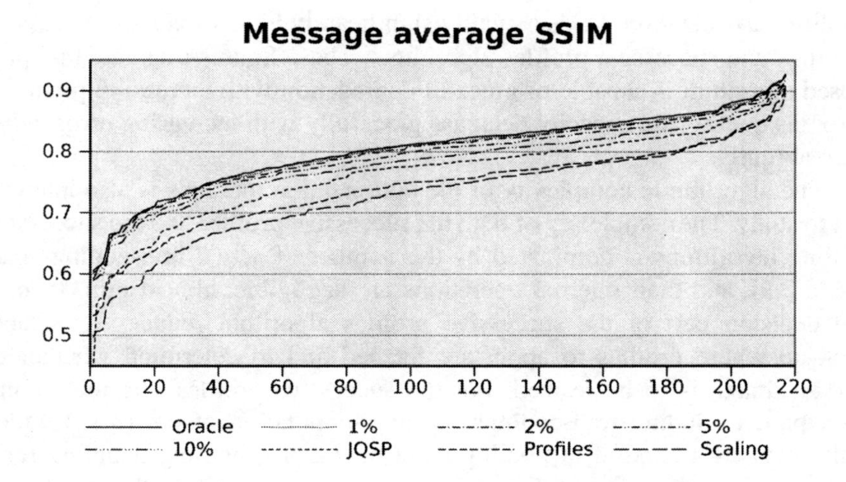

Figure 3.5 Average SSIM by message, by algorithms and predictors.

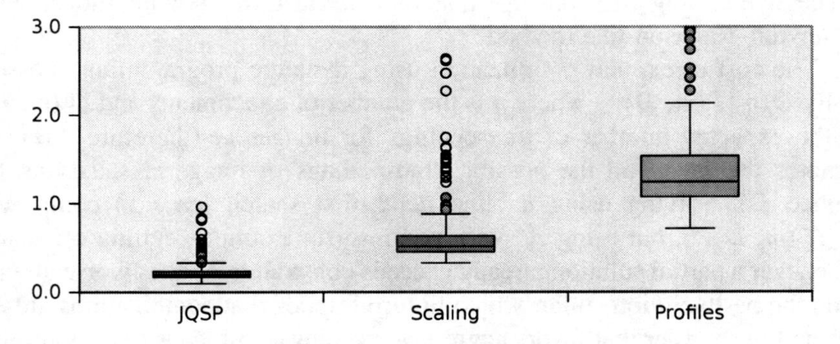

Figure 3.6 Message adaptation times distributions, by algorithm.

ity, merely attempting to do so heuristically, the resulting quality is, in fact, inferior to our proposed method, all predictors considered.

Let us remark that maximizing eq. (3.7), the products of SSIM scores, for a message is not the same as maximizing the average SSIM score for the same message (although as discussed in Section 3.4.1, both are strongly linked); but figures 3.4 and 3.5 confirm that the two are highly correlated, in particular, the ordering of algorithms and predictor pairs is preserved.

Figures 3.2 and 3.5 show that the JQSP predictor behaves close to the 10% relative oracular predictor (probably closer to a 15% relative error predictor), and Figures 3.4 and 3.5 show that the predictor-based method using explicit quality maximization yields a quality significantly higher than the successive scalings and successive profiles algorithms. These figures illustrate that proposed algorithm is capable of graceful degradation with increasing predictor error, as quality and capacity decrease gracefully with increasing error rather than abruptly.

The algorithmic complexity of the optimization methods is also interesting to study. The complexity of both the successive profiles and the successive scaling algorithms is dominated by the number of actual transcodings (see Table 3.3), and their internal operations are negligible, almost null. Indeed, the decision part of the successive profiles algorithm reduces to a table look-up where profiles to apply are fetched and to determine whether or not an image is to be resized. For the successive profiles, the most complex part, excluding transcodings, is the computation of eq. (3.12) which only involves a handful of floating-point operations: nothing daunting for a server-class computer. For the successive profiles algorithm, the number of iterations is upper-bounded by the number of profiles in its table, while the successive scaling will converge quadratically fast, in a few iterations, being a Newton–Raphson-like method.

The cost of explicit optimization using dynamic programming is essentially $O(n\,|T(m, D)|^2)$ where n is the number of attachments and $|T(m, D)|$ is the expected number of transcodings for images m (therefore $T(m, D)$ denotes the set of all the possible transcodings of image m satisfying the device D). Solving using a blind depth-first search leads to complexity $O(|T(m, D)|^n)$, but using A^* with pruning (for example, cutting off search whenever a partial solution already exceeds constraints or does worse already than the best solution so far, which in turn implies that combinations are examined in an order that favors aggressive pruning), still leads to an algorithm with an exponential run-time worst case, but using admissible pruning, it can be made polynomial-time in $O(|T(m, D)|^\delta)$, where δ is a constant that does not depend on n but rather in the average depth explored [45, p. 85].

Figure 3.6 shows that the run-time of explicit maximization is offset by the gain in fewer transcodings, as to be much faster, on average, with an average of 5.55 transcodings per message, than the successive scalings algorithm with an average of 15.02, and than the successive profiles algorithm with an average of 33.36, as shown in Table 3.3. The slight overshooting of the file size prediction for the JQSP predictor keeps the proposed algorithm from achieving maximal capacity, while still producing higher quality messages than the comparative algorithms, with the side-effect of keeping the number of retries lowest after the oracular predictor; a predictor that undershoots significantly would imply far more transcodings before reaching a solution that satisfies the device constraints.

Furthermore, the number of predicted transcoding parameters examined during optimization plays a non-negligible role in the proposed method performance. Even if the optimization algorithm is efficient, a great number of predicted transcoding parameters per image means a larger graph to explore and necessarily increased run-time, even with pruning, as run-time grows (at least) quadratically in the number of predicted transcoding parameters [36]. It then becomes a trade-off between the precision of the predicted transcoding and run-time. The successive scalings and successive profiles algorithms are also subject the speed/quality trade-off, but the results are far less interesting. The successive profiles algorithm could use fewer profiles, but already the resulting quality is inferior to the proposed algorithm. The same is true for the successive scalings method, which could start with an aggressive scaling factor of, say, $\beta_1 = 0.8$, but that would result only in even worse quality.

The predictor-based solution we propose is not limited to JPEG-only messages. It can be generalized to any type of media by introducing new predictors for the desired media types. One could propose predictors for PNG, GIF, voice audio formats, etc., and the general framework of the solution would remain exactly the same. File size and resolution constraints are unchanged, except for the fact that the predictors now accommodate other media types; however, it may be necessary to add other constraints. Certainly one would need to consider the case where the compression format itself is to be changed.

The presentation itself, ignored in this work, can most likely be modeled as supplementary constraints or with minor changes to the optimization algorithm. For example, one could enforce proportionality (i.e., all pictures maintain relative size) by sharing the scaling factor across all images, as the successive scaling algorithm does. If the new constraints are not amenable to dynamic programming (as not all types of constraints or objective functions can be used [11]) one will need to use an A^* search which is, in many regards, a more general solution than dynamic programming.

3.7 Conclusion

In this chapter, we proposed a solution to MMS adaptation where the problem is explicitly modeled as an optimization problem where the goal is to maximize image quality (with MSSIM as a measure of user experience) under the constraints dictated by the receiving mobile terminal. However, the novelty of our solution resides in the fact that we propose the use of predictors to speed-up optimization significantly without jeopardizing the resulting quality of adapted messages. We have shown, also, that the proposed method degrades gracefully with increased predictor error and that, overall, it performs much better, even with rather large predictor errors, than algorithms found in and inspired by the existing literature.

The proposed objective function, based on the structural similarity (MSSIM), is necessarily a simplification of the user experience. Future work could explore how to model user experience more appropriately and determine what objective functions would best replace eq. (3.7) as a measure of the goodness of the adapted message. One could also consider the use of better predictors, for example [40], which was not available at the time of writing [41]. Of course, the more accurately one can predict resulting file size and quality, the best the adaptation can be. Not only can we find better solutions, we can also use this knowledge to further prune optimization and obtain a faster adaptation algorithm.

Acknowledgments

This work was funded by Vantrix Corporation and by the Natural Sciences and Engineering Research Council of Canada under the Collaborative Research and Development Program (NSERC-CRD 326637-05).

References

[1] 3rd Generation Partnership Project. `http://www.3gpp.org/`.
[2] Global System for Mobile Communications. `http://www.gsm.org/`.
[3] ImageMagick and Magick++. `http://www.imagemagick.org/`.
[4] Synchronized Multimedia Integration Language (SMIL 3.0). `http://www.w3.org/TR/SMIL3/`.
[5] The Independent JPEG Group. `http://www.ijg.org/`.
[6] XHTML 1.1 – Module-based XHTML – Second Edition. `http://www.w3.org/TR/xhtml11/`.
[7] Recommendation T.81: Digital Compression and Coding of Continuous-Tone Still Images – Requirements and Guidelines. Technical Report, CCITT, September 1992.
[8] 3GPP TS 23.140 V10.0.0. Multimedia Message Service (MMS) Media Formats and Codecs (Release 10), April 2011. `http://www.3gpp.org/ftp/Specs`.
[9] 3GPP2 X.S0016-000-A. 3GPP2 Multimedia Message System MMS Specification Overview, Rev. A, May 2003. `http://www.3gpp2.org/`.
[10] 3GPP2 X.S0016-330. 3GPP2 MMS MM3 Stage 3 for Internet Mail Exchange, June 2004. `http://www.3gpp2.org/`.

[11] Richard Bellman. *Dynamic Programming*. Dover, New York, 2003.

[12] R.B. Blackman and J.W. Tukey. *The Measurement of Power Spectra, from the Point of View of Communications Engineering*. Dover, 1959.

[13] N. Borenstein and N. Freed. RFC 1521: MIME (Multipurpose Internet Mail Extensions) Part One: Mechanism for Specifying and Describing the Format of Internet Message Bodies. Technical Report, IETF, September 1993.

[14] Dick C.A. Bulterman and Lloyd Rutledge. *SMIL 3.0 Interactive Multimedia for the Web, Mobile Devices and Daisy Talking Books*. Springer, 2008.

[15] Patrick Carey. *New Perspectives on HTML, XHTML, and XML*. Course Technology, 3rd edition, 2010.

[16] Rick C.S. Chen, Stephen J.H. Yang, and Jia Zhang. Enhancing the precision of content analysis in content adaptation using entropy-based fuzzy reasoning. *Expert Systems with Applications*, 37:5706–5719, 2010.

[17] David H. Cooker. RFC 822: Standard for the Format of ARPA Internet Text Messages. Technical Report, IETF, August 1982.

[18] Nokia Corporation. How to Create MMS Services, V 4.0, June 2003.

[19] Stéphane Coulombe and Guido Grassel. Multimedia adaptation for the multimedia messaging service. *IEEE Communication Magazine*, 42(7):120–126, July 2004.

[20] Stéphane Coulombe and Steven Pigeon. Quality-aware selection of quality factor and scaling parameters in JPEG image transcoding. In *Proceedings of IEEE 2009 Computational Intelligence for Multimedia, Signal, and Video Processing (CIMSVP)*.

[21] Stéphane Coulombe and Steven Pigeon. Low-complexity transcoding of JPEG images with near-optimal quality using a predictive quality factor and scaling parameters. *IEEE Trans. Image Processing*, 19(3):712–721, March 2010.

[22] D. Dugad and A. Ahuja. A fast scheme for image size change in the compressed domain. *IEEE Trans. Circuits and Systems for Video Technology*, 11(4):461–474, April 2001.

[23] R. Gellens. RFC 4356: Mapping between the Multimedia Messaging Service (MMS) and Internet Mail. Technical Report, IETF, January 2006.

[24] Richard Han, Pravin Bhagwat, Richard LaMaire, Todd Mummert, Véronique Perret, and John Rubas. Dynamic adaptation in an image transcoding proxy for mobile web browsing. *IEEE Personal Communications Magazine*, 5(6):8–17, 1998.

[25] Frederick S. Hilier and Gerald J. Lieberman. *Introduction to Operation Research*. McGraw-Hill Science, 9th edition, 2009.

[26] Tatsuo Ishihara. The distribution of the sum and the product of independent uniform random variables distributed at different intervals. *Trans. Japanese Soc. for Industrial and Applied Mathematics*, 12(3):197–207, 2002.

[27] J. Dwyer, III. MMS to prosper as mobile marketing becomes mainstream. *Wireless Week*, January 2011.

[28] Tim Kennedy and Mary Slowinsky. *SMIL: Adding Multimedia to the Web*. SAMS Publishing, 2002.

[29] Ihor O. Kirenko and Ling Shao. Local objective metrics of blocking artifacts visibility for adaptive repair of compressed video signals. In *Proceedings of IEEE International Conference on Multimedia and Expo*, pages 1303–1306, 2007.

[30] Gwenaël Le Bodic. *Mobile Messaging, Technologies and Services, SMS, EMS and MMS*. John Wiley & Sons, 2nd edition, 2005.

[31] Rakesh Mohan, John R. Smith, and Chung-Sheng Li. Adapting multimedia internet content for universal access. *IEEE Trans. Multimedia*, 1(1):104–114, March 1999.

[32] Jayanta Mukherjee and Sanjit K. Mitra. Image resizing in the compressed domain using subband DCT. *IEEE Trans. Circuits and Systems for Video Technology*, 12(7):620–627, July 2002.

[33] Open Mobile Alliance. Wireless Application Protocol. `http://www.openmobilealliance.org/Technical/wapindex.aspx`.

[34] Open Mobile Alliance. Multimedia Messaging Services Encapsulation Protocol. March 2005. Open Mobile Alliance OMA-MMS-ENC-V1_2-20050301-A.

[35] Open Mobile Alliance. Enabler Test Specification for (Conformance) for MMS Candidate Version 1.3. Technical Report, October 2010. Open Mobile Alliance OMA-ETS-MMS_CON-V1_3-20101015-C.

[36] Christos H. Papadimitriou and Kenneth Steiglitz. *Combinatorial Optimization: Algorithms and Complexity*. Dover, 1998.

[37] Luca Passani. WURFL – The Wireless Universal Resource File. `http://wurfl.sourceforge.net/`.

[38] William B. Pennebaker and Joan L. Mitchell. *JPEG: Still Image Data Compression Standard*. Digital Multimedia Standards. Springer, 1992.

[39] Steven Pigeon and Stéphane Coulombe. Computationally efficient algorithms for predicting the file size of JPEG images subject to changes of quality factor and scaling. In *Proceedings 24th Queen's University Biennial Symposium on Communications*, 2008.

[40] Steven Pigeon and Stéphane Coulombe. Efficient clustering-based algorithm for predicting file size and structural similarity of transcoded JPEG images. In *Proceedings IEEE International Symposium on Multimedia (ISM)*, 2011 (to appear).

[41] Steven Pigeon and Stéphane Coulombe. Optimal quality-aware predictor-based adaptation of multimedia messages. In *Proceedings of Intelligent Data Acquisition and Advanced Computing Systems (IDAACS)*, Vol. 1, pages 496–499, 2011.

[42] R. Resnik. RFC5322: Internet Message Format. Technical Report, IETF, October 2008.

[43] Soroosh Rezazadeh and Stéphane Coulombe. Novel discrete wavelet transform framework for full reference image quality assessment. *J. Signal, Image, and Video Processing*, pages 1–15, September 2011.

[44] J. Ridge. Efficient transform-domain size and resolution reduction of images. *Signal Processing: Image Communication*, 18(8):621–639, September 2003.

[45] Stuart Russell and Peter Norvig. *Artificial Intelligence: A Modern Approach*. Prentice-Hall, 3rd edition, 2009.

[46] Hamid Rahim Sheikh, Muhammad Farooq Sabir, and Alan Conrad Bovick. A statistical evaluation of recent full reference image quality assessment algorithms. *IEEE_TIP*, 15(11):3440–3451, November 2006.

[47] Yair Shoham and Allan Gersho. Efficient bit allocation for an arbitrary set of quantizers. *IEEE Trans. Acoustics, Speech, and Signal Processing*, 36(9), September 1988.

[48] John R. Smith, Rakesh Mohan, and Chung-Sheng Li. Content-based transcoding of images in the internet. *Proceedings of International Conference on Image Processing (ICIP)*, 1998.

[49] John R. Smith, Rakesh Mohan, and Chung-Sheng Li. Transcoding internet content for heterogeneous client devices. In *Proceedings of International Symposium on Circuits and Systems*, Vol. 3, pages 599–602, 1998.

[50] M.D. Springer and W.E. Thompson. The distribution of products of independent random variables. *SIAM Journal on Applied Mathematics*, 14(3):511–526, May 1966.

[51] M.D. Springer and W.E. Thompson. The distribution of the products of beta, gamma and Gaussian random variables. *SIAM Journal on Applied Mathematics*, 18(4):721–737, June 1970.

[52] Luc Trudeau, Stéphane Coulombe, and Steven Pigeon. Pixel domain referenceless visual degradation detection and error concealment for mobile video. In *Proceedings of International Conference on Image Processing (ICIP)*, 2011.

[53] Zhou Wang, Alan Conrad Bovick, Hamid Rahim Sheikh, and Eero P. Simoncelli. Image quality assessment: From error visibility to structural similarity. *IEEE Trans. Image Processing*, 13(4):600–612, April 2004.

[54] Zhou Wang, Alan Conrad Bovick, and Eero P. Simoncelli. Structural spproaches to image quality assessment. In A.C. Bovik (Ed.), *Handbook of Image and Video Processing*, 2nd edition, pages 961–974. Elsevier Academic Press, 2005.

[55] We-Qi Yan and Mohan S. Kankanhalli. Multimedia simplification for optimized MMS synthesis. *ACM. Trans. Multimedia Computing, Communications, and Applications (TOMCCAP)*, 3(1), February 2007.

[56] Michael Yuen and H.R. Wu. A survey of hybrid MC/DPCM/DCT video coding distortions. *J. Signal Processing*, 70:247–278, 1998.

4

A Comparative Study on Methods for Improved Tracking of Mechanical Movements of IPMC Actuators

Kyriakos Tsiakmakis and Theodore Laopoulos

Physics Department, Electronics Lab, Aristotle University of Thessaloniki, Thessaloniki 54124, Greece; e-mail: {ktsiak, laopoulos}@physics.auth.gr

Abstract

A comparison of three different methods for end-edge detection of the IPMC actuator as it moves is presented in this chapter, using image analysis captured by a CCD camera. The degrees of freedom for actuator move are specific and experiments were realized in underwater conditions when one end-edge is fixed-clapped and the other is free. The tracking techniques of free edge are based on three different approaches while the two recent of these use the new technology of Graphical Processing Units (GPUs) to reduce more the processing time. The results show that the new techniques increased by 47 and 70 times the speed of detection. Comparing the two last methods it can be concluded that they offer satisfactory reduced processing time for the speed of material which can be measured to values up to 10 KHz theoretically. The second improved method uses more calculations regarding position estimation and is slower if material bends out of the way track. The last approach is slower than second because is implemented in the entire area but handles better the bigger bending of the material within shorter execution times. A prerequisite for high speed measurements is the adequate support of the frame rate of camera.

Richard J. Duro and Fernando López Peña (Eds.), Digital Image and Signal Processing for Measurement Systems, 101–133.

Keywords: visual measurements, fast displacement tracking, GPU, CUDA, IPMC actuator.

4.1 Introduction

In the last decades, computer vision and motion tracking have concentrated on the development of general purpose applications [1, 2]. Most of the algorithms were focused on fast and accurate detection of a certain shape or tracking the movement a certain figure [3, 4]. Increasing image resolution gives the ability to increase the accuracy of the above procedures. Improvement and optimization of the algorithms is the main step to achieve better results and more accurate detection or tracking. A key factor to obtain optimum performance in such cases is the processing time of each procedure. Today, a high number of hardware and algorithmic techniques is offering help to improve the processing time [5]. Many developers use Field Programmable Gate Arrays (FPGA) units for the implementation of fast image processing algorithms using multiple processors. Special techniques are then used for the parallel execution of the commands [6, 7].

FPGA can handle communications with other peripherals and also the communications with a computer, if it is necessary. Many researchers prefer to use the computer for image analysis when an external device is unnecessary. Using only the camera connected to the computer all procedures are executed by the CPU. There are many technologies that allow us to develop high speed algorithms: among them MMX, MPI and GPU. Multimedia Extensions (MMX) instructions are a single-instruction multiple-data (SIMD) instruction set. Intel developed a set of few instructions that would significantly accelerate the execution of multimedia applications [8]. The MPI processing technique allows different code routines corresponding to different processes to be executed in multiple CPUs of the computer, without support from hardware-level performance acceleration [9, 10].

In the past few years, many researchers have focused on parallel image processing using the multi-processors of graphics cards. Using the Graphics Processing Units (GPUs) more processing speed can be achieved [11–13]. GPUs, in the beginning, were intended to be used in high speed graphics applications. In the last few years, many developers use this technology in other fields such as machine vision, or non-graphics applications using parallel programming.

Using GPUs was very difficult in terms of programming, since programmers had to learn how to handle the GPU programming model. A certain vendor, NVIDIA, offered a GPU program development with the so-called Compute Unified Device Architecture (CUDA), which is an interface to handle the GPUs easily [14, 15]. CUDA allows the use of a wide-spread programming language for developing algorithms to be executed on the GPU. In the field of computer vision, CUDA technology is used to improve the performance of the algorithms and accelerate the speed of the computations [16, 17]. Edge and corner detection [18] and feature tracking [19] are the major fields of image analysis algorithms that use GPUs.

A number of high-speed image processing applications using CUDA have been reported in the literature. Ogawa et al. [20] recently implemented a CANNY edge detection algorithm on CUDA with a speed-up factor of 61 over a conventional software implementation. Gomez-Luna et al. [21] presented various interesting implementations of well-known algorithms such as the Hough Transform, the CANNY algorithm and cross correlation. These algorithmic approaches are also compared with an OpenMP version of the algorithm, running on two platforms with multiples cores. Xie et al. [22] presented a novel GPU-based scale invariant interest point detector, coined as Harris-Hessian (H-H) and achieved up to a 10–20$\times$ speedup with respect to CPU-based method. Faasold et al. [23] ported the KLT tracker to GPU, using the CUDA technology. The team uses the feature point tracking on a multi-resolution image representation to allow tracking of a large motion. They compare the CUDA implementation with the corresponding OpenCV (using SSE and OpenMP) routines in terms of quality and speed and they achieve a significant speedup of up to a factor of 10. Zhang et al. [24] worked on the SOBEL edge detector and on homomorphic filtering to solve the compute-intensive character of image processing. The two algorithms were implemented on GPU and achieved an acceleration of approximately 25 times and 49 times as fast as those on CPU, respectively. All of these methods are very important in terms of the design and implementation of applications for detecting points-of-interest and for measuring visually the displacement of a certain point of interest.

The methods presented in this chapter are tested in a certain application: a fast and accurate visual measurements system for measuring displacements in underwater environments. A micro electro-mechanical actuator, which is a strip of Ionic Polymer Metal Composite (IPMC), is used as the moving object for all experiments. The information on the motion parameters and the accurate measurement of the position of the edge of the actuator are very important for the development of such systems working in closed-loop configurations [25–27]. The IPMC is a type of electroactive polymer which bends if a voltage is applied across its thickness and vice versa: it generates a detectable voltage if subjected to a mechanical deformation [28, 29]. It can work as a low-voltage activated motion actuator or as a motion sensor or as an energy harvester. The general structure of the IPMC is a polymeric ionomer membrane covered with metal surfaces which are the electrodes. The actuator bends when an electric field is applied between the electrodes. Then a solvent flux of hydrated cations and free water is produced through the cathode. The solvent flux creates an electro-osmotic differential pressure and leads to a bending motion of the anode side of the IPMC.

Different systems are available for measuring displacement characteristics of IPMC actuators in the literature [30, 31]. Most of them use a non-contact distance measurement technique, based on a camera or on a laser-beam. A laser-based measuring instrument offers high resolution (in the order of 2μm) along with the ability to measure small magnitudes of displacements at high frequencies (more than 100 Hz).

Measurement accuracy via the laser-based instruments is obviously very good, but in practical applications of IPMC movements the system presents a lot of problems. Some of them are the large relative angle between the different positions of the IPMC strip surface and the change in the refraction index when working in underwater applications. If the application involves a long IPMC strip (i.e. longer than about 4 cm, which is a very usual case) the laser is not able to track accurately all consecutive positions of the actuator; due to the large bending of the material the free end goes out of the field of view of the laser beam [30].

Figure 4.1 shows the comparison of measurements of displacements versus frequency in the air, using laser and camera systems, showing that there is no significant difference between the two measurement approaches for such applications.

The laser system is based on a laser-optical sensor and the appropriate signal conditioning electronics. It uses the principle of optical triangulation to measure distance to a point of the target. A laser beam is projected onto the target surface and depending on the distance, the diffuse fraction of the reflection of this point of light is focused onto a position sensitive element by the receiving lens. The electronics module provides an analog output voltage proportional to the relative displacement from the sensor to the target. It should be noticed that the laser-system has to be located at a specific reference distance of about 3 cm from the material. The measurement range is limited to about 1 cm relative displacement variation to the reference distance, and

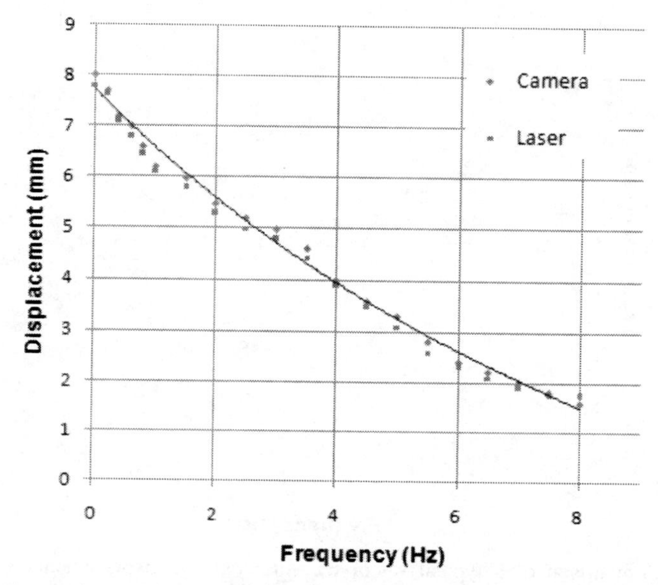

Figure 4.1 Comparison of measurements of displacements vs frequency in the air using laser and camera systems.

the typical error is in the range of 0.5%. The laser system is calibrated by the manufacturer to work in air and has a conversion factor of 1 V/mm.

Significant offsets appear In cases where the laser system is to measure underwater material movements, and recalibration is necessary. The change in the reflection index between air and water causes a change in the position of the beam, and consequently a change in the conversion factor of the laser measuring system is caused by this deviation. Figure 4.2 shows the comparison of the two systems in underwater conditions.

The two systems seem to have an approximately constant difference between their measured values of the displacement.

On the other hand, imaged-based methods for distance measurement by using CCD cameras overcome these problems. The camera system is unaffected by the refraction issues and is more suitable for underwater applications. Nowadays, the improvements in processing techniques and hardware allow handling complicated computations faster. Camera image-based methods are very useful for long strips with large bending, random movements and higher degrees of freedom. A good knowledge of the overall motion parameters and particularly the position of an actuator are certainly essential for

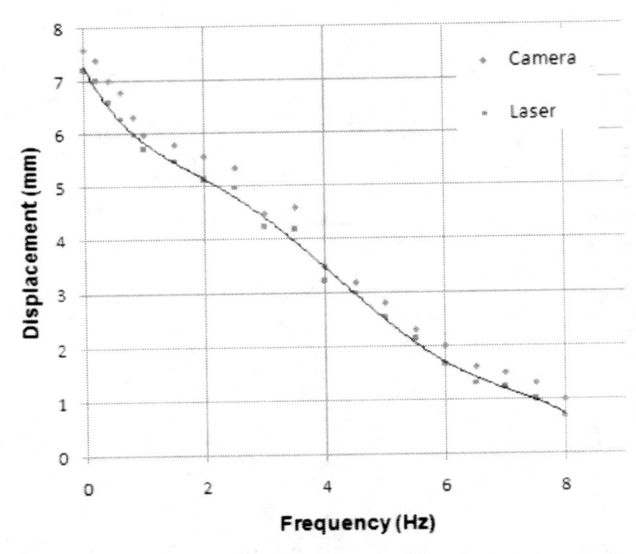

Figure 4.2 Comparison of a typical set of measurements of displacements vs frequency underwater by laser and camera.

the characterization of the electromechanical properties of micromechanical actuators. The three different methods of analysis which are compared in this chapter focus on tracking the end-edge of the actuator under a specific set-up.

All the data required for the analysis of the IPMC position are obtained automatically by the appropriate imaging algorithms using an image sensor. Due to requirements in measurement time and the objective of a real time measuring system there is a need to investigate new image processing approaches for locating the point of interest (end-strip of the actuator). The reduction of the processing time is achieved using faster algorithms and Graphic Processing Units (GPUs), especially CUDA (Compute Unified Device Architecture) technology.

The description of the proposed measuring systems consists of two parts. The first presents the hardware implementation which is based on specific driving circuitry for the IPMC actuator and a simple CCD camera which communicate with computer via a firewire connection. The second part contains the methods for detecting the position of the actuator and the performance improvements using parallel processing.

The major purpose of the development effort for this system is to have a fast and accurate measurement of the displacement of an IPMC actuator.

One end of the actuator is fixed-clamped and when a voltage is applied, the rest of the body bends. We assume that the free-end indicates the position of the actuator where the maximum displacement depends on its length and on the material's characteristics. All experiments have been carried out in underwater environments and consequently the frequency of the movements is slow due to the water.

This chapter presents and compares three approaches to find the free edge of the IPMC strip. The first one considers a simple detection method using a simple scanning of specific points in the material strip. This method shows good results for large displacements with small curvature. The algorithm is executed in the processor of the computer with and without use of the capabilities of the multiple processors of the graphics cards.

The second method uses two improved features that make the detection algorithm much faster, while it searches only in certain small areas and uses parallel processing on GPU units.

The third method uses a new detection method based on grid techniques with the help of histogram analysis and using parallel processing through the multiprocessors of the graphics card. The last two methods accelerate the processing speed by a significant factor.

4.1.1 Graphics Processing Units

The implementation of a real time system for measuring the displacement of a strip, in every received frame, depends on many parameters. In this work the specific experimental setup and the limited degrees of freedom make the tracking easier, especially when the speed and the image resolution are low. In cases where the low frame-rate satisfies the requirements of the application, then a simple modern processor operating at high frequency offers a sufficient performance. The increment of the speed of the material requires a higher frame rate. Between two frames the available time for processing is decreased. Many applications execute many procedures in this time, especially for real time closed loop systems, and they need more time. Such operations are the control signals, the communications with microcontrollers, storing and loading data, execution of complicated calculations and displaying signals in graphs. The image processing procedures require more execution time and it is easier to reduce this amount of the time by using parallel programming. The CUDA technology from NVIDIA is a general purpose parallel computing architecture using graphics processing units (GPUs) where parts of the software may be executed in parallel. Many image processing algorithms include commands that can run independently from each other in separate processors. All operations are controlled and the program flow continues when all complete their execution.

The graphic card used in this work is the NVIDIA Geforce GTS 250 with a Graphics Clock at 738 MHz and a Processor Clock at 1836 MHz, with 128 thread processors, distributed across sixteen multiprocessors. For the implementation of parallel execution each CUDA processor runs one thread independently, allowing thus 128 simultaneous computations. CUDA is a computing engine that makes the use of NVIDIA GPUs easy. Programmers can easily handle the GPUs via commands and programming languages.

Figure 4.3 shows the CUDA programming model used in an algorithm. Usually the algorithm includes mixed code, where the commands executions are performed serially or in parallel. The parallel execution is carried out by the multi-processor of the graphics card by invoking a special function, often called kernel. Parallel execution is considered as an advantage if the calculations need to be executed independently. Few algorithms or procedures offer a significant decrement of execution time due to data transfer and other delays in the GPU cells.

In graphics processing units the process data are divided and stored in memory blocks of the chip where each block is shared by multiple thread processors. The kernel runs several blocks of threads and each thread performs a single computation. The threads are organized into a sequence of grids which have many thread blocks. Hence, a grid of thread blocks consists of a number of blocks that execute the same kernel. The threads in the same thread block can access data from the shared memory which is limited to 16 Kbytes.

4.1.2 Design of the System

A real time experimental system for tracking the moving end-edge of an IPMC actuator is presented in Figure 4.4. This block diagram describes the hardware of the experimental setup for the actuation of the material and the necessary units for the implementation of a real time system.

It should be noted that the camera is placed in a vertical position to the motion of the actuator with a high-contrast background and suitable lighting conditions.

The circuit which has been implemented for the experiment is presented in Figure 4.5. The D/A (MCP4822) is used to drive the power-follower and operates as a function generator taking into account the inputs from ADC channels for the implementation of a closed loop real time system. Three

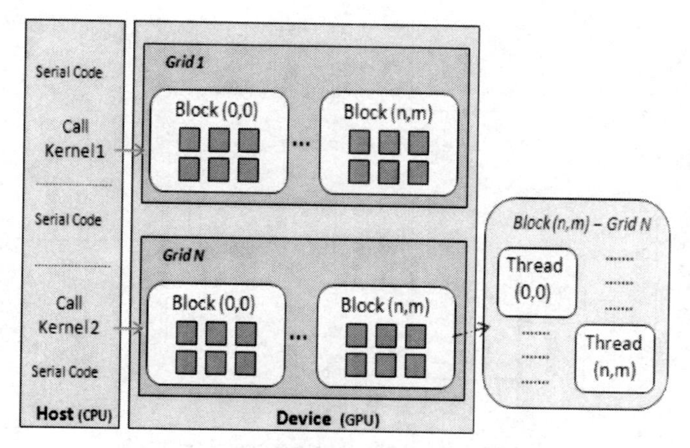

Figure 4.3 Programming model of CUDA.

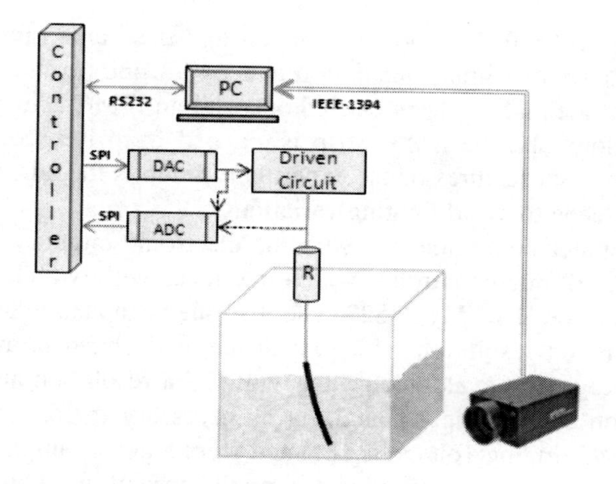

Figure 4.4 Experimental set-up.

channels of A/D (MCP3304) are used to measure all important signals. The resistor R5 is set to 1Ω for measuring the current which goes through the actuator.

The camera model used in the experimental tests was a Unibrain-521b which is connected to the computer via the firewire protocol. The high speed of the firewire communications, the low cost and the image resolution were some of the reasons for selecting this model. The maximum image resolution is 640 × 480 pixels per frame (grayscale mode) and the frame-rate ranges

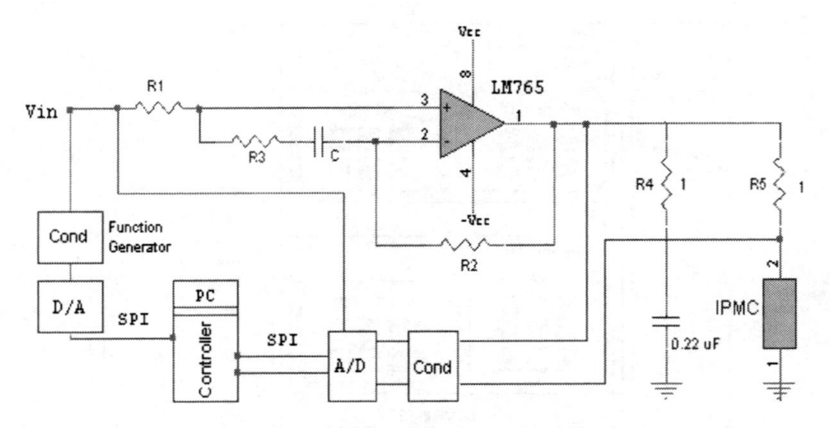

Figure 4.5 Circuit of experimental set-up.

from 3.75 up to 86 fps. It should be noted that the frame rate at 86 fps is available for a smaller image resolution than 640 × 480 pixels.

The experiments have been carried out with an IPMC actuator in underwater conditions and the IPMC strip is clapped from the top end. One of the most important features of the experiment setup is the adjustment of the background scene to avoid lighting variations.

The computer communicates with the electronic circuitry setup via an AVR-ATMega32 microcontroller which interfaces with two chips, the ADC MCP3304 and the DAC MCP4822. The first one is an analog to digital converter with 13 bit resolution with four differential channels available. The last one is a digital to analog converter with 12 bit resolution and is used for the generation of the output signal that is necessary for driving the IPMC actuator. IPMC driving voltage is obtained from a power amplifier (LM675) connected as a voltage follower for driving the current up to 500 mA and its main purpose is to provide the appropriate output current.

The surface resistance of the IPMC is very small and is highly correlated with curvature [32]. Experiments have shown that the amplitude of the applied voltage is increased when the load current gets higher. The A/D unit is used to measure the voltage applied to the IPMC and the load current (current through the IPMC) is monitored by measuring the differential voltage of the R_5 resistor.

Photos of the electronic circuits of the experimental setup are shown in Figure 4.6. In the left image (Figure 4.6a) the circuits of the convertors and the conditioners that adjust the input and the output signals are shown. In the

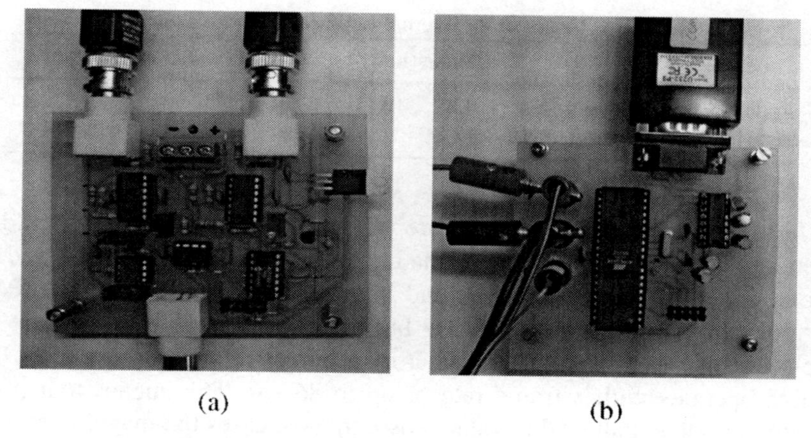

(a) (b)

Figure 4.6 Real photos of the electronic circuits.

right image (Figure 4.6b) the microcontroller and the serial interface circuit are presented.

4.1.3 Calibration of the System and Camera

In this section we present the calibration of the system and the camera for a specific experimental setup. The calibration of the camera is used to find its intrinsic and the extrinsic parameters. Most of them are very useful when the material moves in three dimensions. In this experimental setup and application one needs to know only two of them, as the actuator moves in two dimensions only and at a constant distance from the camera. The three most important parameters that should be calculated are the actual physical focal length, the scale factor (relating pixels to distance) and the radial distortion coefficient. The calibration of the camera follows the technique by Zhang [33], as referred to in his work. A chessboard calibration grid is placed at the same distance as to where the material is moving and the camera is placed in a perpendicular position at constant distance to the layer that the material moves.

The accurate relationship between pixels and millimeters depends on the image resolution and the distance between camera and actuator. The field of view is always adjusted according to the size of the material. For an image resolution of 640×480 and material dimensions of 6.4×4.8 cm the resolution is calculated to approximately 0.1 mm per pixel.

Table 4.1 Values of the internal parameters of the camera.

Parameter	Value (mm)	Standard Deviation Error (mm)
Radial Distortion Coefficient	3.05×10^{-8}	2.1×10^{-9}
Actual Physical Focus Length	12.2	0.26

In the system calibration there are two major parameters that should be taken into account. The first one is the maximum frequency response and the second one is the operation in air and in underwater conditions. The IPMC actuator can operate at over 100 Hz but the camera is able to measure the displacement clearly and with satisfactory accuracy at lower frequencies. The camera operates with a frame rate of up to 86 fps. This means that it can measure less than half of that value. Also, in most cases the magnitude of the displacement decreases as the frequency increases. The results show that this system was capable of measuring at 28 Hz when the frame rate of the camera was 60 fps. In air and underwater experiments the performance of the camera system is the same, unlike the laser measurement system.

For the camera calibration the two main parameters are presented; one is the actual physical focus length and the second is the radial distortion coefficient. The latter is quite low, so an additional algorithm correction is not necessary. The results are shown in Table 4.1.

In this chapter the accuracy of the detector is determined by the localisation and detection accuracy. The first is the position accuracy which means the distance error between the measured (marked) and real corner (point-of-interest, or end of the strip). Here, it is assumed that only one pixel may define the corner point. Consequently, the corner detection is successful if the corner is marked on the real point (pixel). Therefore the accuracy is directly related to the resolution. Experimental results show a one pixel deviation between the marked and real corner which indicates a position error of about ± 0.14 mm for an image resolution of 400×200 pixels.

The detection accuracy is practically representing the number of cases which are detected correctly. Each method presented in this chapter presents a different detection accuracy which is related to different bending conditions of the actuator.

After the camera calibration, the experimental results from the system calibration will be presented. If the lens has a focal length of 12 mm, and a field of view at 400×200 pixels (width $\times$ height) is defined, the theoretical working distance (front of the lens to the object) is calculated to be 75 mm.

Table 4.2 Resolutions for different image sizes and working distance.

Image Size (pixels)	Resolution (mm)	Error (mm)	Working Distance (mm)	Field of View (mm × mm)
400 × 200	0.1	0.002	77	40 × 20
640 × 480	0.11	0.003	123	64 × 48

The experimental calibration results indicate that the value for the actual physical focal length is 12.24 mm. Now, the working distance has to change to the optimum value of 123 mm in order to produce a resolution of 0.1mm approximately. The image resolution of the camera is changed as shown in Table 4.2. For an image size of 640 × 480 pixels, the optimum value for the working distance is 123 mm. If the working distance is decreased, the resolution will be increased, but the actuator will leave the field of view. If the image resolution is increased (for the same field of view), then the conversion of displacement units (mm) to pixels is better, but the size of the image-frame and the processing time are also increased.

4.1.4 Model Identification of the IPMC Using the Camera

The implementation of closed loop control systems for actuators requires the study of the actuator's behavior. In this direction it is necessary to investigate and find the model of the IPMC actuator. The identification of the model it is quite a difficult task especially in underwater environments. The complicated physics acting on the transduction effect of an IPMC actuator makes the identification of a grey or physical box difficult.

Consequently, a black-box approach is preferred when the application is simple. This approach seeks to estimate the values of the coefficients of an "a priory" defined transfer function. The coefficients are experimentally estimated by comparing experimental results to evaluations of the transfer function [34, 35].

The step response of the IPMC based cantilever indicates that two different rising times are present as a cantilever structure presents a resonance frequency. This behaviour appears in some oscillations at the beginning of the step. A possible choice is a model structure such as the one given by the following equation:

$$G(s) = \frac{a_1}{s + p_1} + \frac{a_2}{s + p_2} + \frac{a_3\omega^2}{s^2 + \frac{\omega}{Q}s + \omega^2} \tag{4.1}$$

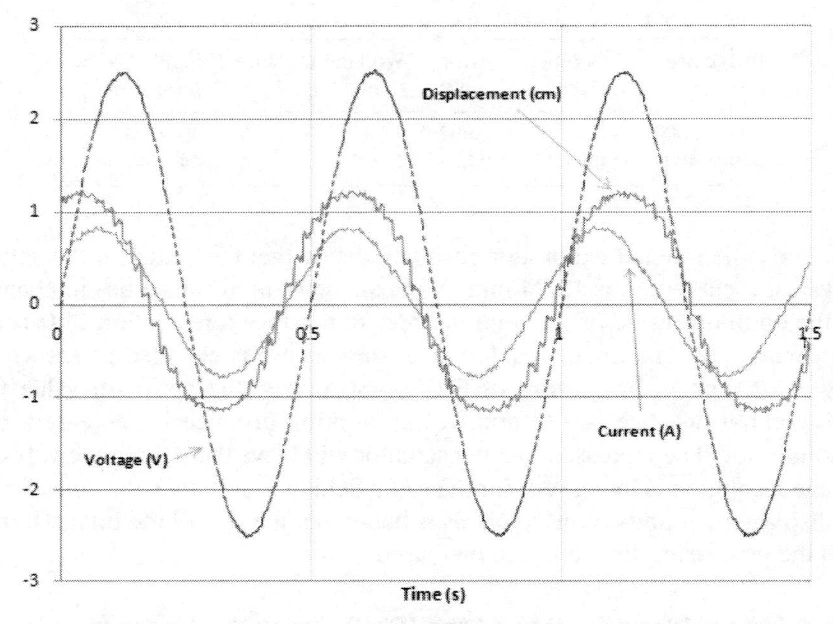

Figure 4.7 Applied voltage, current and curvature k_1 versus time for a typical IPMC strip of 40 mm × 5 mm × 0.1 mm under a frequency of 0.5 Hz.

An identification process consists of two parts, the estimation and the identification of the model from the step response.

The graphical presentation of the input voltage signal applied to the IPMC actuator, and the corresponding current values, are illustrated in Figure 4.7. The same figure also presents the actual displacement of a typical IPMC actuator in response to this voltage and current. A phase difference between the signals is present due to the capacitive behavior of the actuator.

A special algorithm in MATLAB may be used to fit the experimental data to the computed data. Equation (4.2) shows the result of the identification process where a fourth order transfer function has been identified.

$$H(s) = \frac{0.001483s^3 + 0.3665s^2 + 6.244s + 0.0003499}{s^4 + 21.77s^3 + 3544s^2 + 0.000111s + 3292} \tag{4.2}$$

The IPMC has a low pass response. The resonance behavior is observed near 10 Hz. This is a typical response of an IPMC strip where the ionic effect in the transduction reflects the low pass filter, and the mechanical configuration reflects the resonance response.

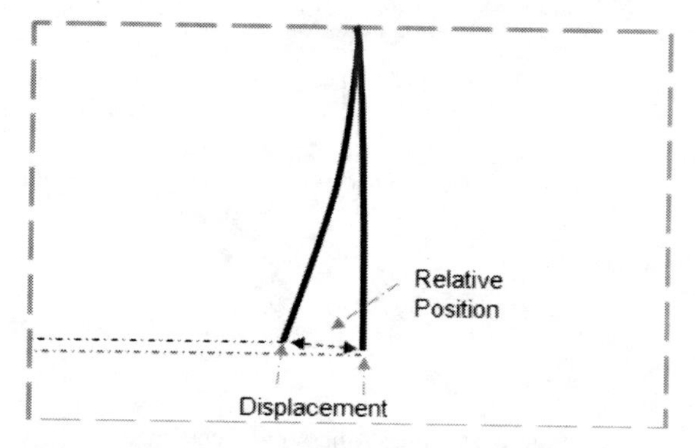

Figure 4.8 Different ways to calculate displacement of an IPMC strip.

Using relatively small approximations the fourth order model may be reduced to a second order model as follows:

$$H_r(s) = \frac{0.00005413s^2 + 0.002358s - 0.000000132}{s^2 + 4.184s + 1.242} \tag{4.3}$$

4.2 First Simple Method

Before analyzing the first method, a detailed description of the characteristics of the measurement method should be given. Figure 8 presents the IPMC actuator in its initial position and in a position when a voltage signal is applied and the actuator bends. The displacement of the actuator is defined in two ways: the edge-end distance between two different positions of the actuator and the distance from the left-hand side of the image. Each technique gives the relative displacement while the second one is recommended for the better comparison with the laser measurement technique.

Figure 4.9 presents a vertical IPMC actuator after the edge detection and thresholding analysis. The central point and the leftmost one are located as shown. The end-point tracking procedure is analyzed comprehensively in a previous work of the same authors [36]. The algorithm detects the bottom of the material shape and seeks to localize the leftmost and the rightmost edge of the bottom of the strip. Once these two points are estimated, the algorithm calculates the center-point.

Depending on the size of the movement, the distance of the two points may be calculated directly (for small displacements). In cases where dis-

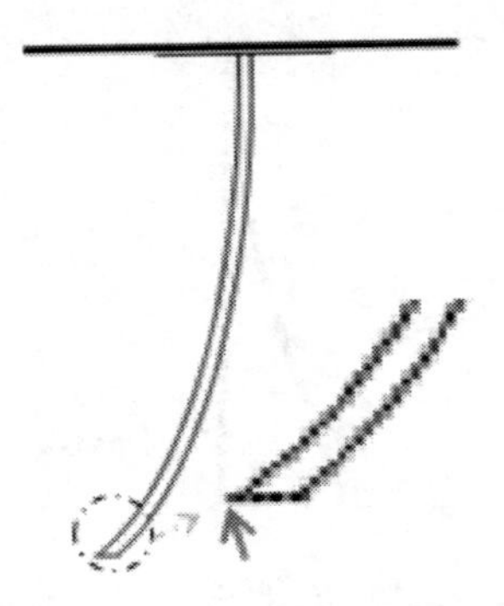

Figure 4.9 Measuring point – end edge.

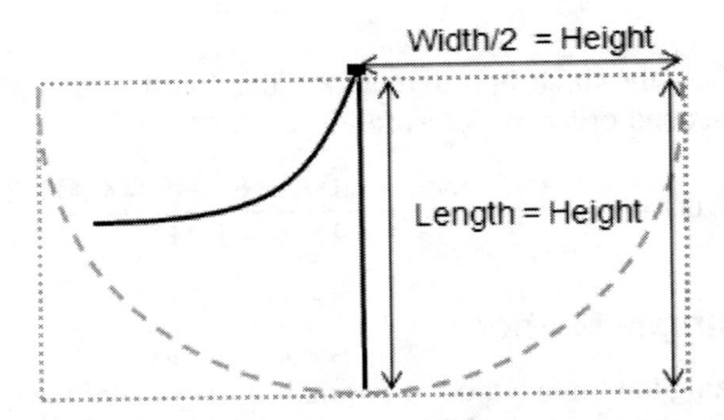

Figure 4.10 The area of interest is limited for best image analysis results.

placement is rather large, the algorithm should calculate the length of the arc in order to obtain a more accurate result.

In order to achieve the best accuracy in calculating the relation between pixels and the real magnitude of the distance in mm, the magnitude of the maximum displacement of the actuator in the field of view is obtained by limiting the vision region of the image as shown in Figure 4.10.

For a better understanding of the feature tracking technique, first a CPU-simple method is discussed. The block diagram of the simple algorithmic process is shown in Figure 4.11.

The first step of each method is the camera activation and parameter initialization. As mentioned above, the camera is calibrated before each specific experiment and for this method has a resolution of 0.1 mm per pixel. While camera frame-rate may be set to over 60 fps, in order to avoid a high number of dropped frames, a value of 60 fps is chosen in all experimental

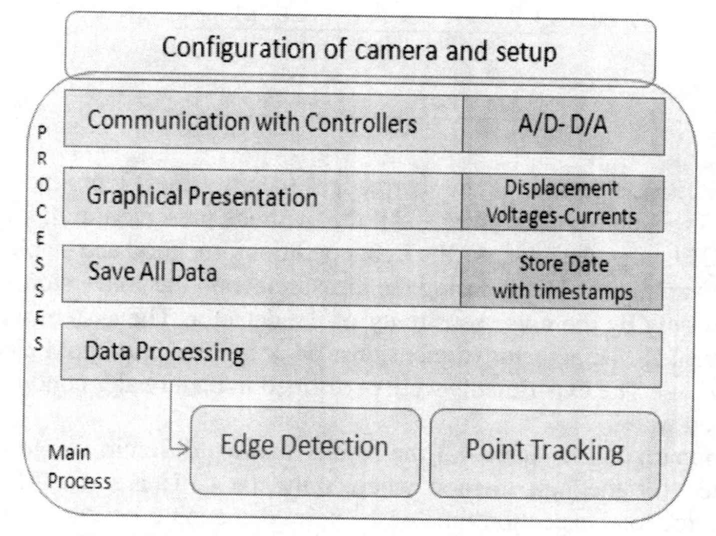

Figure 4.11 Block diagram of a simple algorithm process.

measurements. The algorithm synchronizes video capture communications for receiving images at exactly 60 fps in a separate process without high processing requirements.

The image color depth of each frame is set to grayscale in order to avoid the need for conversion time from a color image. Every captured frame is processed by a separate process which includes edge detection and point tracking methods as shown in Figure 4.11. Every frame is captured directly from a synchronized video stream and is not extracted from a video file.

The algorithm communicates with a microcontroller for the generation of the input voltage of the IPMC actuator and for the measurement of the applied voltage and current (when needed). This procedure has been implemented in a separate process which is synchronized with the main process. Moreover, an additional process handles the graphical display of the signals and the data points of the displacement from the frame processing results.

A SOBEL edge detecting procedure [37] is used to identify the edge clearly and to locate the edge-end points for the analysis of the motion parameters. By contrast a CANNY edge detector [38] may provide better results but increases considerably the execution time. For simplicity in the application and as the main aim is to decrease the processing time, the simple SOBEL edge detector is chosen. One of the most significant characteristics

Table 4.3 Average execution time for a simple method.

Image Size (pixels)	Simple (ms)
640 × 480	9.1
320 × 240	4.45

of the SOBEL algorithm is the sharp edges (depth values) it produces using a pair of 3 × 3 convolution masks. All the methods presented in this work use the SOBEL edge detector. As the light conditions are good and stable (no significant variations of light during the measurements) the conversion is mainly affected only by the noise sensitivity of the detector. The system was tested with a real IPMC actuator (dimensions: 64 × 5 × 0.1 mm, gold electroded, Nafion/Li+). The experiments were performed in underwater conditions with deionised water.

Summarizing, in Table 4.3 the results for two different image sizes are presented for the first method where only the CPU is used. The results concern the total execution time of the point detection procedure.

These results represent the average value of the computation time of processing about 1500 successive images captured by camera during 25 s.

4.2.1 Simple Method and GPU

In the first method, as mentioned above, the meaning of multi-processing is defined as the fast, simultaneous and synchronized execution of all the processes controlled by the main process using only one of the CPU units. In this section we describe the improvement in the execution time achieved for the real parallel execution of the calculations using the multi-processors of GPU units.

The proposed block diagram of the process using CUDA technology is presented in Figure 4.12. This procedure is executed every time an image-frame is received. It is a mixed code where commands are executed in parallel or serially. At the beginning the algorithm checks the synchronization of the video streaming and then starts a loop where all the operations executed.

After a frame is received, the algorithm calls a special function of the kernel which executes in parallel the edge detection and point tracking technique. For the edge detection a thread block size of 16 × 32 = 512 threads was chosen due to the fact that their image height and weight are multiples of 16 and the choice of 32 × 32 = 1024 threads exceeds the limit of 512 threads per block. Due to the dimensional shape of the image the dimension of the Blocks and Grid was chosen equal to 2.

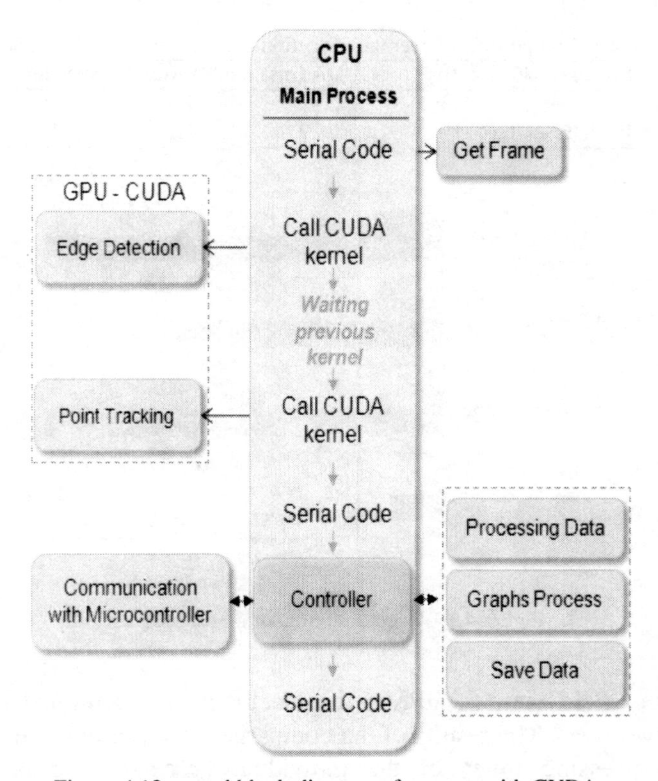

Figure 4.12 posed block diagram of process with CUDA.

Table 4.4 Comparison of average execution time between CPU and GPU.

Image Size (pixels)	CUDA (ms)	CPU (ms)	Speedup
640 × 480	2.5	9.1	3.64
320 × 240	1.9	4.45	2.34

After the edge detection kernel is executed the point tracking algorithm is called. The main process as the controller process controls the synchronization of all operations and calls of the functions.

The measurement technique for the execution time is implemented using time-stamps for both methods. Additionally, CUDA technology provides functions to measure elapsed time using events.

Two image sizes have been tested experimentally in grayscale mode, namely 640 × 480 and 320 × 240. The second one is chosen for larger

Table 4.5 Comparison of average execution time for individual processes.

Process (640 × 480)	CUDA (ms)	CPU (ms)	Speedup
Edge Detection (Sobel)	2.19	7.9	3.61
Point Tracking	0.29	1.1	3.79

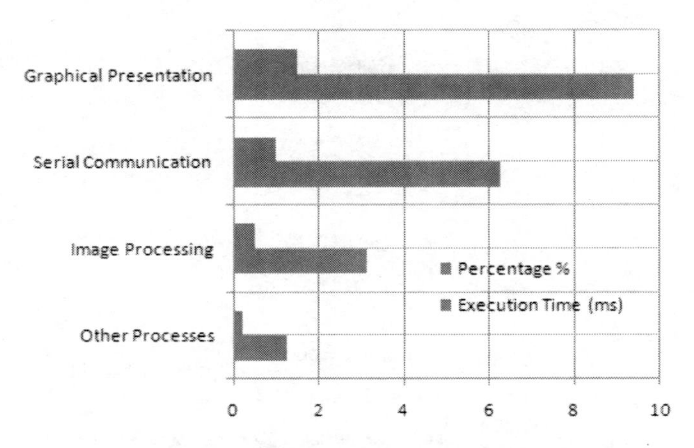

Figure 4.13 Execution time of all procedures.

limitation of field view for IPMC displacement and examined in order to compare two sizes. The results of the comparison are presented in Table 4.4.

The processing times of the individual processes are presented in Table 4.5. The processing time for finding the magnitude of the displacement has been decreased by 72.2%. The second row shows that the point-tracking algorithm has an even better improvement.

Figure 4.13 shows the execution time for each procedure and the corresponding percentage of work in a total time period of 16 ms for a frame rate of 60 fps using CUDA. It should be noted that the graphical presentation of the points can be disabled to reduce the total execution time. The other processes bar stands for data storage and thread handling procedures.

The decrease of the execution time using CPU and GPU satisfies the required specifications of the application. For the frame rate of 60 fps, the image processing time is 9.1 ms which is less than 16.6 ms between two frame receptions. Further decrease of time processing is achieved using CUDA technology.

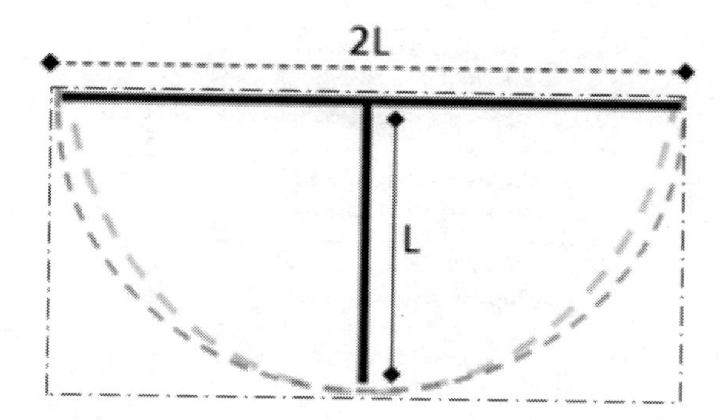

Figure 4.14 Adjusted region of vision on the route of material.

4.3 Small Area Search

The second method improves the processing time by considering a decrease of the size of the area of interest, selecting smaller areas for processing and using GPU acceleration.

Due to the high number of processing pixels of each frame, a smaller area of interest is chosen. There are many factors that must be considered in order to keep the accuracy within satisfactory levels. This is related to the resolution of the image, the distance between the actuator and the camera, and the configuration of the camera. The region of vision is chosen as shown in Figure 4.14. The region width is exactly the length of the actuator in the vertical axis. Concerning the height; it is exactly double the length of the actuator in the horizontal axis.

For a material length of 3 cm an area of 6 × 3 cm is chosen, which corresponds to 600 × 300 (WxH) pixels. In most experiments an image size of 400 × 200 pixels is used. The actuator is not moving out of this region because the bending of the actuator is limited to the top of the horizontal border of the image. More details of the point tracking approach are presented in Figure 4.15.

As mentioned in the previous method, the end-edge of the material is detected after searching within the whole image-frame. To reduce the area for processing, a new algorithm has been developed. The algorithm is based on image-splitting of the moving route of the actuator into orthogonal regions. The vision region is segmented into K overlapped small areas that follow the routing path of the end-strip of the material as shown in Figure 4.16.

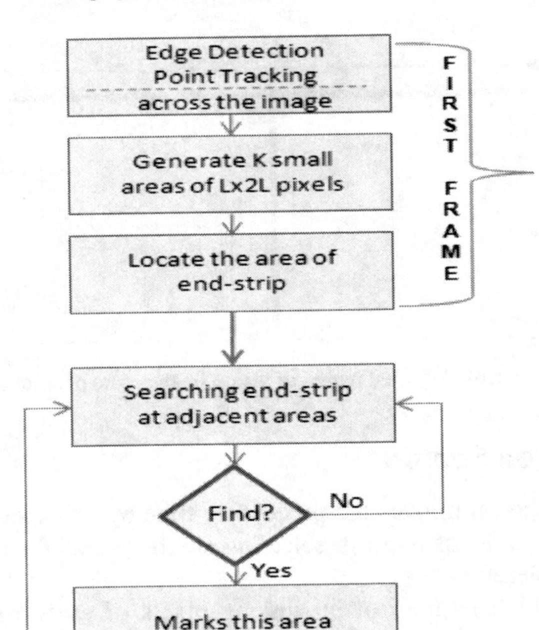

Figure 4.15 Block diagram for second method using small area selection without use of CUDA.

The overlapping area between adjacent small areas is two vertical lines. The size of each dimension for every shape is a multiple of 16 for better adaptation to GPU implemented functions.

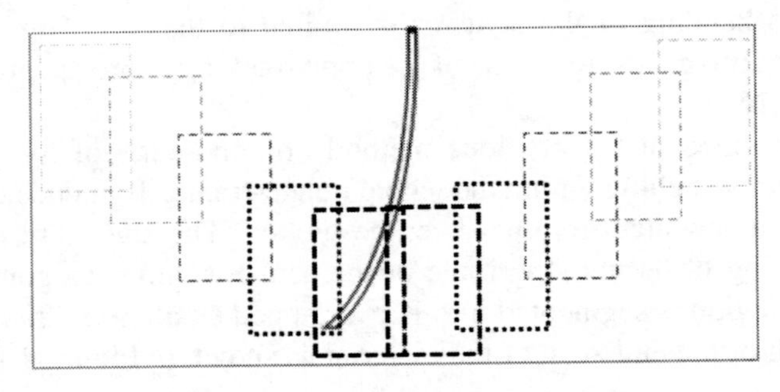

Figure 4.16 Seperation on small areas.

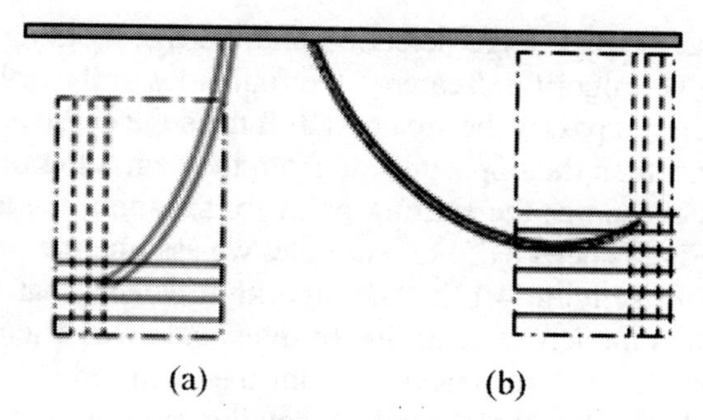

Figure 4.17 Various states of the actuator (a) normal and (b) large bending.

At the beginning the algorithm searches the whole area of the image-frame in order to detect the point of interest, as mentioned in the first method. Then, taking into consideration the arc of the actuator it follows a sinusoidal route and divides the area into small orthogonal areas as shown in Figure 4.16. In the next frame it starts the searching from the small area which is marked at the beginning and continues searching in the adjacent areas until it finds the end of the strip. Again, it marks its area which is the current position of the actuator. Additionally, the algorithm saves the current direction and time slot. In the next frame it uses this information to predict the adjacent area in which the end-point can be found. This procedure is repeated continuously. The edge detection processing is applied in grey-scale image-frames using the SOBEL algorithm.

There are positions of the actuator where the algorithm fails to detect the end of the strip inside the small areas. This happens when the size of the small areas is small or the material presents a large bend because of its non-linear behaviour. Two more considerations are taken into account to solve these problems. The first concerns the increment of the small area size when the algorithm cannot initially detect the end-point. In the second the algorithm executes the search from the beginning covering the whole area of the image. The experiments have shown that the second method provides better results when the material has many random positions.

The material might move on a random bowed track which is specified from the applied input voltage signal, the length and the condition of the actuator. The large bending of the material (Figure 4.17b) should be taken into account in further calculations.

The new improved method for the detection of the end-edge of the IPMC for all bending situations is described as follows.

Initially, the SOBEL edge detection and the thresholding are applied in the small area. The algorithm searches the four sides of the orthogonal area to detect if the material passes the area or not. If the edge of the actuator is inside of the small area then the opposite side from the beginning of the contour of the material is chosen as the starting point for scanning the area using lines until the end-edge is detected. For example, we assume the material is in the situation shown in Figure 4.12b. The algorithm detects that one side of the actuator exists on the left vertical line of the small area. Then, every vertical line at constant intervals is searched from the right side until a significant number of dark pixels are detected. A smaller area around these points is marked for further searching. The algorithm uses the method of matching known patterns for the corner detection. After corner extraction the left-most or the right-most one is chosen. The patterns are applied for each pixel near the end-edge to find a corner.

The position error of this method is ± 0.14 mm for a 400×200 pixel image resolution. In this work the position error is defined as the maximum distance between the measured point-pixel and the real corner-point (point of interest) which is experimentally found as 1 pixel. Another characteristic which is very important in these systems is the detection accuracy, which means the number of corners that are detected correctly. The detection accuracy of the small searching area method is about 90%.

4.3.1 Using GPU When Searching

To further reduce the execution time, CUDA technology is used. The proposed system with a CUDA implementation has two main advantages; it operates in real time and it exhibits quite faster data processing, using parallel executions. The block diagram of the proposed process with CUDA is presented in Figure 4.18. The CUDA approach is a mixed code algorithm, serial and parallel, in the whole area of interest of 400×200 pixels. The two major procedures implemented with CUDA are the "edge detection" and the "end-strip tracking".

At every frame reception the algorithm calls a special function of the kernel which executes the edge detection algorithm in parallel. For this algorithm a thread block size of $16 \times 32 = 512$ threads was chosen due to the fact that the image height and weight are multiples of 16 for 400×200 dimensions. The dimension of Blocks and Grid was set to 2. This part of the algorithm runs for the first time at the beginning of the process.

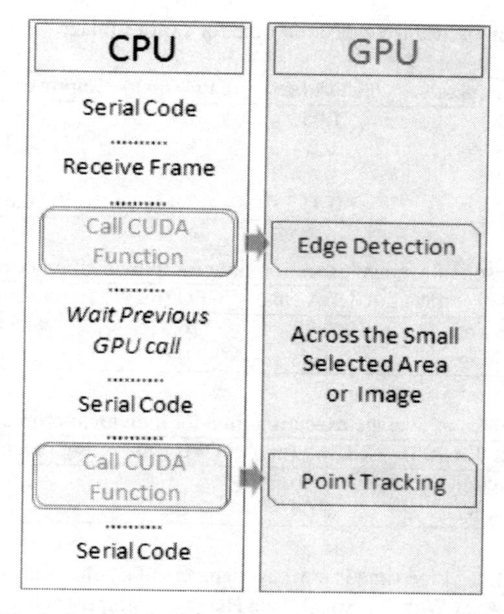

Figure 4.18 Block diagram for the small area seperation method using CUDA.

The data of the image after the edge detection is one-bit black-white and it is easier to find the corner of interest. The algorithm examines side pixels for each pixel-edge and tries to find any matching with known patterns starting from the bottom. The above method can run simultaneously for each pixel-edge since it is independent. The procedure can be fully executed in CUDA.

It should be noticed that the search for the corner of interest is implemented serially. Every point of interest found is marked as a displacement and corresponds to a dedicated time slot which is needed for processing.

The second technique greatly improves the performance of the measurement system in terms of processing speed. The experiments were carried out using GPU and the simple (CPU). The SOBEL filter and the point tracking which is based on the grid and matching approach are implemented with CUDA. Table 4.6 shows that the execution time has been improved for different sizes of images without using the small area selection processing. The algorithm searches the whole image area to find the edge of the strip and then detects the corner of interest using pattern matching.

Table 4.6 Comparison of average execution time for various image sizes without using small area selection.

Image Size (pixels)	CUDA (ms)	CPU (ms)	Improvement factor
640 × 480	1.05	7.1	6.76
640 × 240	0.8	5.5	6.85
400 × 200	0.5	3.2	6.4
200 × 200	0.35	2.4	6.857

Table 4.7 Comparison of average execution time for individual processes for 640 × 480.

Process (640 × 480)	CUDA (ms)	CPU (ms)	Improvement factor
Edge Detection	0.95	6.5	6.84
Point Tracking	0.1	0.6	6

Table 4.8 Comparison of average execution time for individual processes for 400 × 200.

Process (400 × 200)	CUDA (ms)	CPU (ms)	Improvement factor
Edge Detection	0.46	3	6.52
Point Tracking	0.04	0.2	5

Table 4.9 Execution time in various members of small areas (400 × 200).

K	Small Area Width	Small Area Height	Simple (ms)	CUDA (ms)
4	100	200	2.2	0.3
8	50	100	2	0.12
16	25	50	1.8	0.1
20	20	40	1.9	0.2

The most interesting image size is 400 × 200 which is chosen for reducing the processing area. The results of GPU and simple approaches are presented in Tables 4.7 and 4.8 respectively.

It should be noticed that Tables 4.7 and 4.8 show results without the use of the small area selection technique.

Table 4.9 shows the comparison of the small area selection method using the GPU units for a 400 × 200 pixel image size scanning in various K numbers of small areas. It can be observed that a major time improvement appears when K equals 16. For a higher value of K the processing time is the least, but after K surpasses 20 the search becomes ineffective.

Table 4.10 presents the execution time for the detection of the corner-of-interest at various frequencies and actuating signals of the IPMC when the number of small areas is 16. The time is increased for a sinusoidal signal at 7 Hz due to the actuator speed and the fact that the algorithm searches more than one small area to find the correct position.

Table 4.10 Execution time for point of interest detection in various signals.

Signal/Frequency	0.5 Hz	2 Hz	7 Hz
Sinusoidal	0.11 ms	0.12 ms	0.47 ms
Triangle	0.12 ms	0.14 ms	0.165 ms
Square	0.155 ms	0.157 ms	0.157 ms

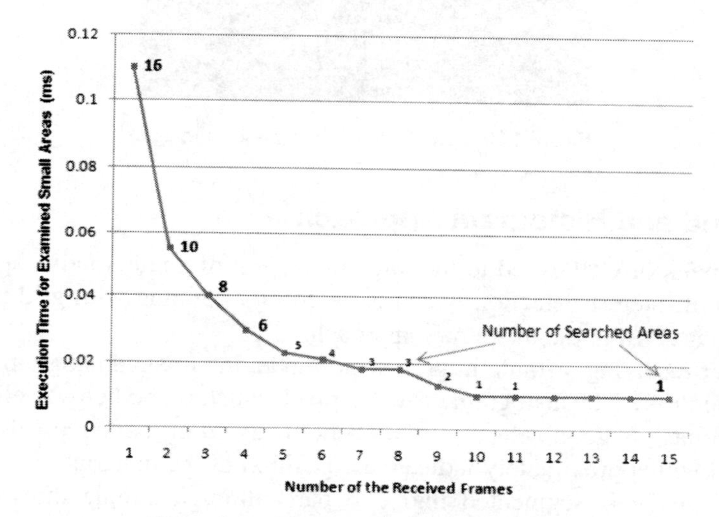

Figure 4.19 Progress for small area tracking.

It is observed that the execution time for a square signal is approximately the same due to more predictable conditions in the actuator position.

Figure 4.19 shows the progress of the small-area tracking approach for the point of interest detection when the number of received frames increases using GPUs, for a sinusoidal signal with the frequency of 1 Hz. At the beginning (first frame), the algorithm searches 16 small areas. First the edge-end is located, then the area is marked and continues by searching in the neighborhood areas. Experiments have shown that the number of searched small areas for tracking is minimized to one only, when the number of the received frames surpasses ten. Therefore, the algorithm can locate the end of the strip very fast in the adjacent areas.

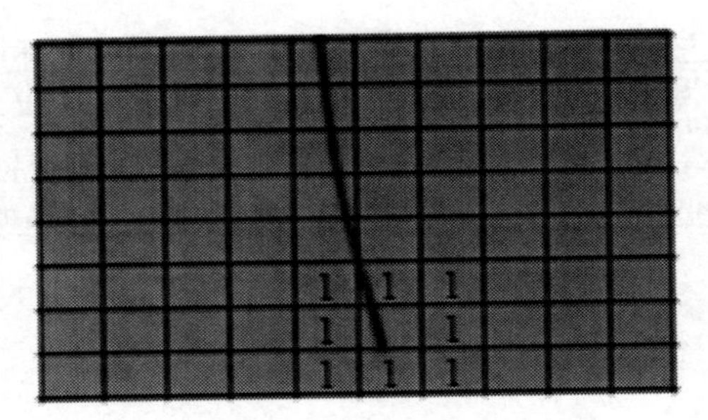

Figure 4.20 Grid approach in grey-scale image.

4.4 Grid and Histogram Approach

The benefits of CUDA led to the implementation of an alternative approach for area of interest detection which is based on searching with grid in grey-scale images using the histogram approach.

After receiving a frame in grey-scale mode, the image histogram can be calculated [39]. The histogram provides pixel statistics and helps to eliminate noise points. A good selection of threshold regarding darker pixels of the material will approximately indicate the position of the material.

The image is segmented into two parts through simple thresholding. The choice of the threshold T is obtained using histogram information. The frame-image consists of the material and the background which have homogeneous intensity. Hence, the histogram has two main peaks. The threshold is the local minimum between the two tops. If the histogram is very noisy a one-dimensional low pass filter is used.

Then the grid search is applied. A fixed number of vertical and horizontal lines are set up at a constant distance as shown in Figure 4.20. For a 400×200 image size 40 vertical and 20 horizontal lines are used with a 10 pixel distance between them. The block diagram of the proposed process is presented in Figure 4.21.

The last horizontal line, where no values above the threshold are detected, is marked. The algorithm calculates where the lower part of the strip is, and proceeds further. If the material is on the left part of the window it marks the leftmost vertical line or the rightmost one respectively.

The algorithm detects and notes how many points (regions of few points) exist with values under the threshold in the first horizontal lines where it is part of the material like the case illustrated in Figure 4.17b. Then the algorithm indicates that the material presents a large bending. If the material is

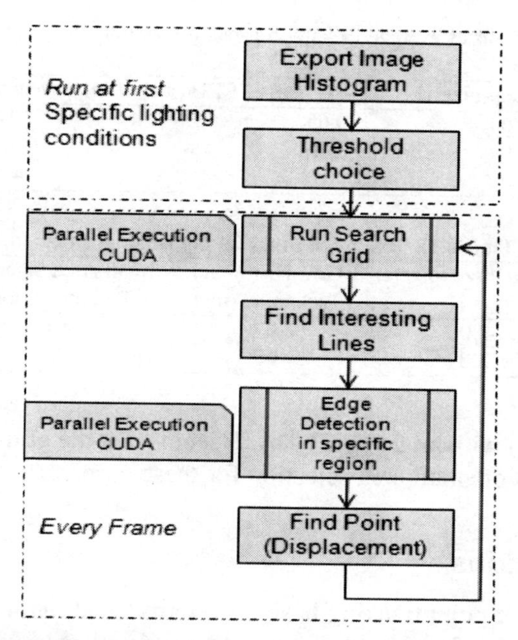

Figure 4.21 Block diagram of point detection using a Grid approach and CUDA.

located on the left side of the image, it takes the first free left line and the first black one with its marked point and marks a small area for processing which has both sizes of dimensions 10 around the centre. In this way the algorithm ensures that the interest point is within the processing area.

The next step is the edge detection in the marked area which is determined by marked lines and is usually a region of 20 × 20 pixels. Hence, the edge detection area is reduced significantly. All the above processes are executed in parallel by CUDA.

Additionally, the histogram extraction is implemented with CUDA and is initially applied in every frame, processing for better threshold detection. Due to the same effect and the good lighting conditions, the extraction procedure is executed only to the top. Next, the interest point detection is implemented in the black-white mode image, as mentioned, using CUDA.

The results of the third method which uses the grid approach and the histogram in grey-scale mode are presented in Table 4.11.

Table 4.12 shows the comparison results (average) in terms of execution time for each method when detecting the point of interest in extreme situ-

Table 4.11 Comparison of average execution times for various image sizes using the Grid-Histogram approach.

Image Size (pixels)	CUDA (ms)	CPU (ms)	Improvement factor
640 × 480	0.63	4.5	7.14
600 × 300	0.4	4	10
400 × 200	0.15	2.2	14.6

Table 4.12 Execution times for big bending situation.

Image Size (pixels)	Grid-Histogram Technique (ms)	Small Area Selection Technique (ms)
640 × 480	3.7	4.1
400 × 200	1.8	2

ations (big bend) without using CUDA. It seems that the grid-method is faster than the method of small area selection for these cases.

4.5 Conclusions

Three methods of point tracking have been introduced in this work, and have been compared for fast image processing of visual measurements of Ionic Polymer-Metal Composite (IPMC) actuator motion parameters.

The first method uses a simple approach for the tracking. A processing time of 9.1 ms for an image size of 640 × 480 can be used for approximately 110 fps if other running processes on CPU are taken into account. The problem of the inefficiency in handling large bending of the moving material and further reductions of processing time are treated by the other proposed techniques.

The second improved method uses a smaller frame, segmentation, edge filter, grid and matching processing. It is based on predictable tracking on small processing areas using motion information such as speed and direction. Results show a total acceleration factor of 70, from 7.1 ms at 640 × 480 to 0.1 at 400 × 200.

The third improved method uses histogram information, grid to grey-scale images and corner searching similar to the previous method. Using only CPU a speedup of factor 1.57 is achieved from 7.1 ms at 640 × 480, using second method, to 4.5 at 640×480 and it is up to 47 times faster for CUDA operation from 7.1 ms at 640×480, while using the second method, to 0.15 at 400×200.

Comparing the last two methods it can be concluded that they offer a significantly reduced processing time for the speed of the movement which can be estimated theoretically to values of up to 10 KHz. A prerequisite for high speed measurements is the adequate support of the frame rate of the camera.

The results have shown that the second method engages a larger number of calculations regarding the estimation of the position and is slower if the material bends out of the track. The third approach is slower than the second one because it is implemented in the entire area, but handles better the larger bending of the material within shorter execution times.

References

[1] C. Steger, M. Ulrich, and C. Wiedemann. *Machine Vision Algorithms and Applications*, Wiley-VCH, 2008.

[2] S. Pankanti, L. Brown, J. Connell, A. Datta, Q. Fan, R. Feris, N. Haas, Y. Li, N. Ratha, H. Trinh. Practical computer vision: Example techniques and challenges. *IBM Journal of Research and Development*, 55(5):1–12, 2011.

[3] M. Vincze and G. Hager. *Vision for Vision-Based Control of Motion*. Wiley-IEEE Press, 2000.

[4] P. Smith, T. Drummond, and R. Cipolla. Layered motion segmentation and depth ordering by tracking edges. *IEEE Transactions on Pattern Analysis and Machine Intelligence*, 26(4), 2004.

[5] R. Weber, A. Gothandaraman, R.J. Hinde, and G.D. Peterson. Comparing hardware accelerators in scientific applications: A case study. *IEEE Transactions on Parallel and Distributed Systems*, 22(1):58–68, 2011.

[6] N. Farrugia, F. Mamalet, S. Roux, Yang Fan, and M. Paindavoine. Fast and robust face detection on a parallel optimized architecture implemented on FPGA. *IEEE Transactions on Circuits and Systems for Video Technology*, 19(4):597–602, 2009.

[7] H. Xinming, L. Cao, and M. Jing. System architecture and implementation of MIMO sphere decoders on FPGA. *IEEE Transactions on Very Large Scale Integration (VLSI) Systems*, 16(2):188–197, 2008.

[8] A. Peleg and U. Weiser. MMX technology extension to the Intel architecture. *Micro, IEEE*, 16(4):42–50, 1996.

[9] W. Wang, X. Zhao, and X. Feng. Parallel wavelet-based image segmentation using MPI. *TENCON IEEE*, 2:80–90, 2004.

[10] M.D. Jones, R. Yao, and C.P. Bhole. Hybrid MPI-openMP programming for parallel OSEM PET reconstruction. *IEEE Transactions on Nuclear Science*, 53(5):2752–2758, 2006.

[11] J.D. Owens, M. Houston, D. Luebke, S. Green, J.E. Stone, and J.C. Phillips. GPU computing. *Proceedings of the IEEE*, 95(5):879–899, May 2008.

[12] Hwu Wen-Mei, C. Rodrigues, S. Ryoo, and J. Stratton. Compute unified device architecture application suitability. *Computing in Science & Engineering*, 11(3):16–26, 2009.

[13] K.P. In, N. Singhal, H.L. Man, C. Sungdae, and C.W. Kim. Design and performance evaluation of image processing algorithms on GPUs. *IEEE Transactions on Parallel and Distributed Systems*, 22(1):91–104, 2011.

[14] NVIDIA CUDA Programming Guide Version 2.2.1, NVIDIA, Santa Clara, CA, May 26, 2009.

[15] R. Cheng, E. Yang, T. Liu, and L. Wu. CUDA-based directional image/video interpolation. In *Proceedings of International Conference on ICALIP*, pp. 125–129, November 2010.

[16] M. Garland, S. Le Grand, J. Nickolls, J. Anderson, J. Hardwick, S. Morton, E. Phillips, Z. Yao and V. Volkov. Parallel computing experiences with CUDA. *Micro, IEEE*, 28(4):13–27, 2008.

[17] R. Arora, R. Tulshyan, and K. Deb. Parallelization of binary and real-coded genetic algorithms on GPU using CUDA. In *Proceedings of IEEE Congress on Evolutionary Computation (CEC)*, pp. 1–8, July 2010.

[18] X. Huang and Q. Wu. Real-time corner detection algorithm based on GPU. *Proceedings of International Symposium on ISISE*, pp. 608–610, December 2010.

[19] E. Ito, S. Saga, T. Okatani, and K. Deguchi. GPU-based high-speed and high-precision visual tracking. In *Proceedings of SICE Annual Conference*, pp. 2151–2154, August 2010.

[20] K. Ogawa, Y. Ito, and K. Nakano. Efficient canny edge detection using a GPU. In *Proceedings of First International Conference on Networking and Computing (ICNC)*, pp. 279–280, November 2010.

[21] J. Gómez-Luna, J.M. González-Linares, J.I. Benavides, and N. Guil. Parallelization of a video segmentation algorithm on CUDA – Enabled Graphics Processing Units. *Euro-Par Parallel Processing* 5704:924–935, 2009.

[22] H. Xie, K. Gao, Y. Zhang, J. Li, and Y. Liu. GPU-based fast scale invariant interest point detector. In *Proceedings of IEEE International Conference on ICASSP*, pp. 2494–2497, March 2010.

[23] H. Fassold, J. Rosner, P.P. Schallauer, and W. Bailer. Realtime KLT feature point tracking for high definition video. In *Proceedings of Computer Graphics, Computer Vision and Mathematics (GraVisMa 09)*, 2009.

[24] N. Zhang, Y.-S. Chen, and J.-L. Wang. Image parallel processing based on GPU. In *Proceedings of 2nd International Conference on ICACC*, Vol. 3, pp. 367–370, 2010.

[25] J. Brufau-Penella, K. Tsiakmakis, T. Laopoulos, and M. Puig-Vidal. Model reference adaptive control for an ionic polymer metal composite in underwater applications. *Smart Materials and Structures*, 17(4).

[26] H.C. Liaw, and B. Shirinzadeh. Robust adaptive constrained motion tracking control of piezo-actuated flexure-based mechanisms for micro/nano manipulation. *IEEE Transactions on Industrial Electronics*, 58(4):1406–1415, 2011.

[27] R.C. Richardson, M.C. Levesley, M.D. Brown, J.A. Hawkes, K. Watterson and P.G. Walker. Control of ionic polymer metal composites. *IEEE/ASME Transactions on Mechatronics*, 8(2):245–253, June 2003.

[28] P. Brunetto, L. Fortuna, P. Giannone, S. Graziani, and S. Strazzeri. Static and dynamic characterization of the temperature and humidity influence on IPMC actuators. *IEEE Transactions on Instrumentation and Measurement*, 59(4), 2010.

[29] M. Shahinpoor, Y. Bar-Cohen, J. Simpson, and J. Smith. Ionic polymer-metal composites (IPMC's) as biomimetic sensors, actuators and artificial muscles – A review. *Smart Materials and Structures*, 7(6):R15–R30, 1998.

[30] K. Tsiakmakis, J. Brufau-Penella, M. Puig-Vidal, and T. Laopoulos. A camera based method for the measurement of motion parameters of IPMC actuators. *IEEE Transactions on Instrumentation and Measurement*, 58(8):2626–2633, August 2009.

[31] Z. Chen, Y. Shen, N. Xi, and X. Tan. Integrated sensing for ionic polymer-metal composite actuators using PVDF thin films. *Smart Materials and Structures*, 16(2):262–271.

[32] A. Punning, M. Kruusmaa, and A. Aabloo. Surface resistance experiments with IPMC sensors and actuators. *Sensors and Actuators A: Physical*, 133(1):200–209, January 2007.

[33] Z. Zhang. A flexible new technique for camera calibration. *IEEE Pattern Analysis and Machine Intelligence*, 22(11):1330–1334, November 2000.

[34] Q.T. Dinh, K.A. Kyoung, N.C.N. Doan, and I.Y. Jong. Identification of a nonlinear black-box model for a self-sensing polymer metal composite actuator. *Smart Materials and Structures*, 19(8), 2010.

[35] C. Bonomo, L. Fortuna, P. Giannone, S. Graziani, and S. Strazzeri. A nonlinear model for ionic polymer metal composites as actuators. *Smart Materials and Structures*, 16(1), 2007.

[36] K. Tsiakmakis and Th. Laopoulos. Image analysis for measuring motion parameters with a CCD camera. In *Proceedings of IEEE International Workshop on Imaging Systems and Techniques (IST)*, pp. 1–6, May 2007.

[37] M. Juneja and P.S. Sandhu. Edge detection techniques: Evaluations and comparisons. *International Journal of Computer Theory and Engineering*, 1(5):614–621, December 2009.

[38] J. Canny. A computational approach to edge detection. *IEEE Trans. Pattern Analysis and Machine Intelligence*, 8(1):679–698, 1986.

[39] N. Ramesh, J.-H. Yoo, and I.K. Sethi. Thresholding based on histogram approximation, Vision. *IEEE Proceedings on Image and Signal Processing*, 142(5):271–279, October 1995.

5

Smartphone-Based Photoplethysmogram Measurement

Yuriy Kurylyak, Francesco Lamonaca and Domenico Grimaldi

Department of Electronics, Computer and System Sciences, University of Calabria, Rende – CS, Italy; e-mail: {kurylyak, flamonaca, grimaldi}@deis.unical.it

Abstract

Smartphones have become one of the widest and often used devices that people bring almost every time and everywhere. Their computational capacities allow their application to many every-day tasks. One of them is health state monitoring. This chapter presents a smartphone-based photoplethysmogram (PPG) acquisition and pulse rate evaluation system. The proposal was designed for different smartphone models, equipped with a LED or not. Different cameras represent the same acquired information in different ways: changes may occur in color saturation, resolution, frame rate, etc. Therefore, several smartphones were used to define the common characteristics of the captured video, and establish proper criteria for PPG extraction. Moreover, the appropriate algorithms were proposed and validated to verify the correct device usage, the system calibration, the PPG and pulse rate evaluation. The experimental results have confirmed the correctness and suitability of the proposed method with respect to the medical pulse measurement instruments.

Keywords: photoplethysmography, pulse wave analysis, pulse rate measurement, smartphone, non-invasive monitoring.

Richard J. Duro and Fernando López Peña (Eds.), Digital Image and Signal Processing for Measurement Systems, 135–164.

5.1 Introduction

Monitoring of vital parameters is very important for timely detection and prevention of any health diseases. The blood pressure, heart rate and their changes are ones of the most important parameters to control.

There are different techniques to control heart activity: electrocardiography, ambulatory blood pressure monitoring, photoplethysmography, etc. When patients are asked to measure their own heart rate, usually the palpation technique is used, however it is not precise. Therefore, individuals should be properly trained on how to measure their own heart rate accurately [1]. To overcome the human factor automatic systems have been proposed.

5.1.1 Electrocardiography

Electrocardiography (ECG) allows evaluating the performance of the cardio vascular system with high accuracy, and it is "a gold standard" for beat-to-beat heart rate (HR) measurements [2]. It is an interpretation of the electrical activity of the heart, detected by the electrodes attached to the skin, and recorded by an ECG machine over a time interval [3]. Each heart beat is represented by a regular sequence of wave patterns (Figure 5.1).

However, the ECG requires attaching and correctly placing multiple electrodes on the body. That limits the device usage to a clinical environment with trained personnel and makes such approach impractical for most individuals interested in monitoring their HR in natural environments [2].

Moreover, many patients are subjects to a so-called "white coat effect". In particular, several studies of white coat effect have confirmed that it occurs in 20% or more of the hypertensive population [4]. The white coat hypertension is defined as the presence of increased blood pressure due to nervousness when undergoing a clinical examination, while at home it remains normal. Indeed, according to a recent research by Kaiser Permanente Colorado in collaboration with the American Heart Association and Microsoft Corp., patients performing self-monitoring of their vital parameters are 50% more likely to

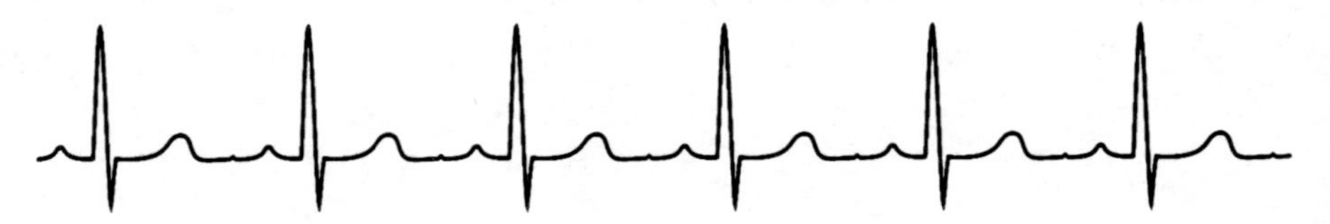

Figure 5.1 ECG wave with detected heart beats.

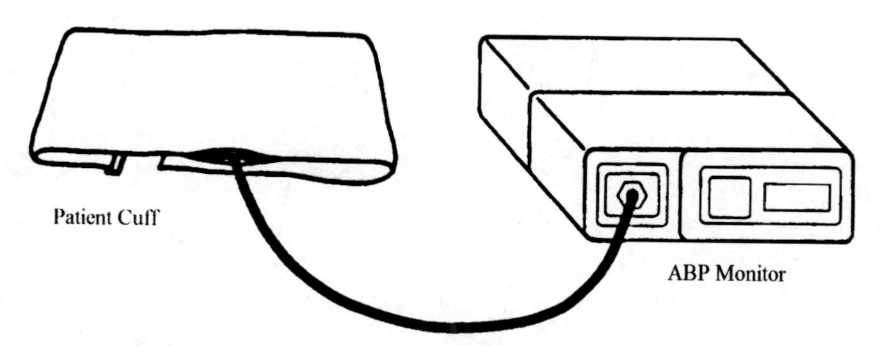

Figure 5.2 Spacelabs 90207 ABP Monitor.

have their blood pressure under control [5]. Therefore, there is a need for low-cost physiological monitoring solutions that are easy to use, accurate, and can be used at home or in ambulatory conditions [6].

There are alterative portable ECG devices such as Holter monitors that allow continuous monitoring of the cardiovascular system. Once the electrodes are attached to the chest, the patient can continue normal activities for 24 hours or more. Then, the cardiologist analyzes the recorded ECG and diagnoses. The main drawback of such solution is that there is no immediate feedback to the user, so there is no possibility to help the patient when the incident occurs [7].

5.1.2 Ambulatory Blood Pressure Monitoring

In a similar way, Ambulatory Blood Pressure (ABP) monitoring devices are used for non-invasive examination of heart activity. They provide continuous 24 hour measurements of the blood pressure and HR at regular time intervals. Having been developed mostly to identify patients with white coat hypertension, they become very useful for the determination of hypertensive end-organ damage risk [8]. For example, a Spacelabs 90207 ABP Monitor (Figure 5.2) [9] is a clinically validated medical device [10–12] tested according to the protocols of the Association for the Advancement of Medical Instruments [13], the American Heart Association [14], and the British Hypertension Society [15].

However, as in the case of portable ECG devices, the ABP monitoring devices are expensive and do not provide the real time measurement results. Patients should visit their doctors for viewing and analyzing the measurements.

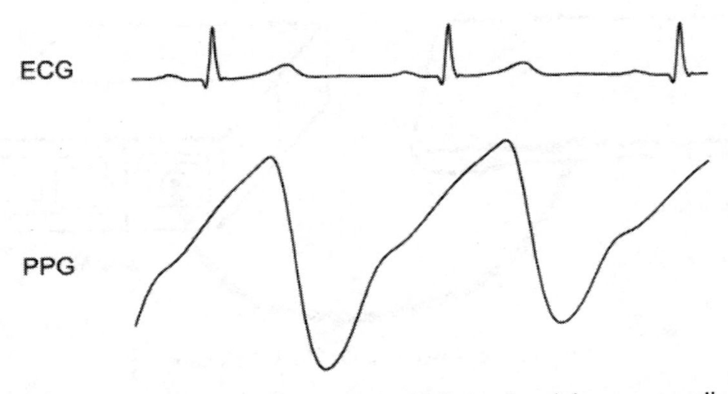

Figure 5.3 The pulsatile (AC) component of the PPG signal and the corresponding electrocardiogram [16].

5.1.3 Photoplethysmography

An alternative non-invasive technique for the detection of blood volume changes during a cardiac cycle is photoplethysmography (PPG). It is a simple and low-cost optical technique that can be used to detect blood volume changes in the microvascular bed of tissue [16]. The technique assumes skin illumination with penetrating optical radiation, usually from a light emitting diode, with a subsequent detection of the signal by a photodetector [17]. Most often the PPG operates at a red or a near infrared wavelength [16].

The PPG has considerable potential for telemedicine including home/remote patient health monitoring. Miniaturization, ease-of-use and robustness are key design requirements for such systems [16]. Clinical PPG applications include monitoring of heart and respiration rate, blood oxygen saturation, pressure as well as detection of peripheral vascular diseases [17].

The PPG waveform consists of a pulsatile ("AC") physiological waveform attributed to cardiac synchronous changes in the blood volume, and a slowly varying ("DC") baseline. The "AC" component has its fundamental frequency typically around 1 Hz, depending on the heart rate. The "DC" component is influenced by respiration, sympathetic nervous system activity and thermoregulation [16]. Figure 5.3 shows the pulsatile component of an acquired PPG waveform and the corresponding electrocardiogram.

As was mentioned, the "AC" component corresponds to the heart beats and can be used for heart activity monitoring. The PPG probe should be held securely in place to minimize the probe-tissue movement artifacts [16].

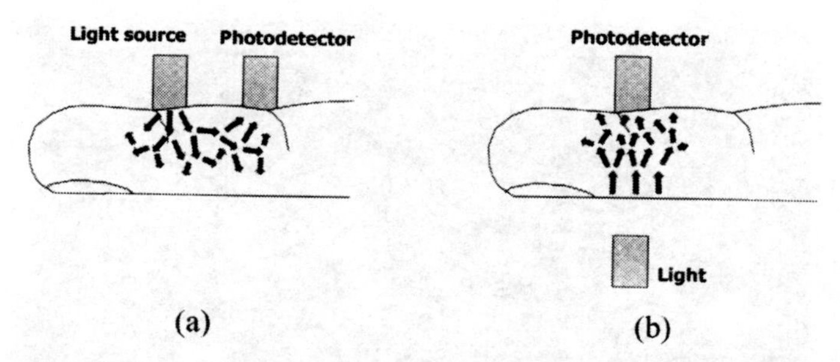

Figure 5.4 Reflection (a), and transmission (b) modes for video acquisition [18].

Figure 5.5 CMS50DL finger pulse oximeter SpO2 monitor.

There are two possible PPG operational modes: (1) transmission, when the tissue sample (e.g. fingertip) is placed between the source and detector (Figure 5.4a), and (2) reflection when the LED is placed next to the detector (Figure 5.4b) [18]. The transmission mode imposes more restrictions than the reflection mode on the body locations available for study [16].

Such photometric-based plethysmogram is normally obtained by using a pulse oximeter (Figure 5.5) [19, 20]. The device is placed on a thin part of the subject's body, usually a fingertip or earlobe. The light with red and infrared wavelengths sequentially passes through the subject to a photo-detector that measures the changes in light absorption [21].

In addition to the PPG waveform, an oximeter evaluates the level of oxygen in blood and computes a pulse rate (PR). Figure 5.6 shows typical information obtained by the CMS50DL oximeter and displayed by the SpO2 software, which comes with the device.

The PPG signal obtained in this way is familiar to clinicians [20]. It clearly shows the pulsatile waveform caused by the pressure wave from the cardiac cycle, and the respiratory sinus arrhythmia induced by breathing [21].

Since the pulse oximeter is non-invasive and relatively inexpensive, in addition to the PR and level of oxygen in blood provided by such devices, much

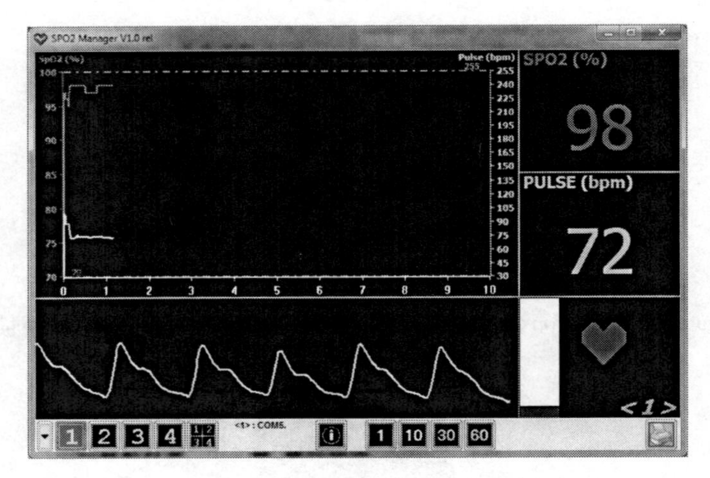

Figure 5.6 Measurement results showed in SpO2 Manager.

research has been carried out in extracting additional biometric information from the waveform. Linder et al. [21] extracted the following parameters from the obtained PPG: the pulse height, peak threshold, cardiac period, full width half max, and peak width (Figure 5.7), and used it to detect changes in posture.

Analysis of the blood volume pulse contour has become important because it contains much information about cardiovascular activity [22]. The final goal is to use the pulse oximeter as a primary sensor in an affordable, wearable health monitoring system [21].

5.1.4 Photoplethysmography Imaging

Replacing the photodetector, used in pulse oximeters, by a video camera enables photoplethysmography imaging. It is an emerging area for research that provides advantages in terms of improved sensitivity, and real-time large surface area measurement [23]. Optical video monitoring of the skin by a digital camera provides information related to the subtle color changes caused by the cardiac signal, and the pulsatile signal [6].

A preliminary CCD camera-based imaging photoplethysmographic system was described in [23, 24]. Fast digital cameras allow the development of PPG imaging, a totally contactless technique for monitoring a larger field of view and different depths of tissue by applying multi-wavelength LEDs. The PPG imaging system can work in both transmission and reflection modes as

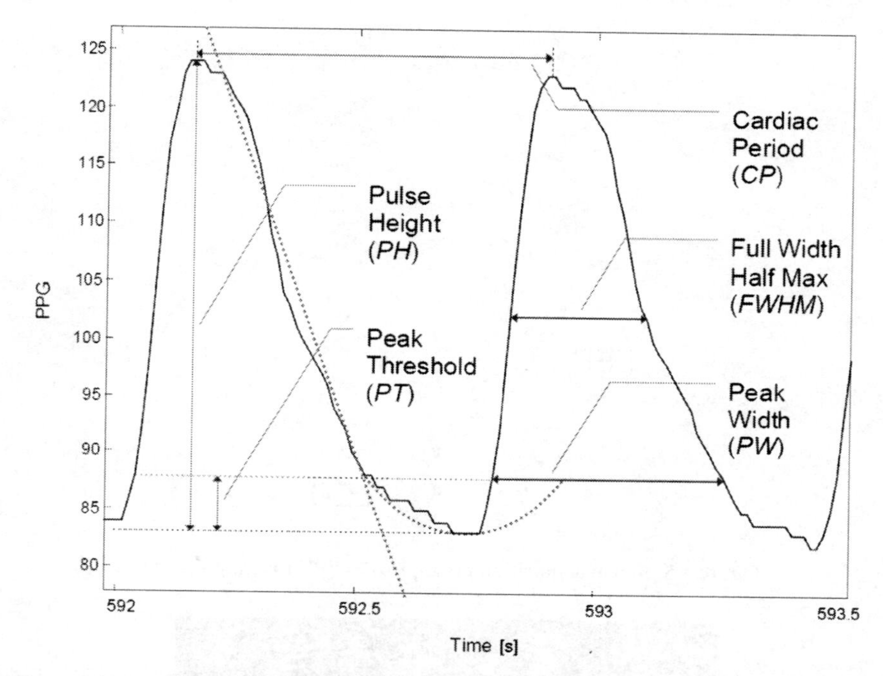

Figure 5.7 The features of the PPG pulsatile component used in [21]: Pulse Height, Peak Threshold, Cardiac Period, Full Width Half Max, and Peak Width.

it is depicted in Figure 5.8. The light intensity that passes through the finger varies with the pulsing of the blood and its plot against time is referred to a PPG signal.

5.1.5 Smartphone-based Health Monitoring Systems

Nowadays smartphones have become one of the widest and often used devices that people bring almost everywhere. In addition, their computational power, possibility of wireless communication as well as their multifunctional user interface allows their usage in very wide spheres.

Smartphones are often used in telemonitoring to receive information from portable medical devices (e.g., blood pressure, glucose and pulse oximeter monitors) and mobile sensors (e.g., physical activity, accelerometer counts, heart rate, respiration rate, pulse pressure, and wireless electrodes) [2]. As an example, the iHealth Lab Inc. has announced the iHealth Blood Pressure Monitoring System for the iPhone, iPod Touch and iPad (Figure 5.9) [25].

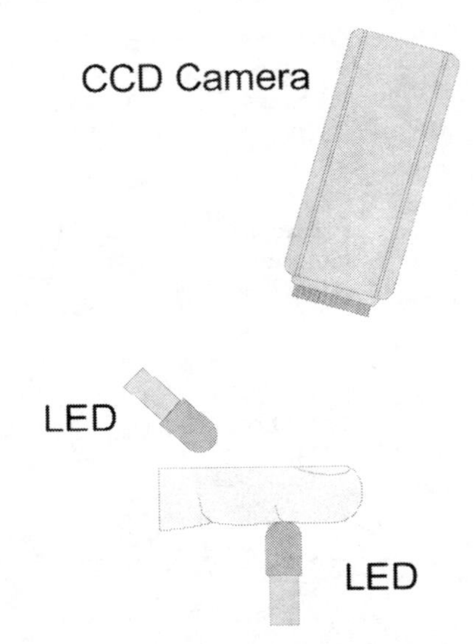

Figure 5.8 Signal acquisition principle of a PPG imaging system.

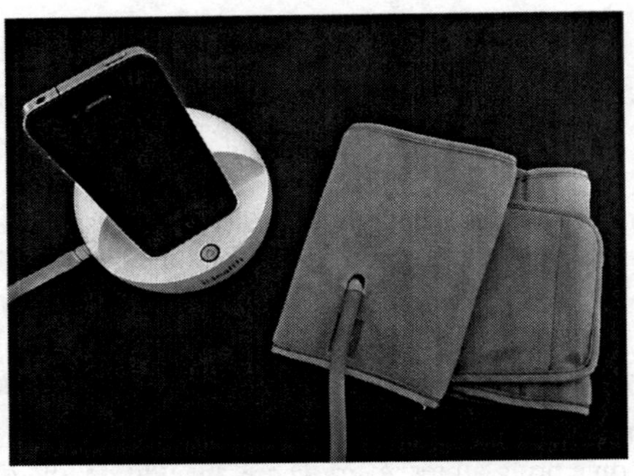

Figure 5.9 The iHealth Blood Pressure Monitoring System.

It consists of a hardware dock, a blood pressure arm cuff and software, and allows the users to self-monitor their blood pressure at home as well as share results with a doctor. There are also pulse oximeters capable of sending the measured results to smartphones using Bluetooth or Wi-Fi connection.

Such devices can be organized then into personalized health monitoring systems (Figure 5.10). The patient fixes sensors (e.g. oximeter) on the

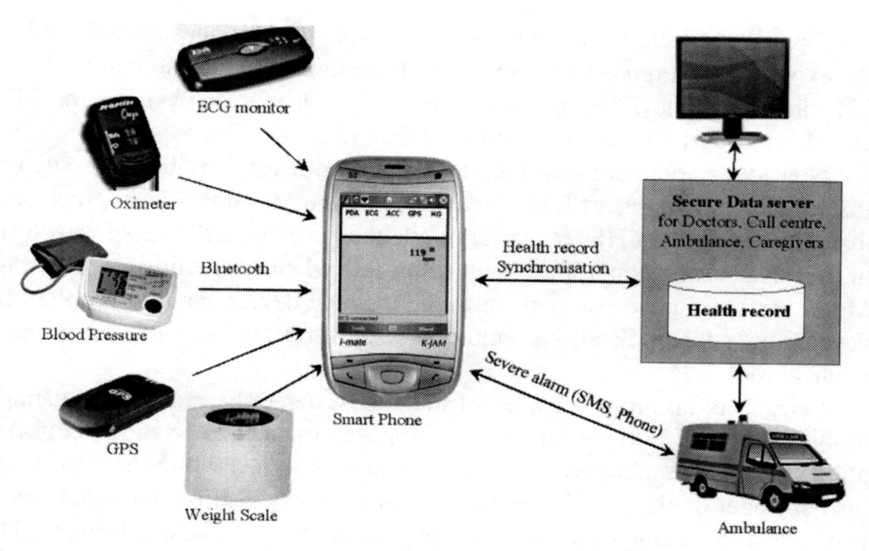

Figure 5.10 Personalized health monitoring architecture [7].

body that communicate with a smartphone sending measurement results. The smartphone then processes the received data and monitors the patient's health. In the case of emergency, it automatically calls an ambulance or sends an SMS to the doctor with the location of the patient and the reason [7, 26].

There are different health and healthcare smartphone applications already available on the market for Android, Apple iOS, RIM BlackBerry, Symbian, Windows Mobile 6.x and Windows Phone 7. As an example we can cite an EU-funded project for older people with multiple chronic conditions eCAALYX (Enhanced Complete Ambient Assisted Living Experiment). The smartphone-based application receives data from the patient-wearable wireless health sensors and communicates over the Internet with a remote server accessible by healthcare professionals who are in charge of the remote monitoring and management of the older patient with multiple chronic conditions [27].

5.1.6 Smartphone-based Photoplethysmography

Most of the current generation cellular phones are equipped with high-resolution cameras, processors and light-emitting diode flashes (LEDs). This is very similar to the PPG imaging technology and, therefore, instead of using a smartphone just as a device for storing and visualizing measured data, they can directly measure some vital characteristics. Smartphones can be used for the express-measurement of such vital characteristics as pulse rate [2, 6, 17, 28], breathing rate [29], as well as providing deeper analysis of the PPG waveform in order to extract additional data [30].

Nowadays, there are smartphone-based commercial applications such as Instant Heart Rate, Heart Rate Tester, Pulse Rate Monitor, Cardiograph, etc. that allow evaluating HR. However, while they provide a PPG-like waveform in the ideal usage conditions, they often fail when something goes wrong. Moreover, there is no comparison to the medical devices and, as reported by developers, such applications should be used for reference only but not as a medical tool.

Pelegris et al. proposed a novel method to detect heart beat rate using a mobile phone [28]. In particular, they proposed to analyze brightness information of the grayscale portion of every captured frame, while the user keeps his/her finger on the lens. To ensure reliability of acquisition, the input signal is matched to a crude heart beat pattern of alternating peaks and troughs. The results were based on the Nokia N95 smartphone, and the authors reported a performance problem of the Android-based smartphone.

Jonathan and Leahy used a Nokia E63 smartphone for pulse rate measurement, and they assessed that the green channel provides a stronger PPG signal than the red one [17, 31]. A central region of interest measuring 10×10 pixels was selected in order to compute the mean intensity value, and a Fourier transform spectral analyses was applied to evaluate the heart rate. The authors reported a possibility to detect changes in HR from rest to after exercise using their approach.

Later, in [2] an Android application was developed and the experimental tests were performed on a Motorola Droid smartphone with a comparison to medical instruments (BioZ ECG and Nonin Onyx II model 9560BT ambulatory finger pulse oximeter). As a result, the validity of HR smartphone measurements was confirmed.

Scully et al. [6] developed a system for physiological parameter monitoring from optical recordings with a mobile phone. The videos were obtained by a Motorola Droid smartphone, and the PPG value was computed at each frame as the 50x50 pixel average of the green channel region. The results for the heart rate were compared to the HP 78354A acquisition system using a standard 5-lead electrode configuration, and the respiration rate was compared with the metronome. In addition, the blue and red channels were used to detect the oxygen saturation and compared to the Masimo Radical SET^{TM}. The high correlation of the results was reported as well.

It is well known that PPG measurements are very sensitive to patient and/or tissue movement artifacts. The automatic detection of such motion artifacts and their separation from good quality signal is a non-trivial task [16].

However, the above research works are based on testing the specific smartphone model for each case and do not refer to the problem of movement artifacts. On the other hand, as was already noted in [32], our tests show that the distribution of the pixels in either green or blue channels is not uniform for different smartphone models, such as HTC, iPhone4, Nokia, or Samsung. The only channel that has similar characteristics is the red one, while the rest can be used to distinguish a normal usage of the system from the abnormal one, when the finger is not located properly or there is no finger at all. Moreover, we noted that the red channel information remains similar even when the smartphone was used without LED, but in a well-illuminated environment.

Therefore, the aim of this research is to develop a method that would address these two problems. In this chapter, we describe a new method to measure the PPG waveform and calculate the pulse rate based on video obtained from a smartphone-based PPG acquisition system. The main emphasis is placed on the development of robust algorithms suitable for different smartphone models.

The rest of the chapter is organized as follows: in Section 5.2 we show a general system overview and the acquisition scheme, Section 5.3 deals with the correct usage assessment procedure, in Section 5.4 we explain the initial system calibration, the PPG evaluation algorithm and pulse computation procedure are described in Section 5.5, while in Section 5.6 the experimental results are presented and we conclude with Section 5.7.

5.2 System Work Overview

The proposed approach utilizes an image acquisition concept similar to the one of a pulse oximeter and PPG imaging. A subject covers with his fingertip a smartphone camera lens, trying to hold the finger steady and pressing without additional force (Figure 5.11). In this case the volumetric variation of blood changes the light absorption that passes through a finger. Such variation of light absorption is registered by a camera, and is used for PPG evaluation.

The measurements are performed continuously and for each new acquired frame the change of color values is computed. The feature of the proposed approach is that both reflection and transition modes of the system usage are

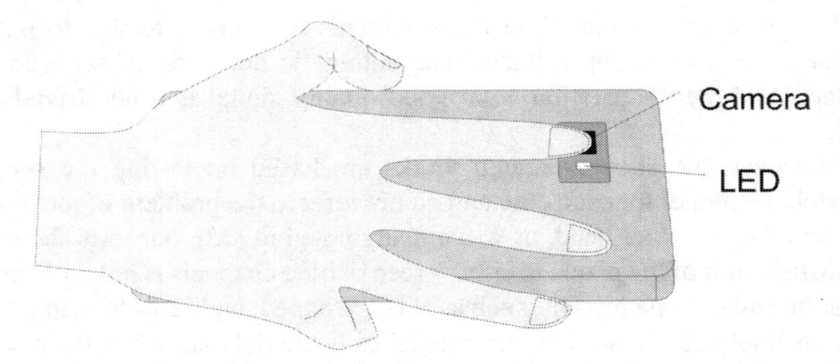

Figure 5.11 General video capturing scheme with a smartphone equipped by LED.

possible. Therefore, to evaluate pulsations it is possible to use smartphones with a LED as well as without it in the case of good lighting conditions.

After obtaining a new frame it is verified for correctness, as shown in the flow chart of the algorithm shown in Figure 5.12. Such verification procedure checks if the system is used in a proper mode: there is a finger in front of the camera and the illumination conditions are sufficient.

Then, there are two stages in the operation of the system: calibration and measurement. In the calibration stage the threshold value is established and the system parameters are updated while in the measurement stage the pulse rate is evaluated based on thresholding results and binary mask analysis. These two stages are explained in detail later in the appropriate sections of this chapter.

5.3 Assessment of Correct Use

When health monitoring is performed in a clinical environment, the medical staff can supervise the whole procedure and detect when it goes wrong. However, when doing self-monitoring, only the person itself can control the correctness of this process. For example, the wrong position of the fingertip on the smartphone optical sensor or its absence, finger movement during the measurement or even changing the force with which fingertip presses the lens may cause wrong results and, as a result, the program gives false alarms or misses a dangerous situation. To prevent wrong health parameter measurement, the program automatically detects all cases of improper usage and instructs the person properly.

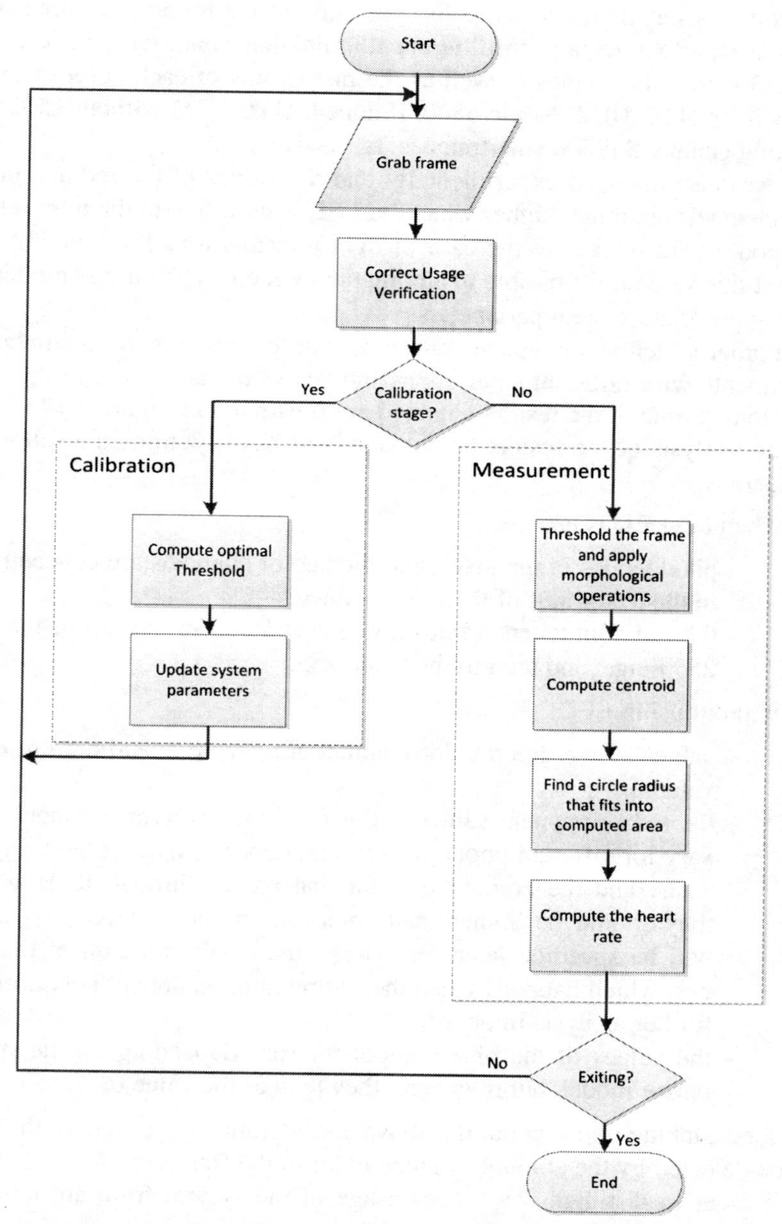

Figure 5.12 The PPG measurement algorithm which includes correct usage verification, calibration and measurement stages.

As it was stated previously, color saturation of the frames, acquired from different smartphones and in different illumination conditions, varies. Figure 5.13 shows the frames as well as the histograms of each color channel, obtained by HTC HD2, Nokia 5800, iPhone4, HTC HD2 without LED and Samsung Galaxy S i9000 smartphones, respectively.

It was also assessed experimentally that the values of the red and green color channels are much higher when the LED is used (i.e. in the light reflection mode) with respect to the case of light transmission. Thus, taking into account this value it is possible to automatically identify the usage mode and select appropriate system parameters.

In order to define the typical color model of the finger image a number of experiments with different smartphone models were carried out in different conditions. Some of the results obtained are illustrated in Figure 5.14.

The analysis of the results obtained (Figure 5.14) permits the following conclusions:

- when the LED is used:
  - pixel values in the green and blue color channels are concentrated in the lower half of their value range;
  - the red component values are concentrated in the top of the 0 to 255 range, and tend to the value 255.
- without using LED:
  - values of the green color channel are very low and are close the value 0;
  - the red component values in this case have no typical range. They vary for different phone models and depend on the patient finger's tissue and the amount of light that passes through it. However, they should be higher than some minimum value R_{NOLEDmin} as will be specified later. Otherwise, the small variation of the values, which happens when the illumination is not sufficient, makes further analysis impossible;
  - the values of the blue component vary depending on the smartphone model, but in general they tend to the value 0.

Hence, taking into account the above considerations, the state of the LED can be detected by the amount of green color in the frame.

In order to distinguish a proper usage of the system from an improper one, the following scheme was applied to each captured frame:

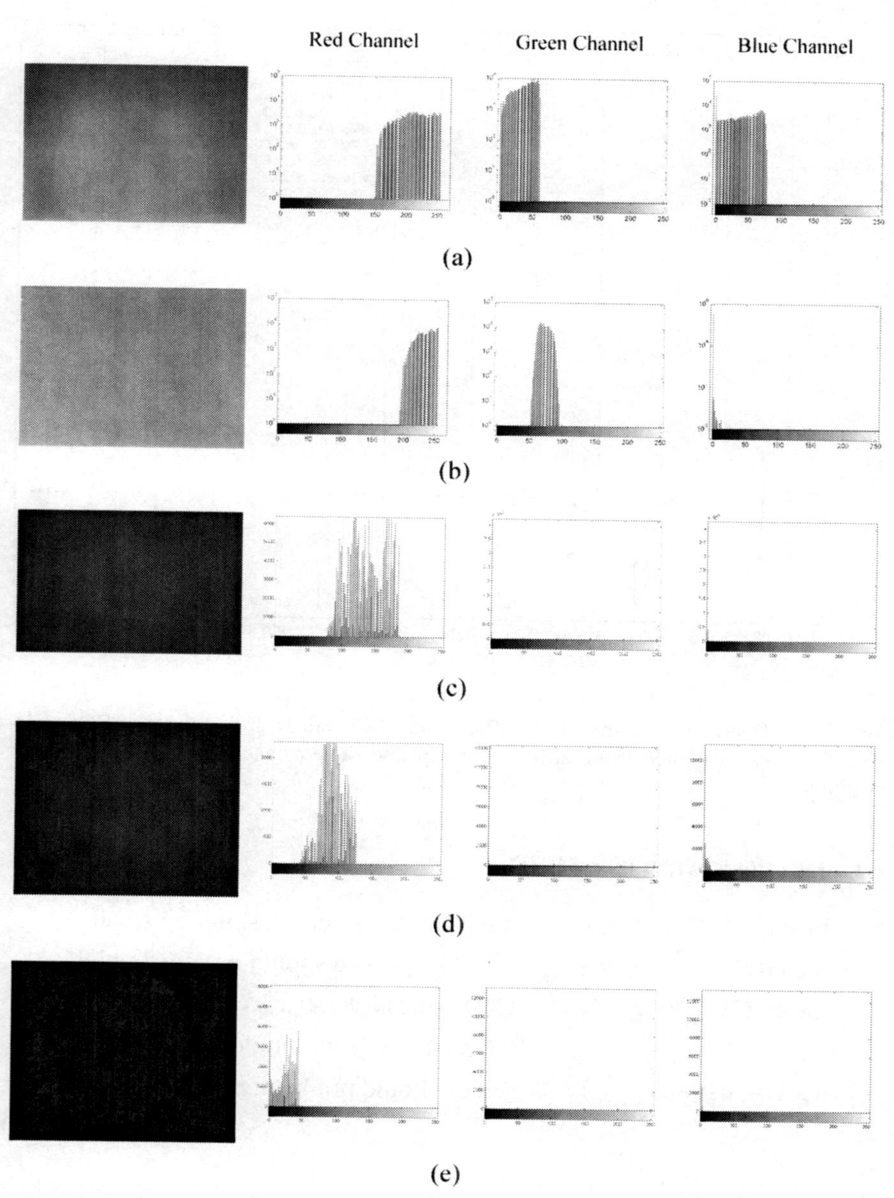

Figure 5.13 Acquired frames and their histograms of the red, green and blue channels for different smartphones and in different lighting conditions: (a) HTC HD2 with LED, (b) Nokia 5800 with LED, (c) iPhone4 with LED, (d) HTC HD2 without LED and (e) Samsung Galaxy S i9000 without LED.

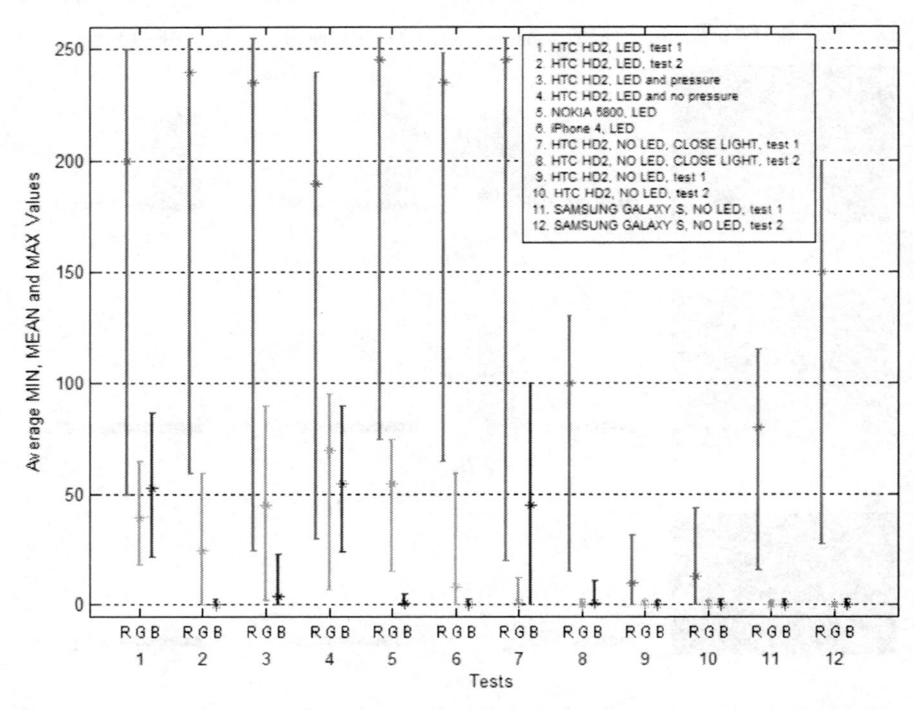

Figure 5.14 Distribution of the MIN, MEAN and MAX values of the pixels in RGB color space for videos captured using different smartphone cameras and under different lighting conditions.

% Color ranges when using LED:

$$
\begin{array}{ll}
\text{mean}(G) + \sigma_G \geq G_{\text{LED}_{\min}} \quad \text{AND} & \text{\% Green must not be small,} \\
\text{mean}(R) - \sigma_R \geq R_{\text{LED}_{\min}} \quad \text{AND} & \text{\% Red should be mostly high,} \\
\text{mean}(G) + \sigma_G \geq G_{\max} \quad \text{AND} \quad \text{mean}(B) + \sigma_B < B_{\max} \quad \text{AND} & \\
& \text{\% Green and Blue should be mostly low} \\
\sigma_R, \sigma_G, \sigma_B < \sigma_{\max} & \text{\% Values should not be distributed too much}
\end{array}
\tag{5.1}
$$

OR

% Color ranges without LED use:

$$
\text{mean}(G) + \sigma_G < G_{\text{NOLED}_{\max}} \quad \text{AND} \quad \text{\% Green must be very small,}
$$

$$\begin{aligned}
&\text{mean}(B) + \sigma_B < B_{\max} \quad \text{AND} \quad \%\ \text{Blue should be mostly low,}\\
&\text{mean}(R) > R_{\text{NOLED}_{\min}} \quad \text{AND} \quad \%\ \text{Red must not be small}\\
&\sigma_R, \sigma_G, \sigma_B < \sigma_{\max} \quad \%\ \text{Values should not be distributed too much}
\end{aligned} \tag{5.2}$$

where mean(R), mean(G) and mean(B) are the mean values of the red, green and blue components, respectively, computed for each captured frame, σ_R, σ_G, σ_B are the standard deviation values, computed for each frame and each color channel, $G_{\text{LED}_{\min}}$ and $R_{\text{LED}_{\min}}$ are the minimum values of the green and red channel, correspondingly, in the case the LED is used, $G_{\max}$ and $B_{\max}$ are the maximum values of the green and blue channels, respectively, $\sigma_{\max}$ is the maximum standard deviation among all color channels, $G_{\text{NOLED}_{\max}}$ is the maximum value of the green channel in the case the LED is not used, $R_{\text{NOLED}_{\min}}$ is the minimum value of the red channel in the case the LED is not used.

The above scheme describes a proper color cluster in the RBG color space of the finger image. For the calibration stage the threshold values in (5.1) and (5.2) are defined based on the analysis of the preliminary experimental results. Thus, they are: $G_{\text{LED}_{\min}} = 10$, $R_{\text{LED}_{\min}} = 128$, $G_{\max} = 128$, $B_{\max} = 128$, $\sigma_{\max} = 40$, $G_{\text{NOLED}_{\max}} = 10$, $R_{\text{NOLED}_{\min}} = 10$, and are used to make sure that the finger is placed on the camera correctly. Such thresholds are valid for different models and used to define if initially the smartphone was used correctly. For the measurement phase, however, the threshold values are updated on the basis of the chromatic parameters of the acquired frames during the calibration stage. This is done to limit the possible color cluster to the characteristics of the current smartphone model, person's tissue and lighting conditions.

The validation step of the correct use is essential for further algorithm execution and quality assessment of the results, especially in the case of health monitoring systems. For the frames with no finger or with a finger in the wrong position, the color distribution in the channels does not fit defined rules, but it is spread out over the whole value range. Therefore, the proposed model allows considering only the case of finger presence, and, as a result, permits validating the correct use.

5.4 Initial System Calibration

As mentioned earlier, the system calibration step is used to adapt the system configuration to the particular smartphone camera and lighting conditions, as well as to the personal characteristics of the finger tissue (skin color, opacity, etc.). There are a few factors that must be taken into account to do that:

(i) different smartphone models lead to different color saturation in the captured frames;
(ii) different fingertip pressure force on the camera lens as well as different features of the tissue change the level of light absorption when it passes through the finger and, therefore, cause different color ratios;
(iii) shifting the finger with respect to the camera lens creates motion artifacts and, as a result, causes wrong segmentation.

Considering the above cases it is clear that a fixed threshold value is not suitable. To compute the PPG signal it is possible to threshold the red components for each frame obtained and compute the number of pixels that surpass the threshold, as was proposed in [32]. The threshold T was established as 95% of the range between the *min* and *max* values during the first 5 s of system operation. That is:

$$T = \overline{\max}(I) - \frac{1}{20}(\overline{\max}(I) - \overline{\min}(I)) \tag{5.3}$$

where $\overline{\max}(I)$ and $\overline{\min}(I)$ are the mean maximum and minimum values, respectively, of the red component for the acquired frames during the first 5 s.

It was confirmed also that acquiring at least three full pulsations is enough, and the number of captured frames is suitable to perform statistical analysis. Such algorithm is reliable and works fine in the case the LED is used. However, if the system works without the LED the range between the *min* and *max* values is small and the number of computed pixels is not enough to make robust measurements. Moreover, it may occur that for some frames the maximum pixel value is lower than the established threshold, and the segmentation does not provide the expected result.

Hence, it was proposed to calculate the T as a mean of such values T_i, computed during the calibration step, at which the number of pixels in the corresponding thresholded image occupy more than θ% of the frame:

$$T = \text{mean}(T_i),\ T_i : \frac{\|\text{val}(P_i) \geq T_i\|}{\|P_i\|} = \theta\%, \quad i = 1, \ldots, N, \tag{5.4}$$

where T_i is the computed threshold for frame i, P_i is the array of red component pixels of the frame i, val(P_i) is the value of each pixel in P_i, $\|\ldots\|$ is the number of pixels in the array, and N is the number of frames for the calibration stage.

Since the finger is not fixed on the camera lens, it can shift during the measurement changing its position as well as its pressure on the lens. Therefore, it is necessary to ensure that the area, which surpasses the established threshold, always fits the image boundaries. Otherwise, the measurement would be incorrect. It was noted also that the pulsating dynamics (i.e. the difference between smallest and largest radiuses) is more for the pixels with high color values. It means that the final result is better if the threshold is high (closer to the max value of the pixels). In this work a value of $\theta = 20$ is used.

Figure 5.15 shows examples of the thresholded image according to (5.4), captured from different smartphones and in different lighting conditions.

As can be seen from Figure 5.15a, the thresholded area contains some artifacts, caused by the close position of the LED and, as a result, high illumination of the pixels. The next section explains how to eliminate such artifacts and extract the proper PPG value.

5.5 PPG Evaluation Algorithm

As discussed previously, only the red component is suitable for PPG measurement since the figure shape remains similar for any smartphone model and any lighting conditions. Normally it has the shape of a paraboloid (Figure 5.16) with the maximum pixel value in the centre.

The shape of the thresholded binary mask depends on how the smartphone is used. Using a smartphone with a LED or specific finger position on the camera this shape can change. As was already mentioned, the mask in Figure 5.15a does not have a circular shape because the frame is acquired with the LED on, and some parts of the finger are better illuminated. Thus, a simple calculation of the number of pixels, as proposed in [32], cannot take into account the above factors.

To overcome this limit, we propose finding the circle that better fits the thresholded image, and use it's radius as the PPG value. In particular, for each captured frame, we first calculate the coordinates C_x and C_y of a centroid of the binary mask as:

$$C_x = \frac{\sum_n x_n}{n}, \qquad C_y = \frac{\sum_n y_n}{n}, \tag{5.5}$$

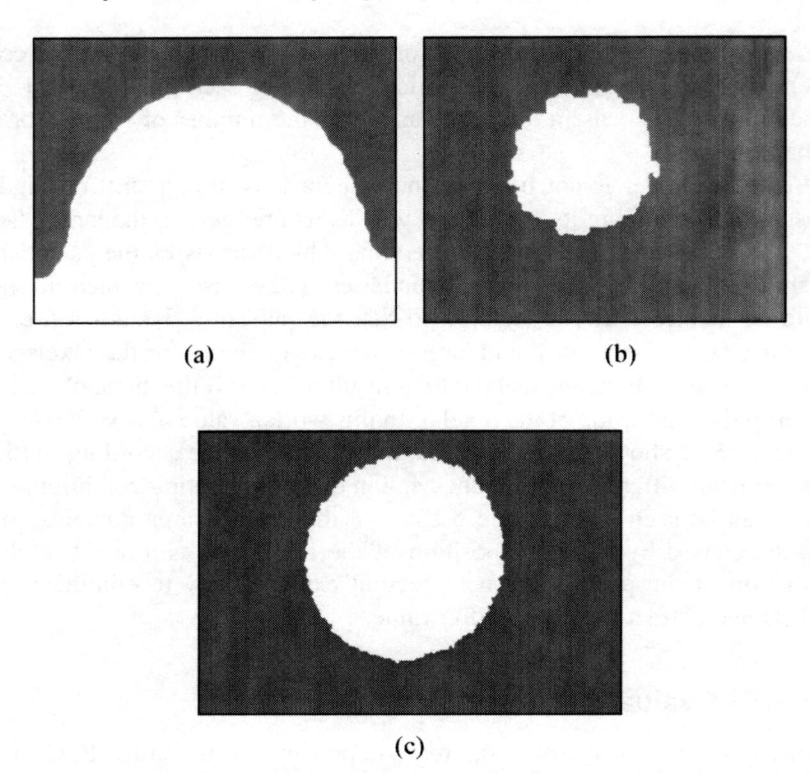

(a) (b)

(c)

Figure 5.15 Computed masks that satisfy the threshold on the frames, captured (a) from HTC HD2 with LED, (b) HTC HD2 without LED, and (c) Samsung Galaxy S without LED.

where x_n and y_n are the coordinates of each pixel with a value 1 on a binary mask, and n is the total number of such values.

Then, the radius of the circle with the centre in the centroid is considered as a photoplethysmogram value (Figure 5.17).

Normally, the Hough transform can be applied to find the circles. However, such approach requires significant computational resources and also does not work well in the case of non-smooth boundaries. Since the computational complexity is important, it was proposed to compute the radius as follows (Figure 5.17):

1. find distance from the centroid to the boundary at positions 0°, 45°, 90°, 135°, 180°, 225°, 270°, 315°;
2. if the length exceeds the distance to the image boundaries, ignore this value;

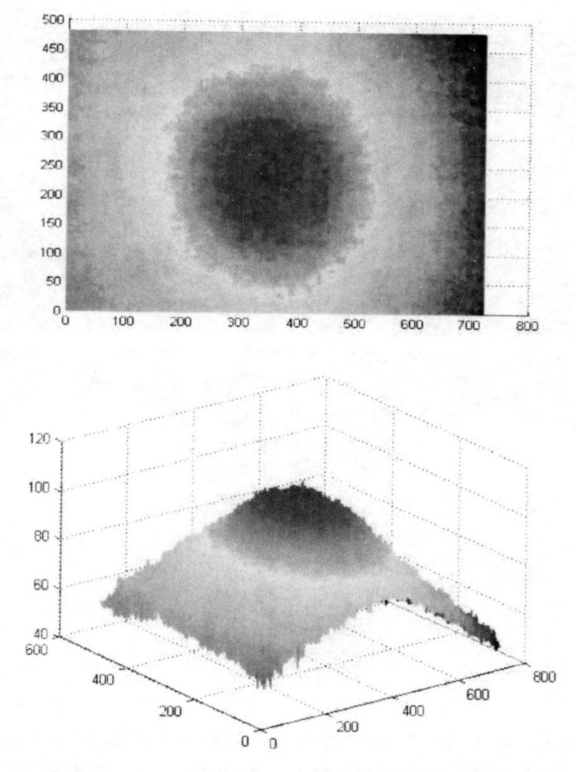

Figure 5.16 The pixels intensity and the surface of the red component for the frame captured from the Samsung Galaxy S smartphone.

3. find the mean value for all the remaining distances, and use it as the radius of a circle.

In this case, if the circle does not fit the image completely because of shifted centre, the radius will still be computed properly.

Computing the radius as described above for a sequence of captured frames gives a photoplethysmogram, where each cardiac cycle appears as a peak. Such waveform is generally referred to as the inverted PPG (Figure 5.18) [33], as the camera corresponds to the received rather than absorbed light intensity [24]. Thus, the final PPG signal is inverted vertically to be used for further processing.

Recovering the PR from such a PPG signal was achieved by passing the computed waveform with a 10s moving window and applying the Fourier

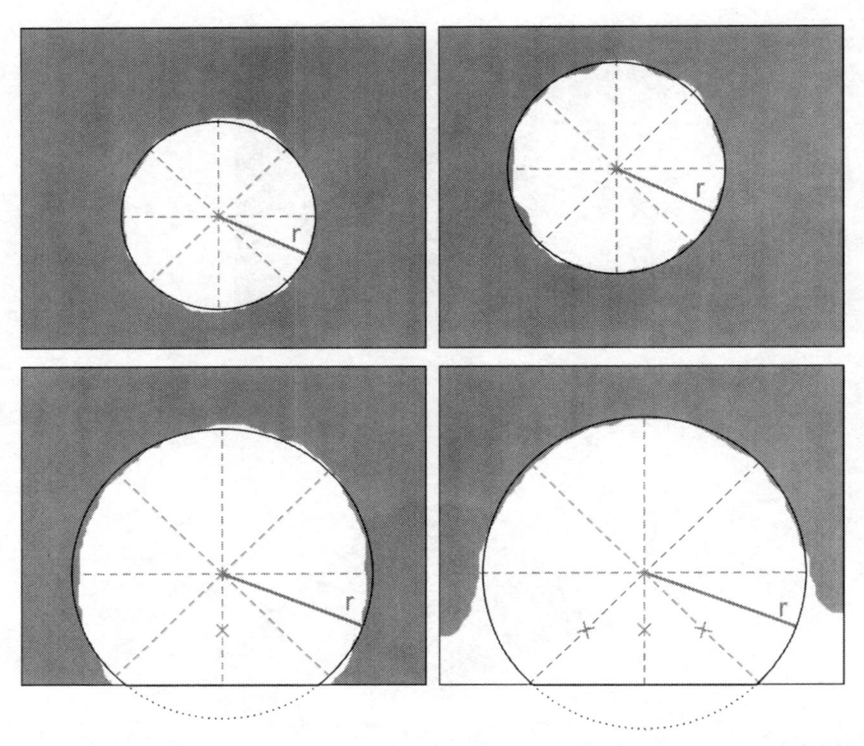

Figure 5.17 Computing the PPG value: white is a thresholded area; dashed lines are the distances from the centroid to the boundaries; crossed out lines are the lines that do not have a boundary on the image and should be skipped; solid bold lines are the radiuses, computed as average values of the above distances, solid lines – circles inscribed into the figures with radiuses r; dotted lines are the parts of the circle that do not fit the picture.

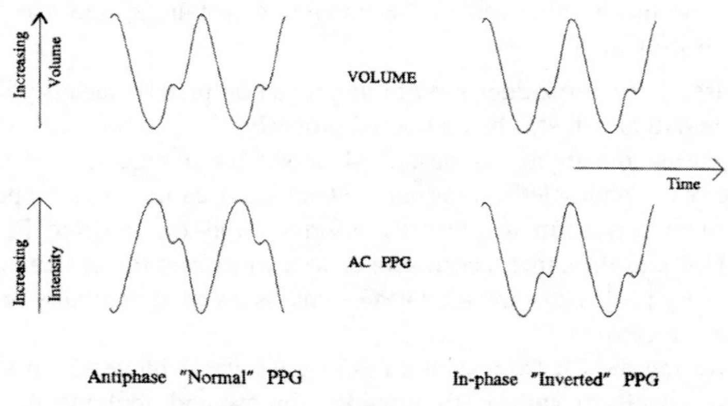

Figure 5.18 "Normal" and "inverted" PPG waveforms [33].

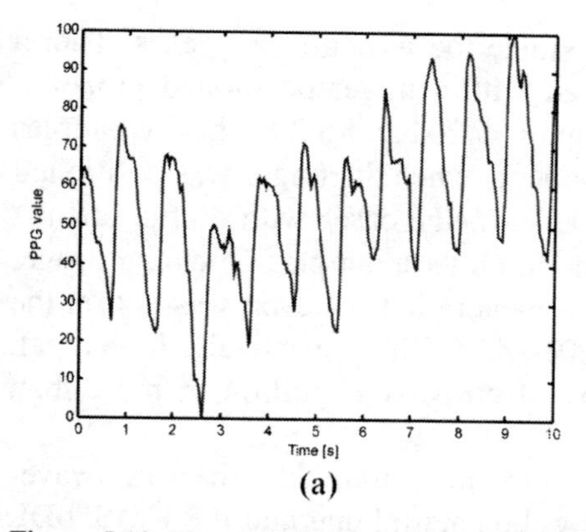

(a)

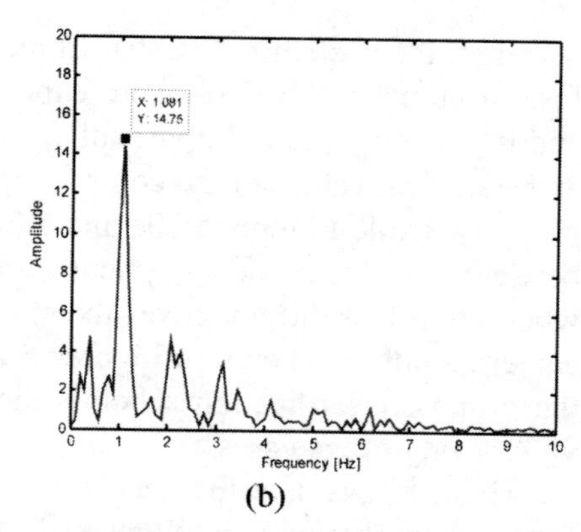

(b)

Figure 5.19 Measured PPG during a time interval of 10 s (a), and a corresponding Fourier spectrum. The evaluated value of PR is equal to 1.081 Hz (about 65 bpm) (b).

Table 5.1 Specification of the smartphones used for experiments.

Device name	Operating System	LED	Video resolution (pixels)	Frequency (fps)
HTC HD2	Windows Mobile 6.5	yes	352 × 288	25
Nokia 5800	Symbian OS v9.4	yes	640 × 480	29
Apple iPhone 4	iOS 4	yes	480 × 272	30
Samsung Galaxy S	Android OS, v2.3	no	720 × 480	30

transformation to each of the windows. The maximal peak on the spectrum near the frequency of 1Hz corresponds to the pulse rate frequency [29].

Figure 5.19 shows an example of computed PPG, inverted and normalized from 0 to 100, and the corresponding Fourier spectrum. The PPG signal itself is unfiltered. As it can be seen from Figure 5.19b, the evaluated PR is equal to 1.081 Hz and corresponds to about 65 beats per minute (bpm).

5.6 Experimental Results

The experimental tests were carried out using different smartphone models, in particular HTC HD2, iPhone4, Nokia 5800, Samsung Galaxy S i9000. Table 5.1 shows their specifications such as version of the operating system, presence of the LED, frames resolution and capturing frequency.

Videos from the smartphones were transferred to a computer, processed and compared to the data obtained from an oximeter. Further processing was done in Visual Studio C++ using the OpenCV library.

First, the system was tested to recognize the wrong usage cases. Therefore, a number of videos were captured with a finger positioned properly, and in a wrong mode. In particular, Figures 5.20a and 5.20b show examples and statistical values of the color components when the finger was positioned properly, while Figures 5.20c and 5.20d show the cases with no full contact between the finger and the phone camera. Other examples of wrong usage, when the finger did not cover the entire camera lens or even was not on the camera at all are shown in Figures 5.20e and 5.20f, respectively. In general, the proposed verification scheme allowed proper recognition of more than 98% wrong usage cases.

Then, to evaluate the accuracy of pulse measurements, the PPG waveforms were obtained simultaneously by the smartphone and the CMS50DL Finger Pulse Oximeter SPO2 Monitor using two fingers of the left hand. Ten subjects participated in the test, going from 26 to 60 years of age. The PPG waveforms obtained by the smartphone were then inverted and normalized from 0 to 100 as explained before for further comparison.

As can be seen from Figure 5.21, which shows the two signals obtained by the smartphone and the oximeter in the normal subject state in the same time period, the peaks and the valleys correspond on both PPGs.

To prove the suitability and the correctness of the proposed method, the above test was repeated again just after squatting for 60 s. In this case the pulse rate changed because of the physical activity. As it is shown in Figure 5.22, the PPG evaluated by the smartphone shows more rapid pulsations and also corresponds to the one from oximeter.

Table 5.2 shows the summary of the tests where the mean PR and standard deviation value were computed for several measurements, performed with the HTC HD2 smartphone using LED and the CMS50DL Finger Pulse Oximeter SPO2 Monitor at respective time periods. The signals were acquired for 60 s and the number of pulsations per minute was computed by applying the Fourier transform to a previous 10 s period.

Then, the mean and standard deviation values were computed. The same tests were performed using other smartphones and the results are shown in Figure 5.23. They confirm that the PPGs obtained from the smartphones are highly correlated to those obtained by a finger pulse oximeter and, therefore, can be used for PR measurements.

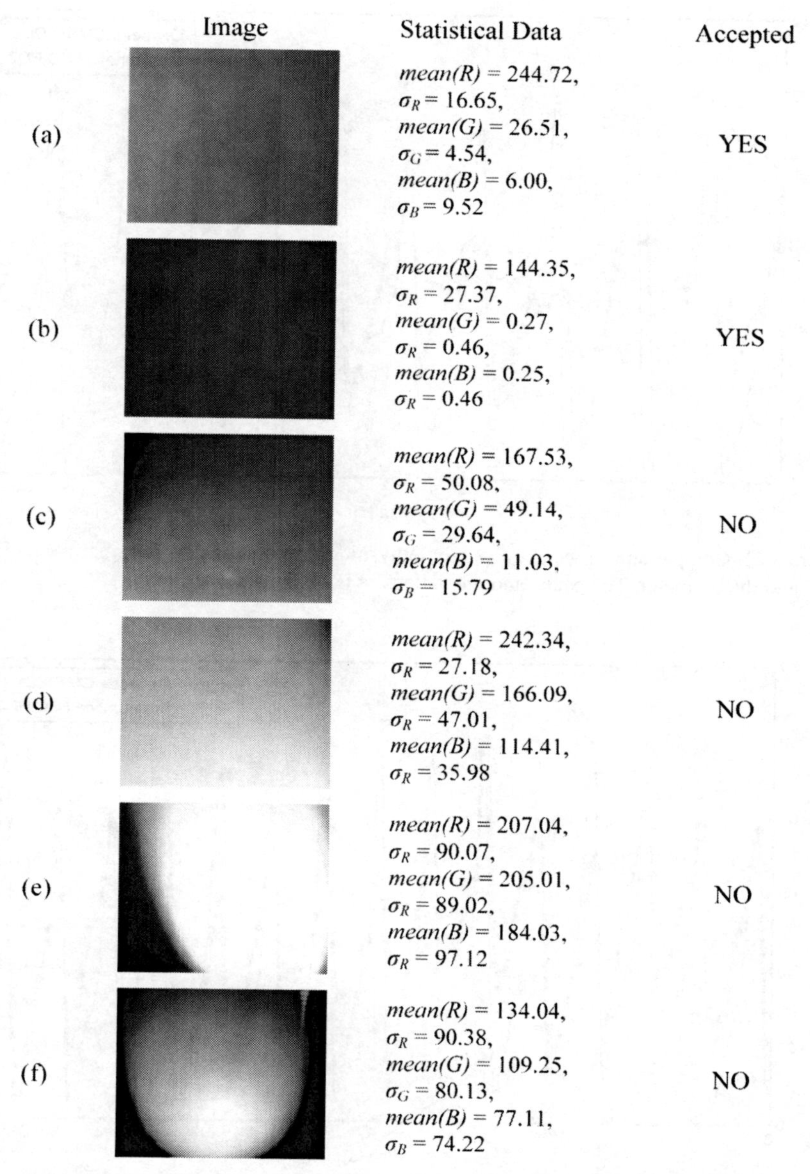

Figure 5.20 Accepted frames (a) and (b), frames recognized by the system as wrong because of not enough pressure of the finger (c), (d), wrong position on the camera (e), and missing of the contact between finger and camera (f).

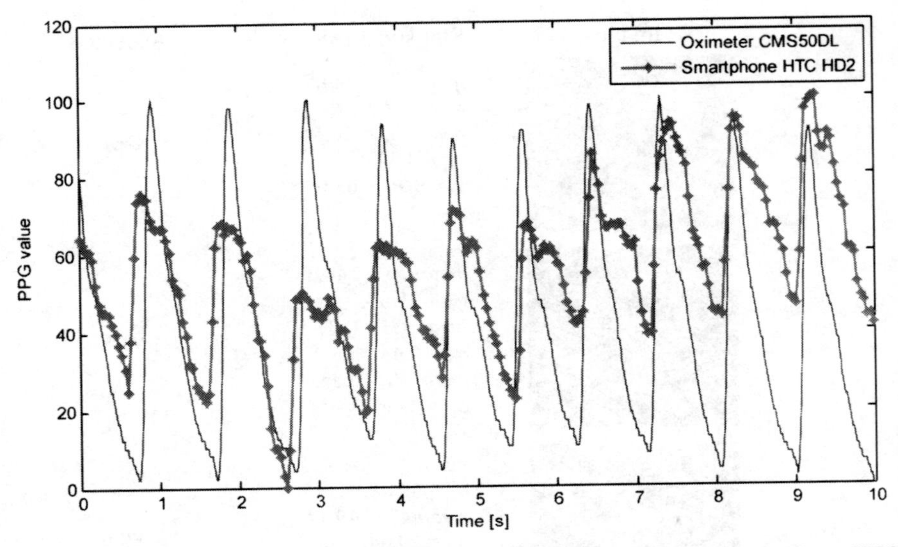

Figure 5.21 Comparison between the photoplethysmograms obtained by the smartphone HTC HD2 and the oximeter. The peaks and the valleys of both signals correspond.

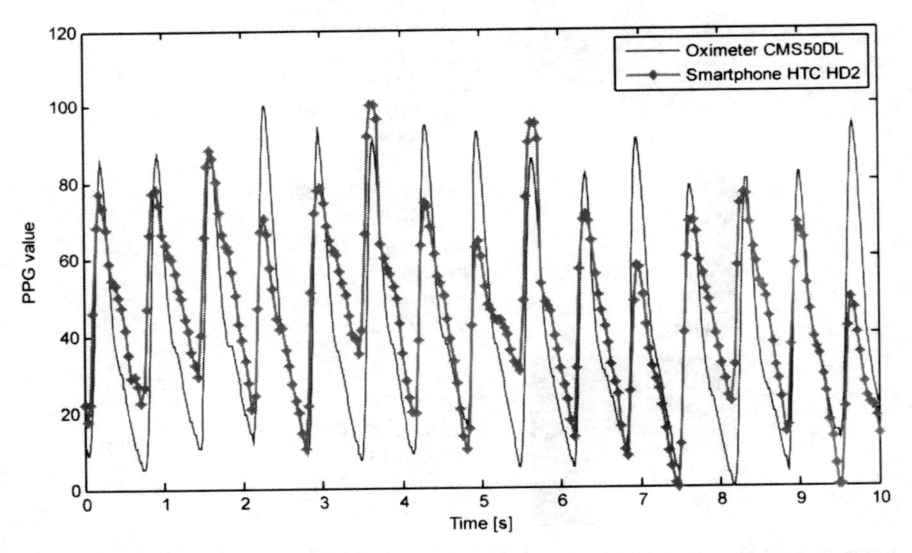

Figure 5.22 Comparison between the photoplethysmograms from the HTC HD2 smartphone and the oximeter after squatting for 60 s. Also in this case the peaks and the valleys of both signals are highly correlated.

Table 5.2 Comparison between the PR evaluated from the HTC HD2 using LED and the CMS50DL Finger Pulse Oximeter SPO2 Monitor.

Test No.	Mean PR from smartphone (bpm)	Mean PR from oximeter (bpm)	Error (%)
Video 1, before squatting	61.46 ± 1.48	62.07 ± 1.34	0.98
Video 2, after squatting	90.15 ± 4.6	91.58 ± 3.29	1.56
Video 3, before squatting	79.42 ± 3.23	79.80 ± 3.14	0.48
Video 4, after squatting	98.86 ± 12.15	97.79 ± 11.17	1.09
Average Error			**1.03**

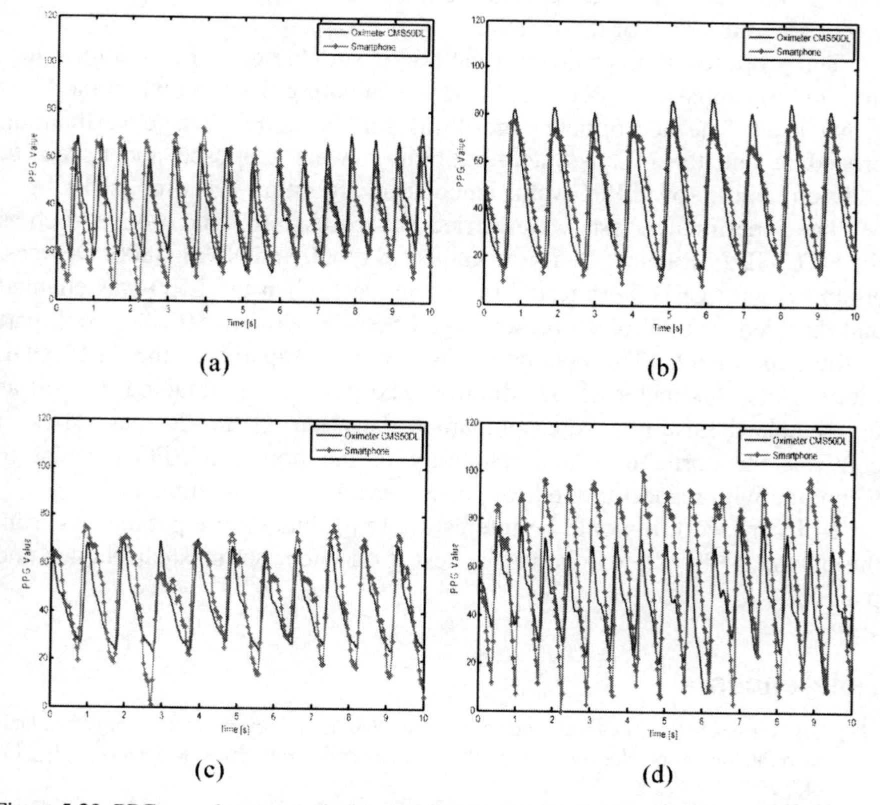

Figure 5.23 PPG waveforms obtained from (a) Nokia 5800 with the LED enabled, (b) iPhone4 with LED, (c) HTC HD2 without LED, (d) Samsung Galaxy S i9000 without LED, and the corresponding waveform, acquired from a CMS50DL Finger Pulse Oximeter SPO2 Monitor.

5.7 Conclusions

This chapter is devoted to the photoplethysmogram measurement by means of a smartphone and integrated camera. Prior work has reported the possibility of smartphone usage for pulse rate measurement. The successful application of the green color component for PPG signal computing was reported.

However, such reports involved a limited number of smartphones and further research has shown that the typical color range varies from model to model, this has been demonstrated in this chapter. In particular, it has been shown that only the red channel has similar characteristics for different models of smartphones while the green and blue may vary dramatically. Although the last two components do not have a fixed color range, they can be used to detect a wrong usage of the system.

The proposed PPG evaluation method is suitable to work in both reflection and transmission modes, and allows evaluating the PPG when the LED is not used. The appropriate algorithms for the correct usage verification procedure and the initial system calibration were proposed and tested. In addition, an improved PPG value calculation algorithm was proposed.

The experimental tests were carried out with smartphone models such as the HTC HD2, iPhone4, Samsung Galaxy S i9000 and Nokia 5800. Devices, equipped with LED were tested in two modes: when the LED was enabled and disabled. A total of 10 persons aged between 26 and 60 years took part in the experiments. The obtained results were compared to the CMS50DL Finger Pulse Oximeter SPO2 Monitor. The pulse rates obtained as well as the signals themselves were comparable between all the devices. Thus, it confirms the correctness and reliability of the proposed PPG calculation technique with respect to medical pulse measurement instruments.

In future work it would be interesting to evaluate more parameters from the obtained PPG signal, as well as test it on wider set of smartphones and patients.

References

[1] D.J. Terbizan, B.A. Dolezal, and C. Albano. Validity of seven commercially available heart rate monitors. *Measurement in Physical Education and Exercise Science*, 6(4):243–247, 2009.

[2] M.J. Gregoski, M. Mueller, A. Vertegel, et al. Development and validation of a smartphone heart rate acquisition application for health promotion and wellness telehealth applications. *International Journal of Telemedicine and Applications*, doi: 10.1155/2012/696324, 2012.

[3] A. Kumar. ECG – simplified, *LifeHugger*, November 2010.

[4] T.G. Pickering. White coat hypertension. *Current Opinion in Nephrology and Hypertension*, 5(2):192–198, March 1996.

[5] A. Smith and T. Ferriss. Home health monitoring may significantly improve blood pressure control. *Kaiser Permanente Study Finds*, May 2010.

[6] C.G. Scully, J. Lee, J. Meyer, A.M. Gorbach, D. Granquist-Fraser, Y. Mendelson, and H. Chon Ki. Physiological parameter monitoring from optical recordings with a mobile phone. *IEEE Transactions on Biomedical Engineering*, 59(2):303–306, 2012.

[7] V. Gay and P. Leijdekkers. A health monitoring system using smart phones and wearable sensors. *International Journal of ARM*, 8(2):29–35, 2007.

[8] M.E. Ernst and G.R. Bergus. Ambulatory blood pressure monitoring. *Southern Medical Journal*, 96(6):563–568, June 2003.

[9] Operations Manual 90207-80/90217-1HABP Monitors, 070-0715-00 Rev. B, Spacelabs Medical, Inc.

[10] A. Shennan, A. Halligan, M. Gupta, D. Taylor, and M. de Swiet. Oscillometric blood pressure measurements in severe per-eclampsia: Validation of the SpaceLabs 90207. *BJOG: An International Journal of Obstetrics & Gynaecology*, 103(7):171–173, 1996.

[11] C.W. Belsha, T.G. Wells, H.B. Rice, W.A. Neaville, and P.L. Berry. Accuracy of the SpaceLabs 90207 ambulatory blood pressure monitor in children and adolescents. *Blood Pressure Monitoring*, 1:127–133, 1996.

[12] P. Iqbal, M.D. Fotherby, and J.F. Potter. Validation of the SpaceLabs 90207 automatic non-invasive blood pressure monitor in elderly subjects. *Blood Pressure Monitoring*, 1:367–373, 1996.

[13] Association for the Advancement of Medical Instrumentation, http://www.aami.org.

[14] American Heart Association, Inc., http://www.heart.org.

[15] British Hypertension Society, http://www.bhsoc.org.

[16] J. Allen. Photoplethysmography and its application in clinical physiological measurement. *Physiological Measurement*, 28:R1–R39, 2007.

[17] E. Jonathan and M. Leahy. Investigating a smartphone imaging unit for photoplethysmography, *Physiological Measurement*, 31(11):N79–N83, November 2010.

[18] D. Damianou. The wavelength dependence of the photoplethysmogram and its implication to pulse oximetry. Doctor of Philosophy Thesis, University of Nottingham, October 1995.

[19] A. Reisner, P. Shaltis, D. McCombie, and H. Asada. Utility of the photoplethysmogram in circulatory monitoring. *Anesthesiology*, 108(5):950–958, May 2008.

[20] P. Leonard, N. Grubb, P. Addison, D. Clifton, and J. Watson. An algorithm for the detection of individual breaths from the pulse oximeter waveform. *Journal of Clinical Monitoring and Computing*, 18(5–6):309–312, December 2004.

[21] S. Linder, S. Wendelken, E. Wei, and S. McGrath. Using the morphology of photoplethysmogram peaks to detect changes in posture. *Journal of Clinical Monitoring and Computing*, 20(3):151–158, June 2006.

[22] U. Rubins. Finger and ear photoplethysmogram waveform analysis by fitting with Gaussians. *Medical and Biological Engineering and Computing*, 46(12):1271–1276, December 2008.

[23] Jia Zheng and Sijung Hu. The preliminary investigation of imaging photoplethysmographic system. *Journal of Physics: Conference Series*, 85, doi:10.1088/1742-6596/85/1/012031, 2007.

[24] Jia Zheng, Sijung Hu, V. Azorin-Peris, Angelos Echiadis, V. Chouliaras, and R. Summers. Remote simultaneous dual wavelength imaging photoplethysmography: A further step towards 3-D mapping of skin blood microcirculation. *Multimodal Biomedical Imaging III, Proc. of SPIE*, 6850, 68500S-1, 2008.

[25] iHealth Blood Pressure Monitoring System, Medisana AG, http://ihealth.medisana.com/en/index.php.

[26] D.L. Carnì, G. Fortino, R. Gravina, D. Grimaldi, A. Guerrieri, and F. Lamonaca. Continuous, real-time monitoring of assisted livings through wireless body sensor networks. In *Proceedings of the IEEE International Conference Intelligent Data Acquisition and Advanced Computing Systems IDAACS'2011*, pp. 872–877, September 2011.

[27] M.N. Kamel Boulos, C. Tavares et al., How smartphones are changing the face of mobile and participatory healthcare: An overview, with example from eCAALYX. *BioMedical Engineering OnLine*, 10(24), 2011.

[28] P. Pelegris, K. Banitsas, T. Orbach, and K. Marias. A novel method to detect heart beat rate using a mobile phone. In *Proceedings of 32nd IEEE Annual International Conference on Merging Medical Humanism and Technology*, pp. 5488–5491, 2010.

[29] W. Verkruysse, L.O. Svaasand, and J.S. Nelson. Remote plethysmographic imaging using ambient light. *Optics Express*, 16(26):21434–21445, 2008.

[30] S.C. Millasseau, J.M. Ritter, K. Takazawa, and P.J. Chowienczyk. Contour analysis of the photoplethysmographic pulse measured at the finger. *Journal of Hypertension*, 24(8):1449–1456, 2006.

[31] E. Jonathan and M.J. Leahy. Cellular phone-based photoplethysmographic imaging. *Journal on Biophotonics*, 4(5):293–296, 2011.

[32] D. Grimaldi, Yu. Kurylyak, F. Lamonaca, and A. Nastro. Photoplethysmography detection by smartphone's videocamera. In *Proceedings of IEEE International Conference Intelligent Data Acquisition and Advanced Computing Systems IDAACS'2011*, pp. 488–491, September 2011.

[33] J. Nijboer, J. Dorlas, and H. Mahieu. Photoelectric plethysmography: Some fundamental aspects of the reflection and transmission method. *Clinical Physics and Physiological Measurement*, 2(3):205–215, 1981.

6

Malaria Classification on Thick Blood Film

Saowaluck Kaewkamnerd[1], Chairat Uthaipibull[2], Apichart Intarapanich[1], Montri Pannarat[1], Sastra Chaotheing[2] and Sissades Tongsima[2]

[1]*National Electronics and Computer Technology Center (NECTEC), National Science and Technology Development Agency (NSTDA), Thailand Science Park, Pathumthani 12120, Thailand; e-mail: saowachuck.kaewkamnerd@nectec.or.th*
[2]*National Center for Genetic Engineering and Biotechnology (BIOTEC), National Science and Technology Development Agency (NSTDA), Thailand Science Park, Pathumthani 12120, Thailand*

Abstract

This chapter describes an automated device for the detection and classification of *Plasmodium* species on thick blood films. An overview of *Plasmodium* parasites that cause malaria, and current diagnosis methods and their limitations are discussed. The chapter concludes with the detailed introduction of an automated device, consisting of an image acquisition unit and its image analysis software module, which could be mounted on a conventional light microscope and operated by a health practitioner/technician with minimal training in the detection and diagnosis of malaria parasites.

Keywords: malaria classification, thick blood film, automatic detection system.

6.1 Malaria Problems

Malaria is one of the devastating infectious diseases that infect people living in tropical and subtropical regions of the world. According to the World Mal-

Richard J. Duro and Fernando López Peña (Eds.), Digital Image and Signal Processing for Measurement Systems, 165–188.

aria Report 2011 published by the World Health Organization, in the year 2010, there were 216 million cases of malaria and 655,000 deaths, mostly children under the age of 5 and pregnant women [1]. Although the burden of infection is highest in sub-Saharan Africa owing to poverty and poor healthcare, countries in Southeast Asia, South America and South Pacific are also categorized as malaria endemic countries. With its fatality and the high prevalence, the disease attributes to great economic impact by causing inability to attend school and work, resulting in reduced productivity that even makes the situation in those endemic countries worse in terms of economic development. With the increase in global warming, more cases of malaria are estimated, not only in the already endemic areas but also in the areas where those tropical diseases have been eradicated [2].

Despite the severity of the disease, the global control of malaria is limited by many factors including the non-availability of an effective vaccine [3] and the acquisition of resistance to currently available antimalarial drugs by the parasites [4]. Parasites with resistance to standard antimalarial drugs such as chloroquine, mefloquine, and pyrimethamine have spread worldwide and are a real threat to the control and treatment malaria disease. The only effective drugs without yet the underlying resistance problems are artemisinin and its derivatives [5]. However a recent report from the Thai-Cambodia border described the possibility of artemisinin resistance developed by the parasites [6]. This was an alarming signal that the parasites will soon adapt themselves to resist all the available antimalarial drugs and soon the control and treatment will be impossible. Hence, the early detection and correct diagnosis of malaria parasite species for proper treatment of the parasite with proper prescribed antimalarial drugs become necessary for the control of malaria disease.

6.2 *Plasmodium* Malaria Parasites

Malaria is caused by parasitic protozoa of the genus *Plasmodium* in phylum Apicomplexa; class: Sporozoa; subclass: Coccidia; order: Eucoccidia; suborder: Haemosporina. More than 100 species of Plasmodium parasites are found in reptiles, birds, and also mammals. Of these, five species of *P. falciparum*, *P. vivax*, *P. ovale*, *P. malariae*, and *P. knowlesi* are reported to cause malaria in humans. Among the five species of human malaria, *P. falciparum* is the most deadly species while *P. vivax* is responsible for more than 50% of all malaria cases outside Africa, and is endemic in the Middle East, Asia and Western Pacific, with a lower prevalence in Central and South America [7].

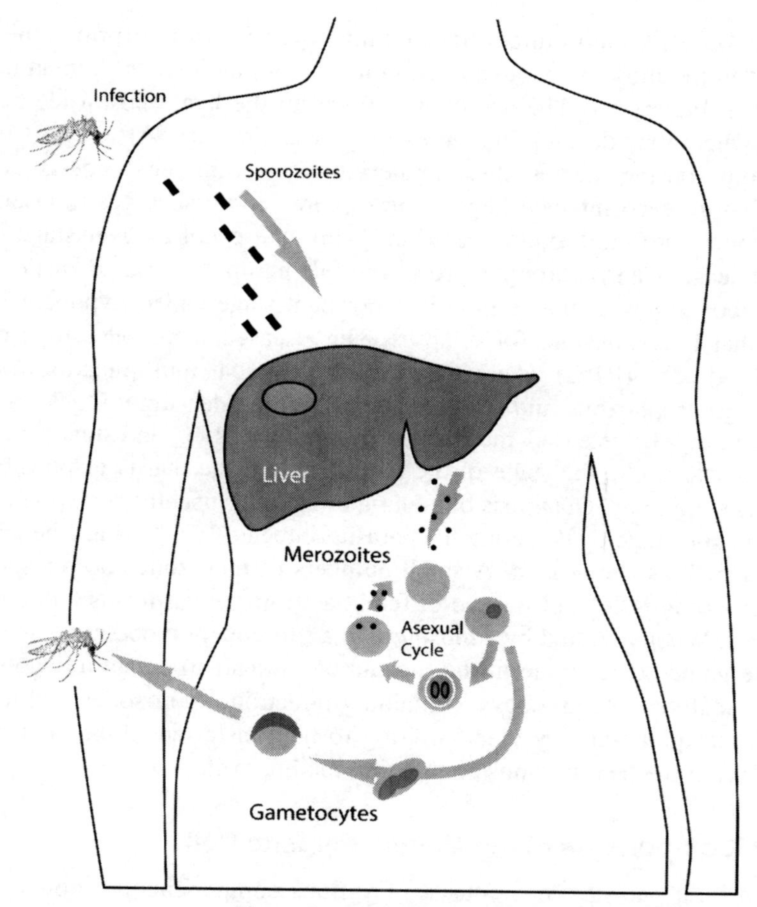

Figure 6.1 Malaria life cycle.

Even though vivax malaria is usually called benign uncomplicated malaria, *P. vivax* infection occasionally results in severe clinical symptoms similar to *P. falciparum* [8]. Interestingly, *P. vivax* and *P. falciparum* often co-exist in many parts of the world.

6.2.1 Malaria Life Cycle

The biology of the five species of human parasites is generally similar and consists of two phases: the sexual and asexual stages, which develop in the mosquito and human hosts, respectively. The infection starts when infected female *Anopheles* mosquitoes transmit infective sporozoites form of the parasite when the mosquitoes take a blood meal from an infected human host as shown in Figure 6.1. The sporozoites travel to the liver and invade hepatocytes, where they develop into an exo-erythrocytic stage. After 7 to 14 days, depending on the species, these exo-erythrocytic schizonts undergo schizogony before each infected hepatocyte ruptures and discharges thousands of merozoites into the blood circulation. While the event of liver stage merozoite release is a synchronous process in falciparum malaria, *P. ovale* and *P. vivax* sporozoites could develop into dormant stage called hypnozoites, the form that is responsible for relapses. The released merozoites then invade red blood cells (RBC) where they undergo asexual multiplication through ring stage, trophozoite and schizont stages before releasing up to 32 progeny merozoites. The released merozoites invade new RBC and thus repeat the erythrocytic cycle, the stage of the parasite life cycle that is responsible for malaria symptoms. Diagnosis of malaria infection is usually performed at this erythrocytic stage by detecting the parasite-infected RBC or parasite-specific factors such as haemozoin. A small numbers of merozoites do not multiply after invading RBC, but instead differentiate into sexual forms termed gametocytes. When ingested by a mosquito in a subsequent blood meal, male and female gametocytes mate in the mosquito's midgut to create a zygote. The zygote matures into an oocyst containing infectious sporozoites that migrate to the mosquito salivary gland waiting to be transferred to the next human host during the feeding, thus completing the life cycle.

6.2.2 Components of the Malaria Parasite Cell

The malaria parasite cells consist of various components as shown in Figure 6.2. The chromatin is the nucleus part of the parasite and is localized inside the nucleus membrane. It has round shape and deep red in color when stained with Giemsa. The cytoplasm is the part surrounding the nucleus. It usually shows blue color, when stained with Giemsa, and has in a number of forms, from a ring shape to irregular shapes.

6.3 Parasite Diagnosis and Problems

Routine methods for diagnosis of malaria parasites are by means of light microscopy and rapid diagnostic tests [9]. The former method involves preparation and microscopic examination of Giemsa-stained thick or thin films of blood from malaria patients. With higher cost, the latter method de-

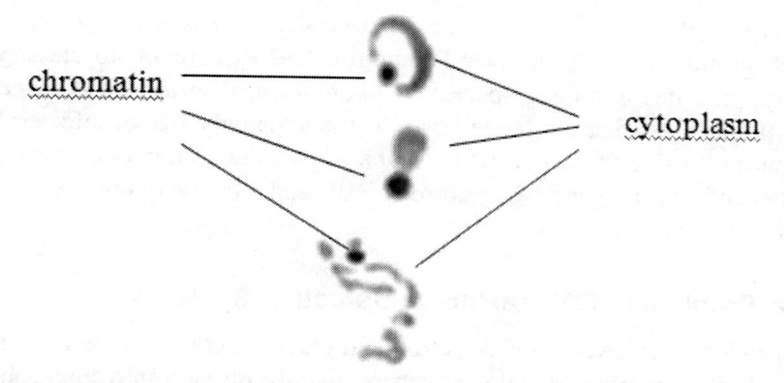

Figure 6.2 Image of malaria parasite on thick blood film, after red blood cell membrane was lysed, showing components inside.

tects parasite-specific antigens or enzymes from blood of malaria patients. Microscopic examination has further advantages since it can be used for identification of species and counting the number of parasites. In particular, this method facilitates treatment of the disease with appropriate drugs and helps control the spread of the disease. However, a vigorously trained technician is required to efficiently and accurately identify malaria and its species for appropriate drug administration. The accurate identification of malaria parasite species is difficult to achieve, because of inherent technical limitations and human inconsistency. A high training budget must be allocated to maintain the number of skillful technicians being familiar with the investigation of malaria parasites.

6.4 Automated Malaria Detection System

6.4.1 Existing Systems

To explore an opportunity to improve the performance of malaria detection and identification, many research efforts have been proposed to automatically detect malaria using digital image analysis. Most of them focused on thin blood smears. Tek et al. [10] used the technique of K-nearest neighbor classifier to identify and classify malaria species. Various image features were utilized in the classification such as the histogram, Hu moments, relative shape measurement, color auto-correlogram. Ross et al. [11] used back-propogation feed-forward neural networks to identify malaria parasites present in thin blood films that classified the species of the infection into four types (*P. falciparum*, *P. vivax*, *P. ovale* and *P. malariae*). Another research group used a Multilayer Perceptron (MLP) network to classify the malaria parasites into three species (*P. falciparum*, *P. vivax* and *P. malariae*) [12]. In this classification process, six features, namely size of infected RBC comparing to size of normal RBC, shape of parasite, numbers of chromatin, numbers of parasite per RBC, texture of RBC and location of chromatin, were introduced.

6.4.2 Problems of Digital Image Detection Systems

The problems in developing detection and classification systems for malaria using digital imaging techniques can be mainly divided into three phases: image acquisition, image segmentation and classification.

6.4.2.1 Image Acquisition

The goal of image acquisition is to acquire an image of sufficiently decent quality with crucial information for further manipulation such as interpretation of the image, i.e., malaria shape recognition. Especially for biomedical images, the lack of crucial information may lead to misinterpretation. Therefore, a high-resolution digital converter is needed to digitize a microscope magnified image (analog form) to a digital image.

During the thick film diagnosis for detection of malaria parasites, a microscopist must operate the fine-focus knob of the microscope to vertically move the lens while viewing the image in order to glean images from various depths of field. Why do we need information from various depths of fields? Figure 6.3 is a cartoon of parasites on a blood film viewed from the side. The dash lines represent the in-focus plane of each depth of field, which presents different in-focus parts of the parasites. Therefore, to see all parts of the parasites, all in-focus parts from various depths of field must be collected. Information from only one field depth such as the top one will lead to miscounting the numbers of parasites because it lacks information of the other parasites that are out-focused. Moreover, the information of only one depth of field is not enough for the estimation of size and shape of a parasite that will directly affect the performance of the species classification process. Unlike a thick blood film, a thin blood film has narrower depth of field due to the preparation process. Thus, detection and species classification of parasites from a thin blood film is easier than those from a thick blood film.

In order to automatically cover all possible areas of thick blood film for the detection of parasites, the designed system has to be able to move freely

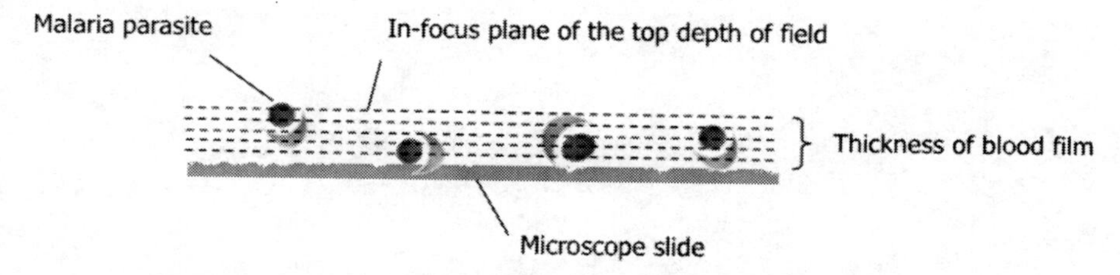

Figure 6.3 Side view of malaria parasite on thick blood film.

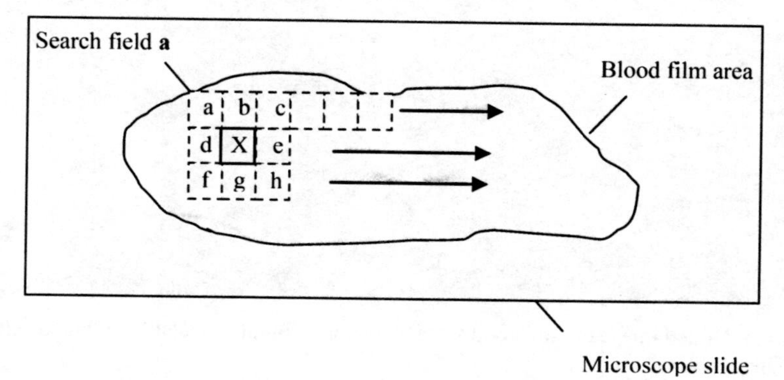

Figure 6.4 The non-overlapping search fields.

both horizontally (on the x and y axes) and vertically (on the z axis). In each microscope field, the vertical movement is needed to cover all in-focus depths of fields by automatically moving fine focus downward and upward in the defined range. One constraint that needs to be addressed is the vertical movement. It is necessary that when the fine focus (z axis) is moved, it must memorize the exact same position before the movement takes place. If the lens could not be controlled back to the start position, the boundary of the in-focus plane will gradually shift in the next microscope fields and will cause the out of focus after a few horizontal movements. The horizontal movement must be set to avoid potential overlaps of adjacent microscope fields during the image acquisition. As illustrated in Figure 6.4, each horizontal movement should not cause the overlap of search field X with other search fields (a-h). If the non-overlapping movement cannot be controlled, it will affect the accuracy of the parasite counting, i.e., the same parasite will be counted twice due to overlapping scans.

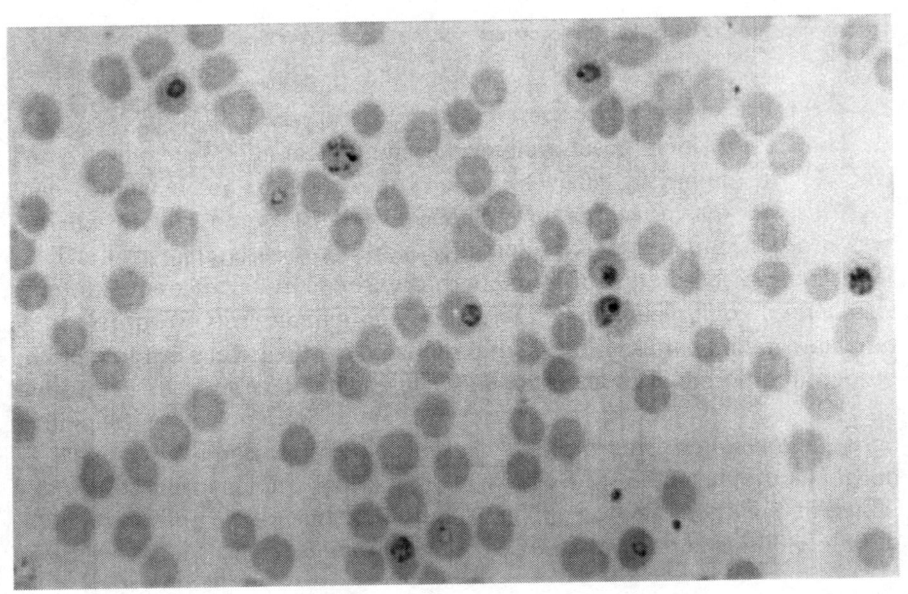

Figure 6.5 Digital image of thin blood smear showing a complete isolation of blood cells from each other.

It is necessary that the lens move in the fashion that it does not cause the overlapping of adjacent microscope fields during the image acquisition and result in the repeated counting of parasites.

6.4.2.2 Image Segmentation

Image segmentation refers to a process that separates the object of interest from other non-interesting objects, e.g., noise or other artifacts. For malaria diagnosis, the objects of interest are malaria parasites and the non-interesting objects are white blood cells, platelets, debris and other artifacts such as residual dye pigments. The segmentation method depends heavily on the characteristics of the image of interest. Therefore, the methods to segment a malaria parasite in the thin and thick blood films are different. Figures 6.5 and 6.7 show the images of thin and thick blood films, respectively. Each type of blood film has different contents and image characteristics that are used to properly address not only the segmentation problem but also the classification problem.

In the thin blood film, the boundaries of both red and white blood cells are clearly visible. The boundary of white blood cells are easier to segment

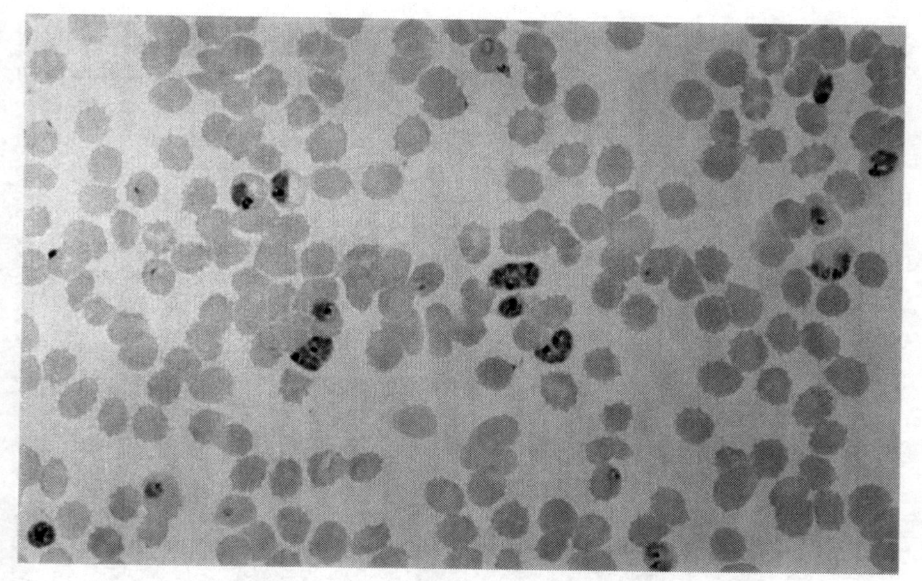

Figure 6.6 Digital image of thin blood film with overlapping red blood cells.

from the background when comparing to the boundary of red blood cells because of their solid color (deep purple or red) and the larger size of cells. The boundary of red blood cells has lighter color, which slightly blends with the background color. In an early state, malaria parasites are embedded inside the red blood cells. To perform the thin film image analysis, the red blood cells must be identified before detecting the parasite in the blood cell. The classical problem for identification of the red blood cell is cell segmentation. In Figure 6.5, the boundary of red blood cells that are completely separate from each other are shown. A conventional edge detection method such as Canny's edge detector [13] can be used to identify the round boundary of the cells. This kind of complete isolation of red blood cells is available only on a well-prepared thin blood smear for a single layer of blood film on a glass slide. The problem of segmentation is more complicated in the images with overlapped red blood cells.

Figure 6.6 shows an image of thin blood film but with overlapping red blood cells. The difficulty for segmentation of this type of image is that the shapes of overlapped cells vary significantly and unpredictably. This causes imperfections in boundary detection, thus a detection method such as Canny's edge detector cannot be used to detect cell boundaries. The cells that do not

Figure 6.7 Digital image of thick blood film showing a complete lysis of red blood cell membrane releasing malaria parasites.

overlap with other cells and stay clearly on the background are much simpler to separate since the contrast between the cells and the background is high. The change in brightness between two cells that overlap each other is not clear. Therefore, a conventional edge detection technique could not be used in this condition since it will recognize both cells as one unit. To separate these connected cells, a more sophisticated detection method is needed.

Figure 6.7 is a digital image of thick blood film. In the thick blood film, it is not possible to visualize the boundary of a red blood cell since the membrane of red blood cells are lysed during the preparation process of thick blood film stained with Giemsa. Only the boundary of white blood cells could be detected. The segmentation process for thick blood film is then to separate malaria parasites and white blood cells from the background. The chromatin of malaria parasites and white blood cells could be easily segmented from the background because of their solid color. Counting malaria parasites from the thick blood film segmentation is much simpler than that of the thin blood film. Using the differences of cell size, a malaria parasite can be distinguished from a white blood cell. On the contrary, for the thin blood smear, the overlapping of malaria parasites in an image is also not a problem since the size of malaria parasite is very small; hence this reduces the chance of overlapping of two parasite cells.

The species of malaria parasites could be classified by observing the differences of shape or pattern of the parasite's cytoplasm, as illustrated in Figure 6.8. However, the malaria cytoplasm is hard to detect even with human eyes, and not all malaria parasites in the thick blood film show clear cytoplasm. The problems could stem from both the blood film preparation process and the light transparent characteristics of the malaria cytoplasm. The change in brightness between malaria cytoplasm and the background is not clear making it very difficult to identify the complete shape of cytoplasm. It is necessary to find solutions that both emphasize the color of cytoplasm in the preparation process and acquire an image with better contrast between the images of malaria cytoplasm and their background during the acquisition process.

6.4.2.3 Classification

The classification step includes the feature extraction and classification processes. These two processes should be considered together since the classification method depends on the features that are extracted from the image. In thin blood film, after red blood cells are segmented, it is possible to extract the shape and color of malaria parasites inside those red blood cells. This makes it easier to distinguish malaria parasites and artifacts, which normally are not inside any cells. In the case of thick blood film, detection of red blood cells is not the concern since they are all lysed during the sample preparation process. The challenging problem is how to classify the early stages of all parasite species, which look very similar to one another. Figure 6.8 shows images of the young trophozoite stages (ring form) of *P. falciparum* (a), *P. vivax* (b), *P. malariae* (c) and *P. ovale*(d).

For the classification problem, the common classical methods are based on structural methodologies. The idea is to extract the parasite's features such as shape, color and some characteristics that are unique to each species and then use these features to distinguish them. In order to extract the structure of a parasite, all elements of the parasite need to be identified. As mentioned earlier, the only parasite element that can be completely isolated is chromatin, which is the first reliable information used for classification. If the aim of the study is to deploy this automatic detection and classification system for use in the malaria endemic areas, the lack of cytoplasm information will be common since the slide preparation facilities are limited in the field. To enhance the feature extraction and classification performance, those three major pro-

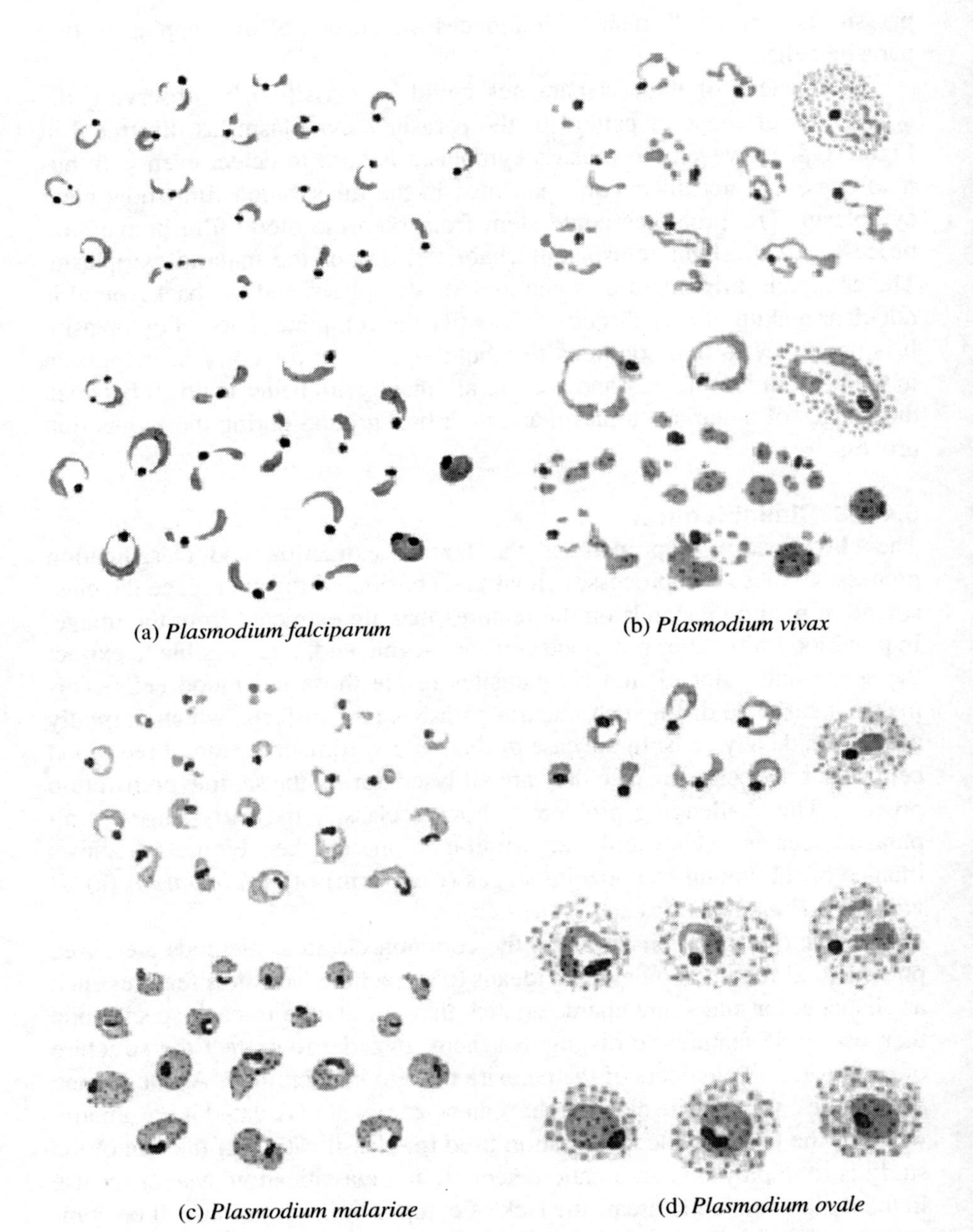

Figure 6.8 Ring forms of four malaria species. Images were modified from [14].

cesses, namely slide preparation, image acquisition and image segmentation are to be improved.

Since the malaria diagnosis using microscopy is the gold standard, to mimic the microscopy diagnosis, prior knowledge and microscopy diagnosis information are adapted for use in the classification process to analyze malarial digital images. For each species, such information is listed as follows:

- *Plasmodium falciparum*
  - The clinical signs appear during days 7–27 after parasite infection.
  - Only young trophozoites (ring form) can be seen on blood film.
  - Gametocytes are rarely observed on blood film.
  - Most mature trophozoites and schizonts are hidden in organs of the body.
  - The trophozoites of *P. faciparum* are smaller than those of other species.
  - Double chromatin can be seen.
  - The ring forms have delicate, thin and smooth ring of blue cytoplasm.
- *Plasmodium vivax*
  - This species can hide in the liver for a long time and may produce relapses of infection repeatedly.
  - All stages can be seen on blood film.
  - The ring forms of *P. vivax* are typically larger than *P. falciparum*.
  - The ring forms are often without a complete circle of cytoplasm.
- *Plasmodium ovale*
  - It is mainly found in Africa.
  - This species is the most difficult to differentiate from other species.
  - All stages can be seen on blood film.
  - The ring forms are similar to *P. vivax*.
  - Schizonts are usually similar to *P. malariae*.
- *Plasmodium malariae*
  - This species have lower parasitaemia than other species.
  - The ring forms may not have a complete circle of cytoplasm.
  - The mature trophozoites may look similar to gametocytes of the same species.
  - The mature schizonts are the most distinctive stage.

6.5 System for Thick-Blood Films

Since microscopic examination of Giemsa-stained thick blood film is mainly used for detection of malaria parasites [15, 16], Frean [17] proposed an automatic parasite counting system on thick blood films using image analysis techniques. The result provided an obvious potential solution for parasite detection. Other work by Frean [18] investigated the use of ImageJ, an open-access Java-based image-processing program, for parasite counting. The results showed that the digital counting method provided higher accuracy as compared to conventional microscopic counting method.

In addition to parasite enumeration, an automated system described here aims for the identification and analysis of parasite species on Giemsa-stained thick blood films by image analysis techniques. The system comprises two main components: (1) image acquisition unit and (2) image analysis module. The image acquisition unit controls the movement of the microscope's stage in 3-directional planes. The captured images are, then, analyzed by the image analysis module to detect and identify malaria parasites. This device is useful for deployment in rural areas where the highly trained healthcare staffs on malaria detection are not available or insufficient to examine large quantity of blood samples.

6.6 Methods

The structure of an automated detection system is illustrated in Figure 6.9. It comprises two main components: (1) image acquisition unit and (2) image analysis module installed inside the processing unit. The image analysis module passes control commands to the image acquisition unit through USB connection A and obtains the captured images from a digital camera mounted on top of the microscope through USB connection B.

6.6.1 Image Acquisition Unit

The image acquisition unit can be mounted on most conventional light microscopes. It automatically controls the movement of the microscope's stage in 3-directional planes. The z-axis vertical adjustment is used for focusing, while the x-axis and y-axis horizontal adjustments are used for repositioning of the thick blood film slide as shown in Figure 6.10. With the motorized controller that is connected to the image analysis module, the module is able to control the vertical and horizontal movements of the microscope's stage.

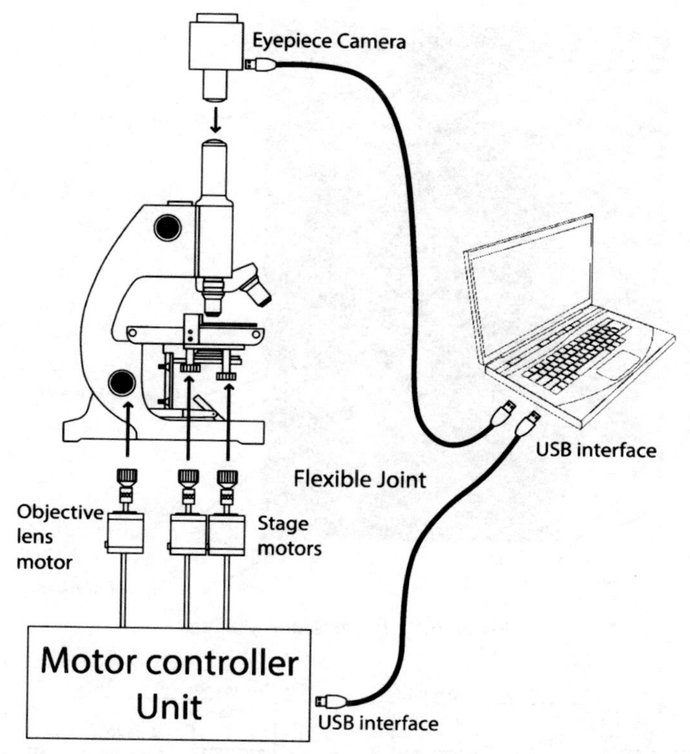

Figure 6.9 Structure of the detection system.

The digital camera that is installed at the top of the microscope captures the 1000× magnified images. Those captured images are sent to the image analysis module, which in turn will detect and identify malaria parasites on each field of the thick blood film. The analysis process of the film will be completed when the preset numbers of field are analyzed.

The image acquisition board consists of a micro-controller, a USB port connected to the processing unit and control ports connected to three motorized units. The z-axis motorized unit attached to the fine-tune knob of microscope operates in the precision of 9 nanometers using a 4-phase stepping motor. To achieve the nanometer scale, each step of the stepping motor is divided into 16 sub-steps. For the x- and y-axes, the precision scale is 0.5 to 1 mm which is a proper scale for moving the stage to the adjacent microscope fields.

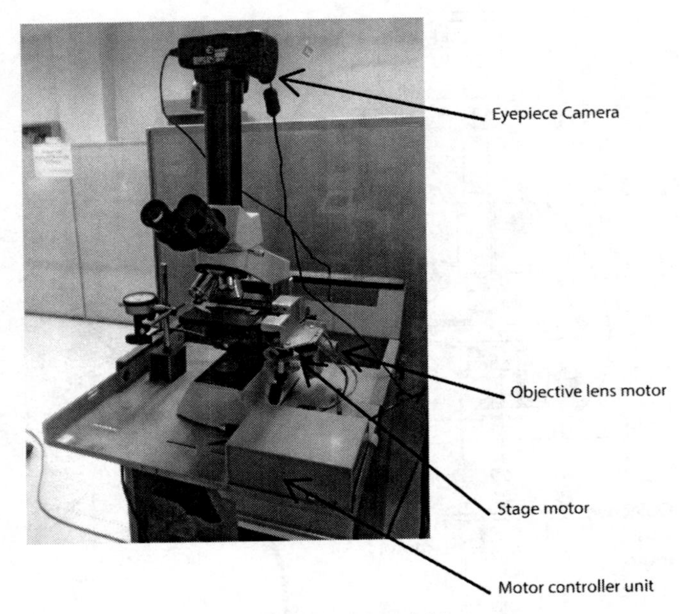

Figure 6.10 Image acquisition unit.

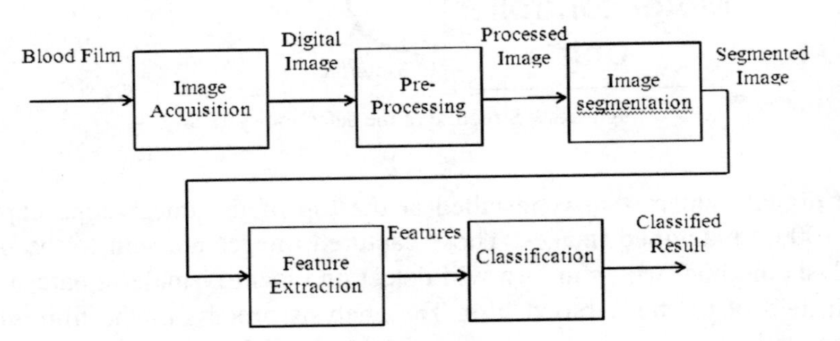

Figure 6.11 Image analysis module.

6.6.2 Image Analysis Module

The image analysis module is composed of five processes (Figure 6.11): image acquisition, pre-processing, image segmentation, feature extraction and classification processes. First, the blood film is placed on the microscope stage. The position and focus of the first field are set before initiating the analysis processes.

For the image acquisition process, the motorized controller controls the movement of the microscope stage by sending the movement command to the individual motorized unit that is attached to the microscope knob. The digital snapshot of one microscope field sample is automatically obtained with the digital camera and sent to the image pre-processing module. The work of the pre-processing process is combined with the image acquisition process. By controlling vertical movement, the system is able to capture images at different depth of fields. This benefits the system to improve the quality of image by merging in-focus information over a range of images to generate one entirely in-focus image. The in-focus regions are identified based on their sharpness. The Laplacian technique is used to perform edge detection for in-focus pixel positions of each depth of field image because of its high accuracy and speed. The $L(x, y)$ is defined as the Laplacian value at the x and y positions, which ranges from 1 to N, $f(x, y)$ denotes the value of the image, X and Y are the width and length of the image, respectively.

$$L(x, y) = \sum_{X}\sum_{Y}\big\{|2f(x, y) - f(x-1, y) - f(x+1, y)| + |2f(x, y) - f(x, y-1) - f(x, y+1)|\big\}. \tag{6.1}$$

Laplacian values of each depth of filed image are computed using Equation (6.1). Figures 6.12a–e illustrate each depth of field of the same microscope field. It could be noticed that each image represents a different in-focus area. Figure 6.12f represents quality improvement by merging in-focus information over various depths of fields.

After the pre-processing process, the in-focus image is converted into HSV (Hue-Saturation-Value) color format. The value component of HSV image is employed for segmentation that is divided into three steps.

Step 1: Find histogram of value components and extract non-background objects (white blood cells, malaria parasites and possible Giemsa stain-derived artifacts). These are extracted using an adaptive threshold found according to information of the histogram. Figure 6.13 shows the histogram of the value component from a HSV image. The small inserted box represents the histogram of the foreground of interest. The foreground component is clearly separated from the background component.

Step 2: After discarding the background as shown in Figure 6.14, the image is divided into small windows of size 300 by 300 pixels and searched for connected regions. Each region is labeled with an identification value for future reference.

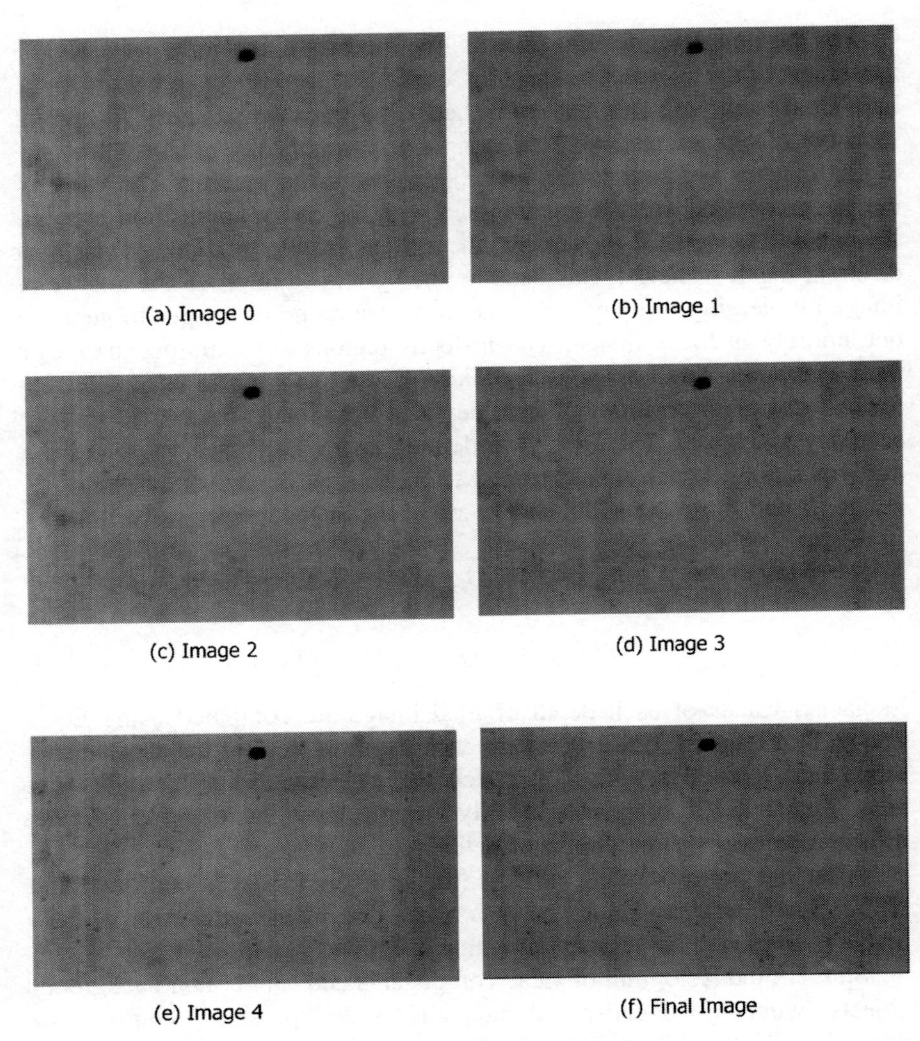

Figure 6.12 Various depths of field images.

Step 3: In each window, malaria parasites are distinguished from white blood cells according to size.

The above processes (steps 2 to 3) are repeated until all the parasites are discovered and labeled.

The labeled regions that may contain the parasites are further processed. As the hue values represent different physical components of the parasites,

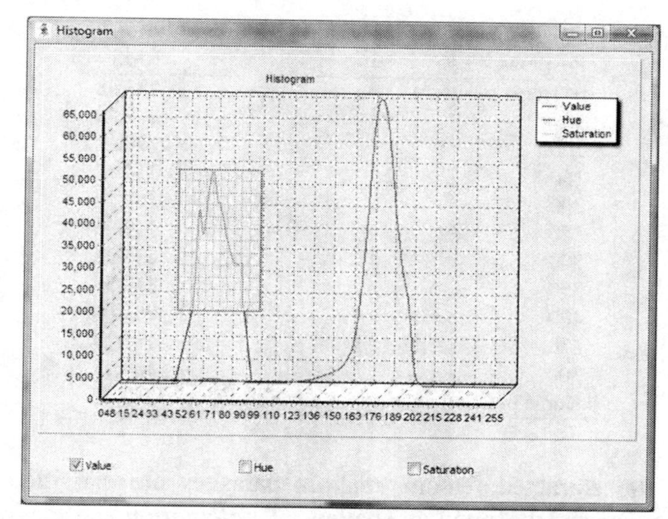

Figure 6.13 Histogram of value component from HSV image.

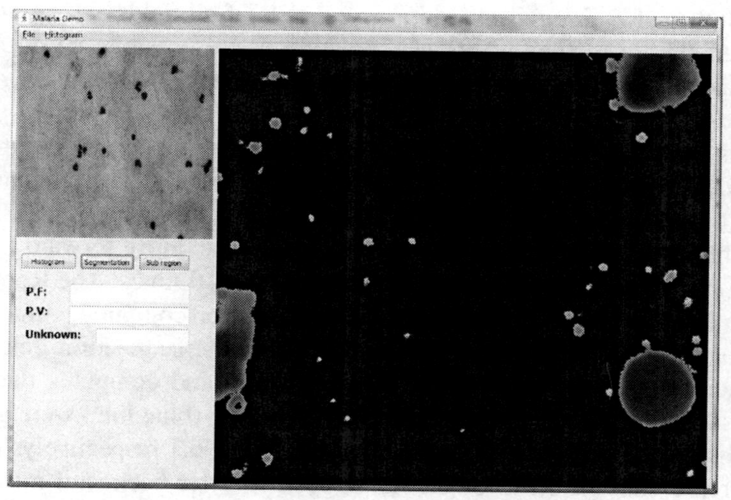

Figure 6.14 Processed image after pre-processing and segmentation processes.

the hue histogram of the HSV image is obtained. The chromatin size represented by the number of red and magenta pixels in the hue histogram of each region are used for distinguishing parasite species in the classification process.

Table 6.1 RMSs of 50-step movements.

Positions	RMS
0 (Home position)	283.48
50	282.49
100	278.1
150	274.82
200	260.74
250	247.96
300	235.25
350	219.5
400	201.04
450	182.29
500	169.41
Home position after backward movement	283.41

Using the extracted feature, malaria parasites are classified into two species that are mostly found in Thailand, *P. falciparum* and *P. vivax*. *P. falciparum* parasites have a smaller size of chromatin than the *P. vivax* parasites. The number of *P. falciparum* and *P. vivax* cells of all fields are counted and recorded. Those regions where classification is not possible are designated as unknown, and the user is alerted that the sample may contain malaria parasites.

It is necessary for the movement of microscope's stage to be precisely controlled in order to acquire high quality digital images. The efficiencies of image acquisition in controlling all 3-directional planes were tested. For the z-axis movement test, the microscope stage was set to move forward from the start point (set as home position) for 50 steps and 10 times. The Root Mean Square (RMS) shown in Equation (6.2) is used for computing image contrast for each 50-step-forward movement. Then, the acquisition unit moves the microscope stage back to the home position and computes the image contrast. The RMSs of 50-step-forward movements (blue line) were graphed and tabulated as shown in Figure 6.15 and in Table 6.1 respectively. The red circle represents the RMS of the home position after backward movement. The comparison of image qualities at the home position before forward and after backward movements is shown in Figure 6.16.

$$\mathrm{RMS} = \sqrt{\frac{1}{MN}\sum_{i=0}^{N-1}\sum_{M-1}^{j=0}(I_{i.j} - \bar{I})}, \quad \bar{I} \text{ is the average intensity of pixels.} \tag{6.2}$$

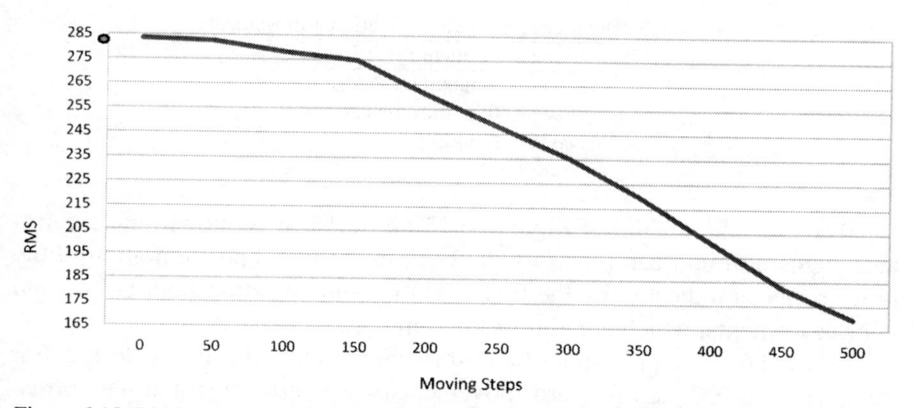

Figure 6.15 RMSs of 50-step-forward movements and RMS of home position after backward movement.

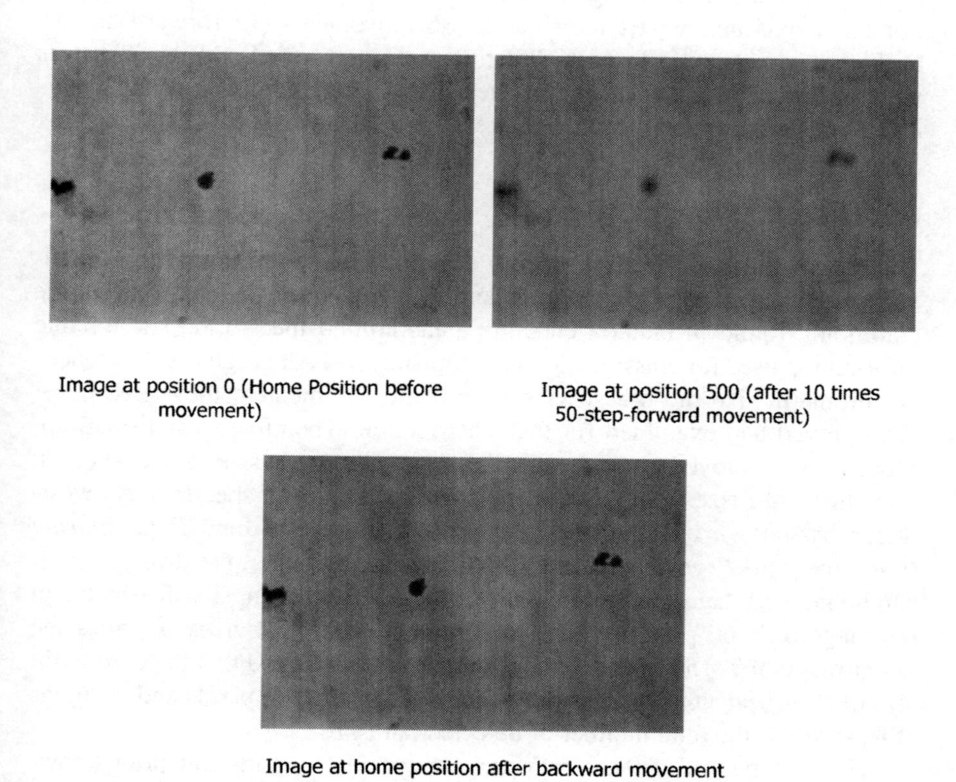

Figure 6.16 Comparison of image qualities at home position.

Table 6.2 The precision of all 3-directional movements.

Axis	Average Precision
z-axis	9 nanometers
x-axis	0.29 millimeters
y-axis	0.2 millimeters

Based on the results of Figures 6.15 and 6.16, it is shown that the image acquisition unit can precisely get back to the home position after all the movements as indicated by the RMSs at the home position both before and after movements, which are almost the same.

In Table 6.2, the precision of all 3-directional movements was tested. For the z-axis, 10,000-step-forward movements were performed and the corresponding distances were measured in a unit of micrometer. This process was repeated 5 times. The average precision of the z-axis is about 9 nanometers. For the x-axis and y-axis movements, 36 movements (18 forward and 18 backward movements) in each direction were performed and the distances were measured in a unit of micrometer. The average precision of the x-axis is 0.29 millimeters and the average precision of the y-axis is 0.2 millimeters.

6.7 Discusssion

The aim of the classification process described here is to identify two malaria parasite species, *P. falciparum* and *P. vivax*, since the two species are commonly found in malaria endemic areas around the world. The reliable information used for classification is chromatin size. Although the differences of chromatin size in both species have been documented, they have never been proved and examined for their distribution. Therefore, a statistical approach was employed to see the distribution of chromatin size of both species. The chromatin size from 2,000 samples of each parasite species was investigated. As shown in Figure 6.17, the chromatin size of both *P. falciparum* (blue line) and *P. vivax* (red line) distribute in the range of 7–150 pixels. It can be noticed that the chromatin size of *P. falciparum* mostly distributes in the range of 7–60 pixels, while the chromatin size of *P. vivax* occupies the wider range of 7–150 pixels. In addition, it was also found that the chromatin size of *P. falciparum* distributes in a range higher than 75 pixels and occupies at least 5% of the total number of detected parasites.

The distribution of these two parasite species supports our prior knowledge that, for *P. falciparum*, only young trophozoites (ring form) can usually be seen on blood films and are smaller than other species. This is why the

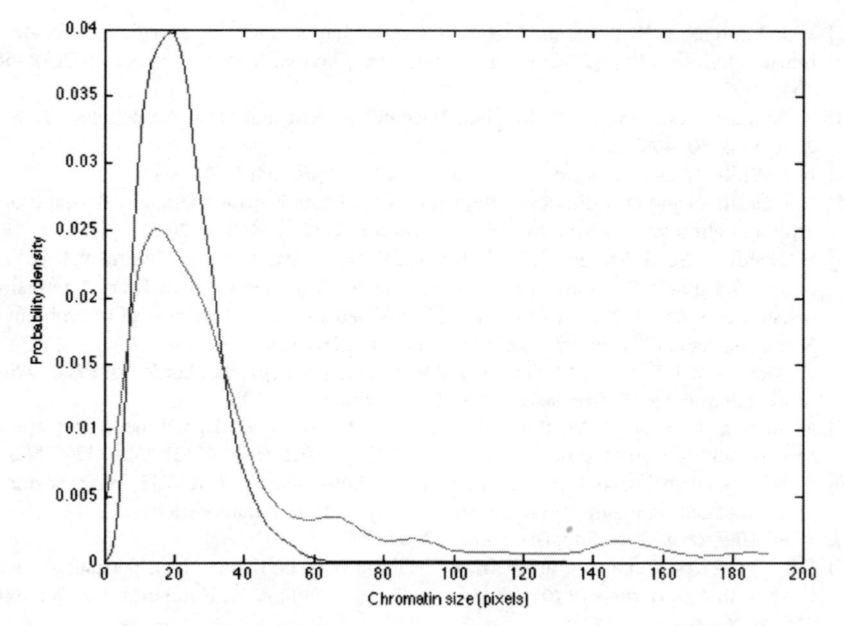

Figure 6.17 Distribution of chromatin size of *P. falciparum* and *P. vivax*.

distribution of their chromatin size is narrow. For *P. vivax*, all stages of parasites including mature trophozoites, schizonts and gametocytes can be seen on blood films and the chromatin sizes are normally bigger than those of *P. falciparum*. Consequently, the chromatin could be used to distinguish these two species.

Using only the chromatin size feature may lead to low classification rate. To enhance the performance of the classification rate, other features can be used. However, with the constraint that the quality of the slides prepared in the field could vary. It needs efforts and cooperation of health officers working in the field to prepare decent quality thick blood films. Therefore, the automated system could be used to extract a quality image and obtain information on parasite infection that aids proper treatment and control of the deadly malaria disease.

References

[1] World Health Organization, World Malaria Report 2011, Geneva, WHO, 2011.

[2] W.J. Martens, L.W. Niessen, J. Rotmans, T.H. Jetten, and A.J. McMichael. Potential impact of global climate change on malaria risk. *Environ. Health Perspect.*, 103(5):458–464, 1995.

[3] K. Matuschewski and A. K. Mueller. Vaccines against malaria – An update, *FEBS J.*, 274(18):4680–4687, 2007.

[4] N.J. White. Drug resistance in malaria. *Brit. Med. Bull.*, 54(3):703–715, 1998.

[5] R.T. Eastman and D.A. Fidock. Artemisinin-based combination therapies: A vital tool in efforts to eliminate malaria. *Nat. Rev. Microbiol.*, 7(12):864–874, 2009.

[6] H. Noedl, Y. Se, S. Sriwichai, K. Schaecher, P. Teja-Isavadharm, B. Smith, W. Rutvisuttinunt, D. Bethell, S. Surasri, M.M. Fukuda, D. Socheat, and L. Chan Thap. Artemisinin resistance in Cambodia: A clinical trial designed to address an emerging problem in Southeast Asia. *Clin. Infect. Dis.*, 51(11):e82–89, 2010.

[7] K. Mendis, B.J. Sina, P. Marchesini, and R. Carter. The neglected burden of *Plasmodium vivax* malaria. *Am. J. Trop. Med. Hyg.*, 64(1–2 Suppl):97–106.

[8] M.A. Beg, R. Khan, S.M. Baig, Z. Gulzar, R. Hussain, and R.A. Smego Jr. Cerebral involvement in benign tertian malaria. *Am. J. Trop. Med. Hyg.*, 67(3):230–232, 2002.

[9] C. Wongsrichanalai, M.J. Barcus, S. Muth, A. Sutamihardja, and W.H. Wernsdorfer. A review of malaria diagnostic tools: Microscopy and rapid diagnostic test (RDT). *Am. J. Trop. Med. Hyg.*, 77(6 Suppl):119–127, 2007.

[10] F.B. Tek, A.G. Dempster, and I. Kale. Malaria parasite detection in peripheral blood images. In *Proceedings of British Machine Vision Conference*, Edinburgh UK, pp. 344–356, 2006.

[11] N.E. Ross, C.J. Pritchard, D.M. Rubin, and A.G. Duse. Automated image processing method for the diagnosis and classification of malaria on thin blood smears. *Medical and Biological Engineering*, 44:427–436, 2006.

[12] N.A. Seman, N.A. Mat Isa, L.C. Li, Z. Mohamed, U.K. Ngah, and K.Z. Zamli. Classification of malaria parasite species based on thin blood smears using multilayer perceptron network. *International Journal of the Computer, the Internet and Management*, 1:46–51, 2008.

[13] Rafael C. Gonzalez and Richard E. Woods, *Digital Image Processing*, Prentice Hall, 2002.

[14] Basic Malaria Microscopy, Part 1, Learner's Guide, WHO 1991.

[15] World Health Organization. *Policy and Procedures of the WHO/INICD Microbiology External Quality Assessment Programme in Africa.* Geneva, WHO, 2007.

[16] World Health Organization. *Basic Malaria Microscopy*, Geneva, WHO, 2001.

[17] J.A. Frean. Improving quantitation of malaria burden with digital image analysis. *Trans. Royal Soc. Trop. Med. Hyg.*, 102:1062–1063, 2008.

[18] J.A. Frean. Reliable enumeration of malaria parasites in thick blood films using digital image analysis. *Malaria J.*, 8:218, 2009.

7

Face Detection and Tracking Framework for Video Processing

Ihor Paliy[1], Anatoly Sachenko[1] and Ognian Boumbarov[2]

[1]*Research Institute of Intelligent Computer Systems, Ternopil National Economic University, Ternopil, Ukraine; e-mail: ipl@tneu.edu.ua*
[2]*Technical University of Sofia, Sofia, Bulgaria*

Abstract

In this chapter we describe the face detection and tracking framework. Face detection is based on the combined cascade of neural network classifiers. A cascade of weak classifiers is used to localize the face structural features. Tracking is performed using a Kalman filter. The framework was experimentally explored on a test video sequence and adjusted to obtain the high processing speed.

Keywords: face detection, face tracking, combined cascade of neural network classifiers, convolutional neural network, Kalman filter.

7.1 Introduction

There are many modern applications, such as a face recognition, videoconferences, content-based image retrieval, video surveillance, human-computer interface, analysis of face expressions, visitor counting, access control, etc., where detection is the first and the most important stage of any face processing [1]. *Face detection* (FD) is the determination of the face(s) coordinates on an arbitrary image. It is quite simple for humans, but it becomes a serious problem for computer systems due to many factors which

Richard J. Duro and Fernando López Peña (Eds.), Digital Image and Signal Processing for Measurement Systems, 189–200.

affect the face appearance such as in-plane and out-of-plane rotations, face expressions, occlusions, lighting conditions, etc. [1].

Nowadays appearance-based methods are more preferable providing face detection [1] in terms of detection validity and speed. They are based on the scanning of an input image at some scale levels by a fixed-size window in order to find faces in different positions and scales. Each window is handled using the two-class (face/non-face) classifier presented by a neural network [2–5], support vector machines [6–9], Bayesian classifier [10], etc. After the input image has been processed, all face detections are grouped, and the areas with a number of multiple detections are accepted as faces only.

In 2001, Viola and Jones [11] presented one of the fastest appearance-based FD approaches able to process the input image in near real-time mode. The novelty of their approach comes from the integration of a new image representation (integral image) and a learning algorithm (based on AdaBoost) and a method for combining classifiers (cascade of weak classifiers). Besides they used a set of rectangular Haar-like features instead of raw pixel information as an input for weak classifiers. Lienhart [12] extended the Haar-like features set with rotated ones and Li et al. [13] proposed a novel learning procedure called FloatBoost.

One of the highest validities is demonstrated by the FD approach of Garcia and Delakis [5]. They used a *convolutional neural network* that processes an input image in two stages: coarse and fine detection. During the coarse detection, the convolutional neural network handles an image of any size at once.

According to the validity evaluation results on the Carnegie Mellon University (CMU) test set [3] the FD approaches with monolithic classifiers outperform methods with cascade classifiers [14]. However, at the same time the cascade structure makes the classifier faster.

Face detection may be followed by the localization of the face structural features (eyes, nose, and mouth) which is useful in some applications, like face recognition, gaze tracking, expression recognition, etc. The localization assumes that there is one face only on the image region that has been already detected at the previous stage, and its main task is to find the coordinates of structural features.

In its simplest form, a tracking can be defined as a problem of estimating the object trajectory in the image plane as it moves around a scene. In other words, a tracker assigns consistent labels to the tracked objects in different frames of the video. Additionally, depending on the tracking domain, a tracker can also provide object-centric information, such as the orientation, area, or the object shape. Tracking objects can be complex due to: loss of information caused by projection of the 3D world on a 2D image, the noise in images, complex object motion, nonrigid or articulated nature of objects, partial and full object occlusions, complex object shapes, scene illumination changes, and real-time processing requirements.

All existing object tracking methods can be divided into the following three classes [15]:

- Point Tracking (MGE tracker, GOA tracker [16], Kalman filter, JPDAF, PMHT, kernel tracking, silhouette tracking).
- Kernel (mean-shift [17], KLT, layering [18], eigentracking, SVM tracker [20]).
- Silhouette Tracking (state space models, variational methods [20], heuristic methods [21], Hausdorff [22], Hough transform, histogram [23]).

This chapter presents a face detection and tracking algorithm using a *combined cascade of neural network classifiers* for detection and *Kalman filters* for tracking. We consider both techniques as a base to build a fast face detection and tracking framework able to process the video flow with high speed and validity.

7.2 Face Detection Method

Here we propose joining the rapid cascaded classifier with the accurate monolithic one within a two-level combined cascade of classifiers instead of using them independently. The first level of the combined cascade is represented by the Haar-like features' cascade of weak classifiers (multilayer perceptron), which is responsible for face-like object detection, and the second level is represented by the convolutional neural network for the objects' verification.

The Haar-like features' cascade of weak classifiers [11] allows us to detect the face candidates very quickly. It consists of levels which include one or more weak classifiers. The weak classifier's input is represented by Haar-like features with a value [11]

$$Feat(x) = s_w \times SUM_w + s_b \times SUM_b, \qquad (7.1)$$

where x is the input image's sub-window, s_w and s_b are the whole rectangle's and its black part's weights accordingly, SUM_w and SUM_b are the whole rectangle's and its black part's sum of pixels. A weak classifier's output value

$$h(x) = \begin{cases} 1, & \text{if } Feat(x) < \theta \\ -1, & \text{if } Feat(x) > \theta, \end{cases} \tag{7.2}$$

where θ is the weak classifier's threshold. The cascade of weak classifiers is a linear combination of weak classifiers [11]:

$$H(x) = \sum_{t=1}^{T} \eta_t \times h_t(x), \tag{7.3}$$

where T is the weak classifiers' number and η_t is the t-weak classifier's weight. The AdaBoost algorithm [24] is used for training the cascade of weak classifiers as well as a selection of the most important Haar-like features.

The verification stage of the combined cascade of neural network classifiers uses the convolutional neural network (CNN) [25] which is more robust to the input image's deformations than other known classifiers. The output value of a neuron with the bipolar sigmoid transfer function and the coordinates (m, n) of p-plane and l-layer [14]

$$y_{m,n}^{l,p}(x) = \frac{2}{1 + \exp(-WSUM_{m,n}^{l,p}(x))} - 1, \tag{7.4}$$

where x is the input face candidate's image and $WSUM$ is the neuron's weighted sum [14]

$$WSUM_{m,n}^{l,p}(x) = \left(\sum_{k=0}^{K-1} \sum_{r=0}^{R-1} \sum_{c=0}^{C-1} y_{2m+r,2n+c}^{l-1,k}(x) \times w_{r,c}^{l,p,k} \right) - b^{l,p}, \tag{7.5}$$

where K is the input planes' number (as well as convolutional kernels), R and C are the convolutional kernel's height and width, $w_{r,c}^{l,p,k}$ presents the synaptic weight with coordinates (r, c) in the convolutional kernel between k-plane of the $(l-1)$-layer and p-plane of the l-layer, and $b^{l,p}$ is the neurons' bias of the p-plane and l-layer.

We used the sparse structure of the CNN instead of the full-connected one as well as the decreased number of layers by performing convolution and subsampling operations per each plane simultaneously [26] in order to increase the neural network's processing speed.

Face-candidate verification and active training set generation methods were proposed for CNN to obtain the high validity and speed of the whole combined cascade of neural network-classifiers [14].

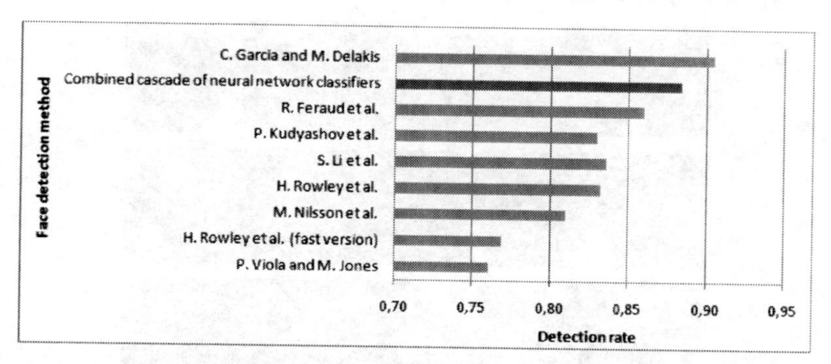

Figure 7.1 Comparison of the FD approaches' detection rate with false alarms rate of 10^{-8} using the CMU test set.

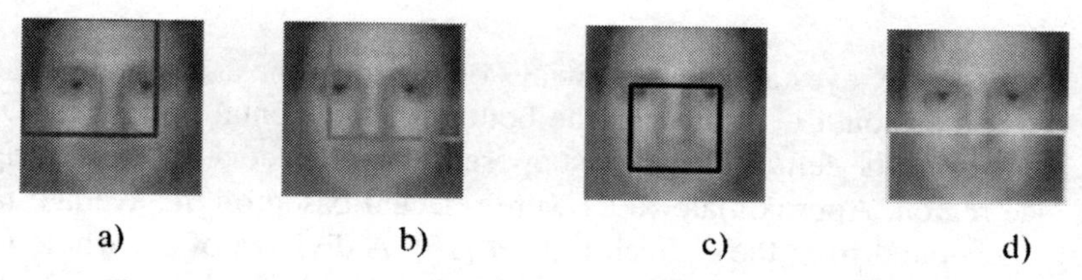

Figure 7.2 ROIs for left eye (a), right eye (b), nose (c) and mouth (d).

The experimental explorations using the CMU test set showed that the combined cascade of neural network classifiers demonstrates one of the best validities as compared to known FD approaches (Figure 7.1). It yields only to the Garcia and Delakis method [5], approximately by 2% in detection rate, but it is 8 times faster. The developed FD system handles 15–20 frames (sized 320×240 pixels) per second using the Logitech QuickCam Messenger web camera connected to an Intel Celeron E1200 Dual-Core 1.6 GHz PC with 1 GB RAM.

A cascade of weak classifiers based on Haar-like features [11] was used to detect such face elements as eyes, nose and mouth. This cascade detects objects very quickly rejecting the majority of the background in the early stages. The output values of Haar-like features, the weak classifier and cascade may be calculated using (7.1), (7.2) and (7.3) correspondingly.

The face features' search is based on regions of interest (ROI) which are indicating based on knowledge of where these face elements are usually located (Figure 7.2).

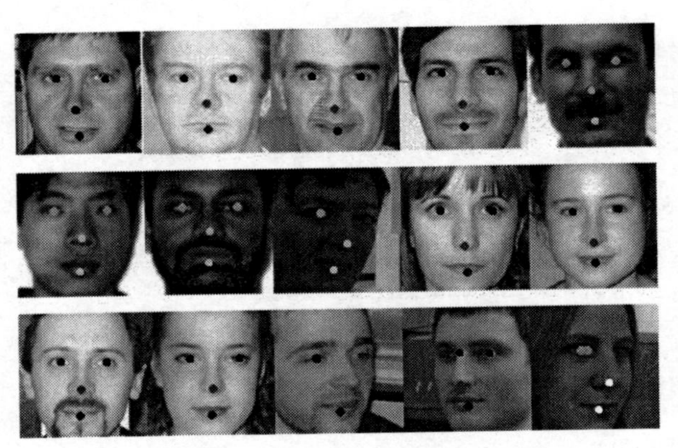

Figure 7.3 Results of face features detection.

For example, eyes are located always on the top, a nose is situated in the middle and a mouth is located on the bottom of the frontal upright face. At the same time the left/right eye is supposed to be detected on the left/right top face region. Approximate ROIs were selected based on the average face image generated from the Caltech test set [15]. A division of the whole face image was made to increase the detection validity and speed.

The experimental exploration performed on several test sets including the Caltech, AT&T, Yale and our own set showed the algorithm failed sometimes on half-profile faces and faces with hard shadows, but it works well with frontal faces (Figure 7.3).

It is possible to determine many other key points for the face orientation estimation based on the coordinates of the detected face features.

7.3 Face Tracking Algorithm

The face tracking algorithm uses a Kalman filter to predict a position of the detected face on the next frame of a video flow. The filter estimates the face coordinates in time moment t based on the history of coordinates for the previous frames [16]:

$$x_t = F_t x_{t-1} + B_t u_t + w_t, \tag{7.6}$$

$$y_t = F_t y_{t-1} + B_t u_t + w_t, \tag{7.7}$$

where F_t is the state transition model applied to the previous coordinate x_{t-1}; B_t is the control-input model which is applied to the control vector u_t; w_t

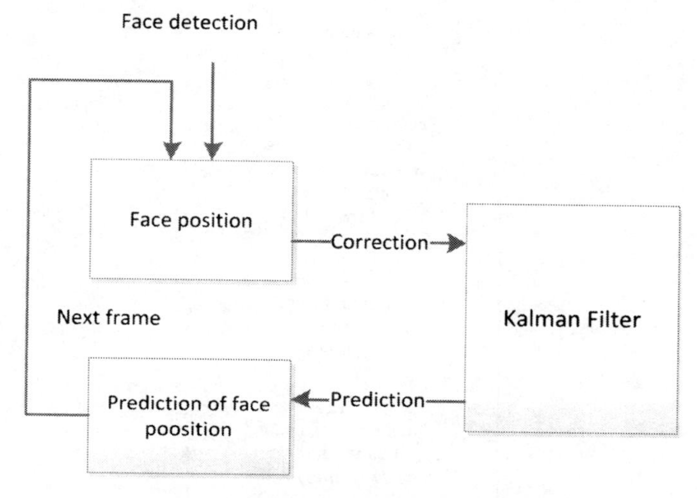

Figure 7.4 Face tracking algorithm based on Kalman filter.

is the process noise which is assumed to be drawn from the zero mean multivariate normal distribution. At the time moment t for the real state x_t an observation (or measurement) [16]

$$z_t = H_t x_t + v_t \tag{7.8}$$

where H_t is the observation model between x_t and z_t; v_t is the observation noise which is assumed to be zero meaning the Gaussian white noise.

The Kalman filter uses the real face coordinates to predict the face position on the next frame (Figure 7.4). Then the parameters of the Kalman filter are corrected using predicted coordinates to ensure handling of the next frames. The filter should be corrected once per some number of frames with the real face coordinates to reduce the prediction error.

7.4 Face Detection and Tracking Framework

The general face detection and tracking framework consists of the following steps (Figure 7.5):

- Step 1: the biggest face is detected on the whole frame by the combined cascade of neural network classifiers.

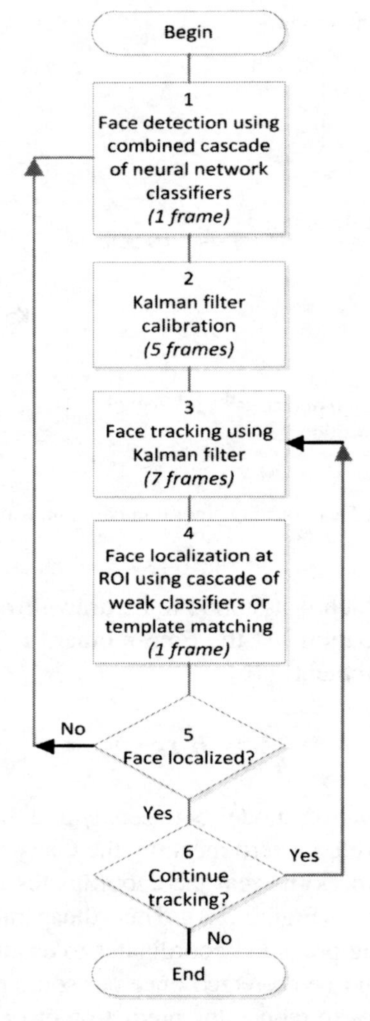

Figure 7.5 General face detection and tracking framework.

- Step 2: the Kalman filter is calibrated on 5 frames using real face coordinates obtained by a cascade of weak classifiers or template matching in the same region of interest (ROI) in order to gather the motion history.
- Step 3: the Kalman filter is corrected using predicted face coordinates. Real values are used only after step 6 to reduce prediction error.

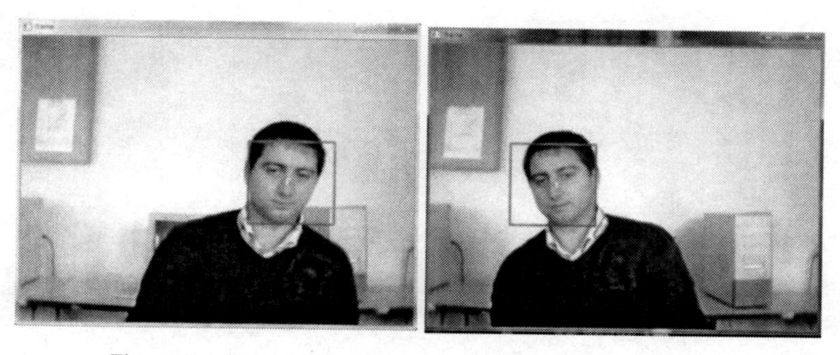

Figure 7.6 Test video sequence frames with tracking results.

- Step 4: a cascade of weak classifiers or template matching is used to localize a face in the ROI.
- Step 5: if the face is missed then step 1 should be executed again to find the next biggest face for tracking. If a face is localized successfully then real coordinates are used to update the Kalman filter in step 3.
- Step 6: it contains conditions to complete the run of the algorithm.

The framework was implemented using the C++ programming language and the Intel Open Computer Vision library [27].

A test video sequence of 500 frames was created to perform the experimental exploration of the developed framework. The sequence contains one face with different motion trajectories (Figure 7.6).

Several experiments were performed to adjust the tracking parameters. For example, two experiments on speed and validity (Figure 7.7) were run to find the rational value for the face tracking period using the Kalman filter without real coordinates in step 3. According to these experiments the results of a correction period of seven frames was selected.

The developed framework allows increasing the video processing speed to 30–50 frames per second (depending on the Kalman filter correction period) in comparison with handling each frame by the combined cascade of neural network classifiers (Figure 7.8).

7.5 Conclusions and Future Research Directions

In this chapter a face detection method and a tracking algorithm are developed. To detect a face we propose joining a rapid cascaded classifier with

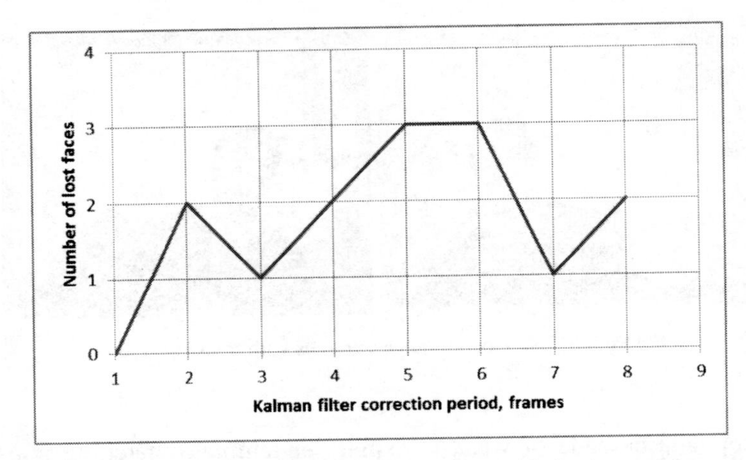

Figure 7.7 Dependence between Kalman filter correction period using real face coordinates and number of missed faces on the test sequence.

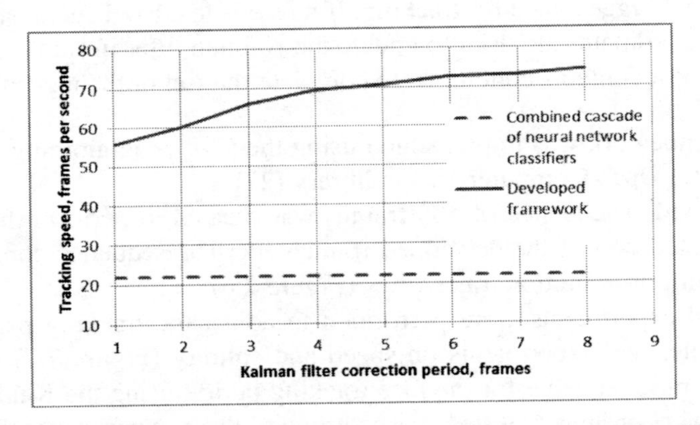

Figure 7.8 Comparison of the combined cascade of neural network classifiers and the developed framework's video processing speed on the test set.

an accurate monolithic one within a two-level combined cascade of classifiers instead of using them independently.

The face tracking algorithm uses a Kalman filter to predict a position for the detected face on the next frame of a video flow.

A general framework for the face detection and tracking procedures was created and explored. The experimental results confirmed this framework allows increasing the video processing speed by 30–50 frames per second (depending on the Kalman filter correction period) as compared to handling each frame using the combined cascade of neural network classifiers. Moreover the high processing speed of 60–70 frames per second enables implementing a framework for further face processing operations like recognition, emotion analyses, age and sex estimation, etc.

Furthermore, a further exploration of the framework by tracking multiple faces simultaneously is planned. Some parallel techniques like MPI and CUDA are expected to be used there.

References

[1] M. Yang. *Recent Advances in Face Detection*, IEEE ICPR 2004 Tutorial, Cambridge, United Kingdom, 2004.

[2] T. Poggio and K. Sung. Example-based learning for view-based human face detection. *IEEE Transactions on Pattern Analysis and Machine Intelligence*, 20(1):39–51, 1998.

[3] H. Rowley, S. Baluja S., and T. Kanade. Neural network-based face detection. *IEEE Transactions on Pattern Analysis and Machine Intelligence*, 20:22–38, 1998.

[4] M. Yang, D. Roth, and N. Ahuja. A SNoW-based face detector. In *Proceedings of Advances in Neural Information Processing Systems 12 (NIPS 12)*, pp. 855–861, 2000.

[5] C. Garcia and M. Delakis. Convolution face finder: A neural architecture for fast and robust face detection. *IEEE Transactions on Pattern Analysis and Machine Intelligence*, 26(11):1408–1423, 2004.

[6] E. Osuna, R. Freund, and F. Girosi. Training support vector machines: An application to face detection. In *Proceedings of IEEE Conference on Computer Vision and Pattern Recognition*, pp. 130–136, 1997.

[7] S. Romdhani, P. Torr, B. Schlkopf, and A. Blake. Computationally efficient face detection. In *Proceedings of ICCV*, Vol. 1, pp. 695–700, 2001.

[8] B. Heisele, T. Serre, S. Prentice, and I. Poggio. Hierarchical classification and feature reduction for fast face detection with support vector machines. *Pattern Recognition*, 36(9):2007–2017, 2003.

[9] T. Vetter, M. Rä tsch, and S. Romhani. Efficient face detection by a cascaded support vector machine using Haar-like features. In *Proceedings of the 26th German Association for Pattern Recognition Symposium (DAGM'04)*, Tü bingen (Germany), pp. 62–70, 2004.

[10] H. Schneiderman and T. Kanade. Probabilistic modeling of local appearance and spatial relationships for object recognition. In *Proceedings of IEEE Conference on Computer Vision and Pattern Recognition*, pp. 45–51, 1998.

[11] P. Viola and M. Jones. Robust real-time face detection. *International Journal of Computer Vision*, 57(2):137–154, 2004.

[12] R. Lienhart and J. Maydt. An extended set of Haar-like features for rapid object detection. In *Proceedings of the IEEE International Conference on Image Processing*, Vol. 1, pp. 900–903, 2002.

[13] S. Li and Z. Zhang. FloatBoost learning and statistical face detection. *IEEE Transactions on Pattern Analysis and Machine Intelligence*, 26(9):1112–1123, 2004.

[14] I. Paliy, Y. Kurylyak, A. Sachenko, S. Sokolov, and O. Boumbarov. Combined approach to face detection for biometric identification systems. In *Proceedings of the IEEE*

Fifth International Workshop on Intelligent Data Acquisition and Advanced Computing Systems: Technology and Applications (IDAACS'2009), Rende (Italy), pp. 425–429, 2009.

[15] A. Yilmaz, O. Javed, and M. Shah. Object tracking: A survey, *ACM Comput. Surv.*, 2006.

[16] G. Welch and G. Bishop. An introduction to the Kalman filter. University of North Carolina at Chapel Hill, http://www.cs.unc.edu/welch/kalman/∼, pp. 1–11, 2006.

[17] D. Comaniciu, V. Ramesh, and P. Meer. Kernel-based object tracking. *IEEE Trans. Patt. Analy. Mach. Intell.*, pp. 564–575, 2003.

[18] H. Tao, H. Sawhney, and R. Kumar. Object tracking with Bayesian estimation of dynamic layer representations. *IEEE Trans. Patt. Analy. Mach. Intell.*, pp. 75–89, 2002.

[19] S. Avidan. Support vector tracking. In *Proceedings of the IEEE Conference on Computer Vision and Pattern Recognition (CVPR)*, pp. 184–191, 2001.

[20] M. Bertalmio, G. Sapiro, and G. Randall. Morphing active contours. *IEEE Trans. Patt. Analy. Mach. Intell.*, pp. 733–737, 2000.

[21] R. Ronfard. Region based strategies for active contour models. *Int. J. Comput. Vision*, :229–251, 1994.

[22] D. Huttenlocher, J. Noh, and W. Rucklidge. Tracking nonrigid objects in complex scenes. In *Proceedings of IEEE International Conference on Computer Vision (ICCV)*, pp. 93–101, 1993.

[23] J. Kang, I. Cohen, and G. Medioni. Object reacquisition using geometric invariant appearance model. In *Proceedings of International Conference on Pattern Recongnition (ICPR)*, pp. 759–762, 2004.

[24] Y. Freund and R. Schapire. A decision-theoretic generalization of on-line learning and an application to boosting. In *Computational Learning Theory*, pp. 23–37, Springer-Verlag, 1995.

[25] Y. LeCun, L. Bottou, and Y. Bengio. Gradient-based learning applied to document recognition. In *Intelligent Signal Processing*, pp. 306–351, IEEE Press, 2001.

[26] P. Simard, D. Steinkraus, and J. Platt. Best practices for convolutional neural networks applied to visual document analysis. In *Proceedings of Seventh International Conference on Document Analysis and Recognition (ICDAR'03)*, Vol. 2, p. 958, 2003.

[27] OpenCV library: http://sourceforge.net/projects/opencv/

8

Special Area Detection and Recognition on Agricultural Fields Images

Valentin Ganchenko[1], Rauf Sadykhov[2], Alexander Doudkin[1], Albert Petrovsky[1] and Tadeusz Pawlowski[3]

[1]*United Institute of Informatics National Academy of Science, Minsk, Belarus; e-mail: ganchenko@lsi.bas-net.by*
[2]*Belarusian State University of Informatics and Radioelectronics, Minsk, Belarus*
[3]*Industrial Institute of Agricultural Engineering, Poznan, Poland*

Abstract

Problems of practical using of aerial photograph processing methods based on textural and fractal characteristics of images are considered in this chapter (on an example of agricultural field images). ASM, Contrast and Entropy are used as textural characteristics. Calculation of fractal signature is based on fractal dimension. The results of joint segmentation based on proposed methods are tested at processing potato field aerial photographs. Furthermore, computational speedup problem using MPI are considered.

Keywords: aerial photograph, textural characteristics, fractal characteristics, Mass-Parallel Processing.

8.1 Introduction

Every year a need for information obtained using remote sensing data (RSD) grows. RSD are used in problems of cartography and land cadaster [1], in agronomy and precision agriculture [2–7], forestry [8–11], development of water systems [12], environmental monitoring [13], etc. Constantly grow-

Richard J. Duro and Fernando López Peña (Eds.), Digital Image and Signal Processing for Measurement Systems, 201–233.

ing requirements for perfect data processing systems are increasing, because information is a key element in decision-making, and the amount of information of different degrees of complexity increases. One of the major problems arising in connection with the creation of modern information systems is the automation of the processing of raw data presented as images.

One of the most important areas of image processing is the precision agriculture area. Efficient processing of raw data allows reducing material and other costs in problems associated with crop cultivation and forecasting, monitoring of the level of crop germination and many other applications. The solution of such problems involves using geographic information systems (GIS), which combine the necessary techniques for image processing. A number of institutes and companies around the world deal with research in this area, which include the "ScanEx" Research and Development Center (http://www.scanex.ru/), the ESRI company (http://esri.com/), the ERDAS Inc. company (http://erdas.com/). There are also a number of research centers that solve the problem of precision farming such as the Australian Centre for Precision Agriculture (http://www.usyd.edu.au/su/agric/acpa/pag.htm), the Centre for Precision Farming (http://www.silsoe.cranfield.ac.uk/cpf/), the Ohio State University Precision Agriculture (http://precisionag. osu.edu/) and others.

8.1.1 Thematic Processing

Remote sensing methods allow effectively detecting field areas that are infected by plant diseases. Detection and recognition of an infection in the early stages of its development reduces costs in plant protective measures. There are two main approaches for the detection of the infected areas: spectrometric and optical or visual [14–18]. The spectrometric approach allows detecting a number of infections in very early development stages. For example, a change of the reflective features of potato plants in the infrared area allows identifying phytophthora even before the appearance of visual features [14, 17]. In spite of that fact, the development of optical methods for infection detection takes place both for independent systems and for spectrometric ones increasing a quality of disease recognition of the affected areas of agricultural fields. They include the following:

- methods and algorithms for preprocessing and selection of features of the objects in agricultural field images based by combining a spectral approach and a separation in color space coordinates;

- artificial neural network (ANN) models for fuzzy data clustering and classification methods.

The main stage of thematic image processing is image segmentation (groups of pixels), i.e., a separation of the image into homogeneous color or spectral characteristics areas using any criterion of homogeneity (similarity), and assigning them certain pre-defined classes [19–24].

The segmentation of image $f(x, y)$ by a predicate L_p is a partition $S = \{S_1, S_2, \ldots, S_k\}$, which satisfies the following conditions:

(a) $\bigcup_{i=1}^{k} S_i = X$;

(b) $S_i \bigcap S_j = \emptyset$ for any $i \neq j$;

(c) $L_p(S_i)$ *is* true for any i;

(d) $L_p(S_i \cup S_j)$ takes a false value for any $i \neq j$.

Note that k resulting areas are often grouped into m classes, where $2 \leq m \leq k$.

A majority of the known image segmentation methods, as well as edge detection methods can be divided into two main groups: regional-oriented methods, which directly build S_i, and boundary-oriented ones, which determine boundaries of S_i [25–29].

In methods of the first group, the criteria for the homogeneity of the areas in L_p is based on common features of image: an intensity or a chroma, a texture type, spectral properties of the image, etc. This group includes methods for threshold separation, region expansion and area separation.

$$L_p(S_i) = |\|\nabla f(x, y)\| - \|\nabla f(x', y')\|| \leq T;$$

$$|\phi\nabla(x, y) - \phi\nabla f(x', y')| \leq A;$$

where ∇ is a gradient of function f; φ is a direction of the gradient; T and A are thresholds.

Local analysis operators, such as Sobel, Roberts, or Kirsch operators, are used for the calculation of the gradient. In turn, convolution operators allow to directly select pixels that may, for example, belong to straight line of a particular orientation and a width. The choice of a segmentation method depends on the ultimate goal of the image processing operation, the image type and the available computing powers. The most difficult task is to construct an algorithm for segmenting agricultural field images that are very noisy. As a rule, a combination of some approaches provides a full solution to the problem of land cover recognition, especially using multi-temporal data.

The method of building areas combined with using deformable models is proposed to process digital aerial images [19]. This hybrid approach is called "a method of competition areas". Deformable models are represented as a set of flexible curves that adapt to the contour vectorization of the segmented area minimizing energy dynamically. The method includes the best features of the build-up method and the method of deformable models.

ANNs are actively used for processing RSD of agricultural lands. For instance, the ARTMAP neural network [30] based on adaptive resonance theory is used for mapping vegetation according to spectral imaging and landscape data. The approximate fidelity of ARTMAP is about 80%.

Modular ANNs are proposed for finding plants or areas of crops, contaminated by biological agents [31]. They reduce the influence of confusing factors and provide a classification of the spectral curves of chemical components. An evaluation of the vegetation state, i.e., detection of chemical components, is based on processing the images of plant leaves. Modular ANNs can provide better results than traditional ANNs.

A mathematical morphology method is proposed for the segmentation of gray-scale agricultural landscape images [20]. It is based on a watershed transformation and halftone pseudo skeleton building operations and is characterized by the non-use of binarization operations that can significantly improve the quality of the segmentation. The operation of the halftone pseudo skeleton building technique enables performing grayscale image segmentation with a significant reduction of the effects of quantization and noise errors on the segmentation result. This feature helps to significantly reduce segmentation time both in an analysis phase and in area association. For increasing sensitivity a number of morphological gradients computed for different structuring elements are proposed.

The accurate herbicide dispersion system [32] uses capture and image processing stages based on the fuzzy logic apparatus. Weeds are sought on the images and marked in shades of green. The system automatically controls herbicide application for effective removal of the weeds, reducing the cost of work and minimizing contamination of the soil and water. Fuzzy logic membership functions are easily modified and allow to quickly create control instructions for the system.

A technology for automatic rice field detection is proposed in [33]. The main feature of the technology consists in the application of regional-oriented classification based on geographical data and a set of image regions of interest are obtained during period of study. The normalized difference vegetation index over time for detecting rice fields is studied.

The Advanced Spaceborne Thermal Emission and Reflectance Radiometer (ASTER) [23] is used to locate and display agricultural crops, soil and some types of a land cover. Four attributes of arable land are considered: planted culture, stages of yield growth, color and texture of soil. The user (operator of the system) carries out a classification using a minimum average distance.

Thus, there are a large number of algorithms that can be used for vegetation RSD processing. However, these algorithms are quite highly specialized and designed for specific tasks that can greatly reduce their results in application to the images of agricultural fields. Use of widespread algorithms is complicated by the complexity of the research object. Sometimes image processing is very difficult because of noise on the images (foreign objects, sun glare on the leaves of plants, etc.).

In this context a development of feature extraction algorithms is required. They can rely on expert data (for example, the expert can specify the different characteristics of the vegetation depending on the illumination). One should bear in mind that sometimes only gray-scale images are available. Hence, the algorithms providing additional features should be based only on these data.

Ultimately, a segmentation algorithm is needed which can be applied to both conventional spectral data and additional data (e.g., a texture).

8.1.2 Data Processing System in a Vegetation Monitoring Task

The basic concept of precision agriculture is the fact that a vegetation cover is not uniform within a single field. Up-to-date technologies are used to evaluate and detect these irregularities: global positioning systems (GPS, GLONASS), special sensors, aerial photographs and satellite imagery, as well as special software systems based on GIS. RSD are used for a more accurate evaluation of the seeding density, calculation of application rates and crop protection, more accurate prediction of yield and financial planning. Also, it must take into account local peculiarities of soil and climatic conditions. In some cases it may make it easier to determine the reasons for the deterioration of vegetation [7].

Sometimes precision agriculture is associated with the desire to maximize profits applying fertilizers only on those portions of fields where fertilizers are needed. Following this, agricultural producers use variable or differential fertilization technologies in those areas of the field, which are identified with the help of GPS-receivers and where requirement for a certain rate of fertilizers is identified using yield maps. Therefore, the rate of application or spraying is less than an average in some areas of field, and a redistribution of fertilizers takes place in favor of areas where the rate should be higher. The application of fertilizers is thus optimized.

Precision agriculture can be used to improve the state of fields in several directions:

- agronomical, taking into account the real requirements of the crop in terms of fertilizers;
- technological, leading to better planning of agricultural operations;
- environmental, reducing negative impacts of an agricultural production on the environment, for instance, a more accurate estimation of the nitrogen fertilizer requirements of the crop leads to a restriction in the use and spread of nitrogen fertilizer or nitrates;
- economic, increasing the efficiency of agriculture, including reducing costs for nitrogen fertilizers.

Other benefits for agriculture may be found in having an electronic recording and a storage of field work history and harvest, which may help in subsequent decision-making processes and in the preparation of special reports on production cycles.

8.1.3 Existing Systems

There are a number of systems that are intended for commercial and research tasks for precision farming.

MARS (the Monitoring of Agriculture with Remote Sensing; the Joint Research Centre of the European Commission's monitoring of agricultural land, http://mars.jrc.it/) is intended to manage and control agriculture: crop forecasting, agricultural insurance, the interaction with environment. The system is based on an analysis of crop models using weather and GIS or GPS data, as well as RSD.

VESPER (Variogram Estimation and Spatial Prediction plus Error, the Australian Centre for precision farming, www.usyd.edu.au/agriculture/acpa/software/vesper.shtml) was designed for spatial prediction based on kriging, i.e., on optimal statistic interpolation [34]. It is intended to generate maps of a crop yield [35] based on digital relief models and the evaluation of salinity in the upper layer of soil [36]. The system also allows performing kriging variograms of fields with an ability to manually adjust and fit them. Its friendly interface allows to define boundaries of fields and to generate grid interpolation.

Ag Leader Insight (Ag Leader Technology Ltd., www.agleader.com/) is a complete package of tools for precision farming, from planting to harvesting (from development of a project on a desktop of computer to its implementation in the field). It uses one of the most popular technologies in the world for grain yield monitoring and allows to create and view maps of the crop yield and the moisture content during harvest, get instantaneous information on how the state of the field effects the yield. SeedCommand provides complete control over sowing operations, which varies depending on the requirements of crops, using AutoSwath functions, which turn on and off sections of drills based on the map of the field and already planted areas. It reduces cross-seeding saving 3–12% of seeds and increases potential yield at same time.

Topcon Positioning Systems (Topcon Positioning Systems Company, Inc., www.topconpa.com/) is a major producer of software and hardware control and monitoring systems for agriculture. Topcon produces equipment to optimize spraying, planting and fertilizing.

AGRO-NET NG ("GEOMIR" Engineering Center, www.geomir.ru/) is a GIS-based software and provides a GIS database for each field, a review of cultivated crops and an agronomic analysis. The GEOMIR company also manufactures mobile GIS electronic data acquisition systems for agricultural needs and electronic circuits for statistical and thematic analysis of agricultural fields.

8.1.4 Research Object

Agricultural field color images are the object of our research in this chapter (Figure 8.1). Rectangles show the same area of a field captured from 5 m height.

The purpose of the work consists in the development of an effective method for processing vegetation cover color images produced with help of high resolution digital camera, and also its realization as a software platform for computer vision systems. In this case, the spatial resolution of the image refers to the size of the square of the original object contained in one pixel. A lower value for this quantity equals a higher spatial resolution for the image. In this article, if the side of the square is less than 0.6 cm, the spatial resolution is considered high, otherwise it is taken as low.

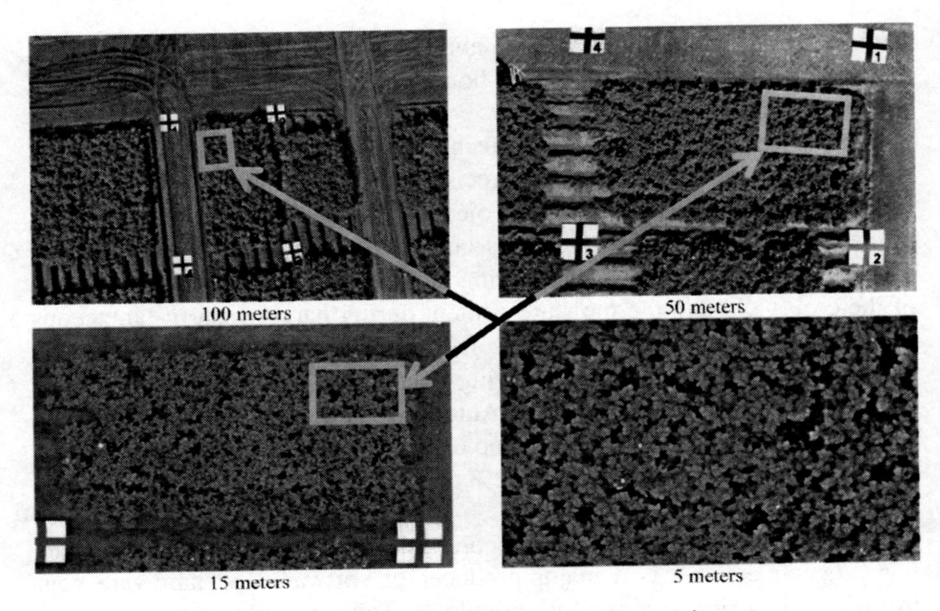

Figure 8.1 Examples of initial aerial photographs.

Color, fractal and textural features of the images are used as a basis for the area detection method under consideration. These features have been successfully used in several investigations related to image processing [16, 18, 37–39].

After the detection of special areas we need to solve the problem of the recognition required for mapping a disease. This can be done by recognizing the initial image or by recognizing the special area received. For the image recognition stage we used histogram analysis of RGB- and HSV-color features.

8.2 Special Area Detection

8.2.1 Textures

Heterogeneity or frequency of small fragments is called digital image texture. A characteristic feature of the texture is uniformity at a neighborhood or a local level, i.e., at the level of groups of adjacent pixels with different brightness [40]. There are two main approaches to the definition of textures:

- a texture is a set of repeated basic primitives (elements) with a different orientation in space;
- a texture is regarded as a kind of anarchic and homogeneous aspect, not having pronounced edges.

The textures are subdivided into fine-grained, coarse-grained, smooth, granulated and undulating in accordance to used base attributes and interactions between them. In view of the interaction degree of the base elements, the structures are subdivided into strong (the interaction follows to some rule) and weak (the interaction has a casual character).

A texture is one of the major features used for the identification of areas (objects) on the image. It represents a two-level structure [40]:

- at the top level – a set of base elements connected by some spatial organization;
- at the bottom level – base elements representing casual aspects.

There are several methods to calculate textural characteristics [40]:

- methods based on measuring spatial frequency (high spatial frequencies dominate in fine-grained textures, and low spatial frequencies in coarse-grained textures);
- methods based on calculating a quantity of brightness jumps per unit of area (the quantity is small in coarse-grained textures and it increases with decreasing grain of the texture);
- methods using the brightness adjacency matrix (change in brightness distribution is much slower in coarse-grained textures than in fine grained ones with increasing distance between evaluated points);
- methods describing textures by run lengths (the lines of points with constant brightness are longer on coarse textures than on finer ones). The basic textural features are given below [41]:

$$\mathrm{ASM} = \sum_{i=1}^{N_g} \sum_{j=1}^{N_g} \left(\frac{P(i,j)}{R} \right)^2; \tag{8.1}$$

$$\mathrm{Contrast} = \sum_{n=0}^{N_g-1} n^2 \left(\sum_{\substack{j=1 \\ |i-j|=n}}^{N_g} \sum_{j=1}^{N_g} \left(\frac{P(i,j)}{R} \right) \right); \tag{8.2}$$

$$\mathrm{Entropy} = -\sum_{i=1}^{N_g} \sum_{j=1}^{N_g} \left(\frac{P(i,j)}{R} \right) \log \left(\frac{P(i,j)}{R} \right). \tag{8.3}$$

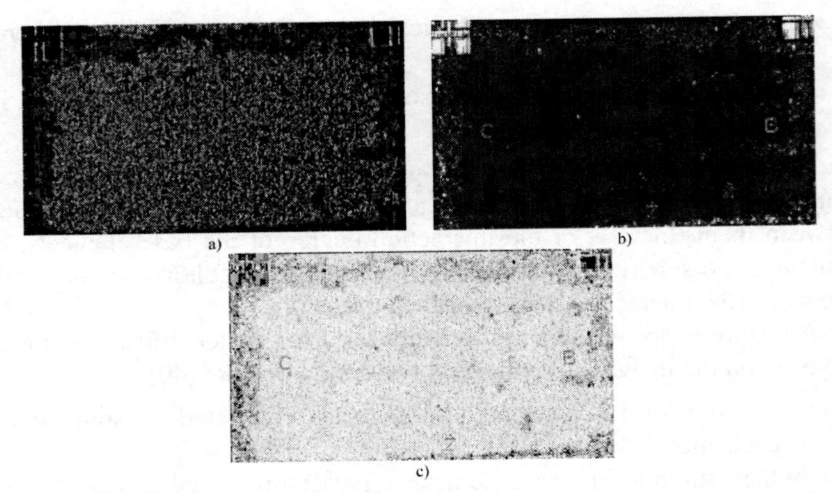

Figure 8.2 Result of textural features calculation for a field aerial photograph taken from a height of 15 meters: (a) contrast; (b) ASM; (c) entropy.

The essence of the proposed method for textural feature calculation consists in the calculation of separate channel image signatures with their subsequent association using factors whose values depend on vegetation type and condition. An example of textural feature calculation is represented in Figure 8.2, where the visualization of the calculated values is presented. Contrast approaches 1 for small variations of the initial data, and it decreases at greater variation. ASM approaches 1 in homogeneous initial data values. Entropy decreases to 0 in objects of homogeneous color.

8.2.2 Fractals

In most cases natural objects on the Earth's surface represent a fractal formation [42]. It is possible to determine qualitatively whether an object is fractal by considering it at different scales. There is no natural length scale in fractal objects, and they look similar for different image enlargements [43, 44]. For example, a tool for investigating the electrical properties of natural objects comprises the external electromagnetic fields at different frequencies. Each section of a surface has its conductivity at different frequencies. However, the fractality of objects means that the boundary of each conducting area is similar to the boundary of any other area at different frequencies in a statistical sense.

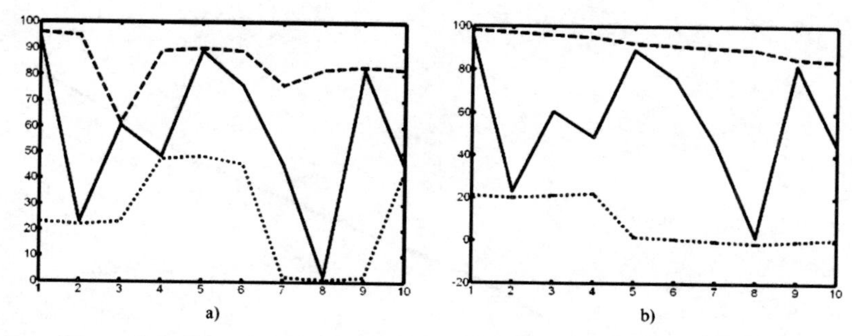

Figure 8.3 Construction of top and lower surfaces: (a) $\varepsilon = 1$; (b) $\varepsilon = 3$.

Fractal theory can be used for the segmentation of RSD, when the fractal (fractional) dimension D and the fractal signature are used as quantitative estimated parameters for segment description [43].

Fractal signature calculation is based on the fact that quantified intensity values of two-dimensional signals are located between two functions named the top and bottom surfaces [42]. The top surface U contains a set of points whose values always exceed the intensity of the initial signal. The bottom surface L has values of points which are always lower than the initial image (Figure 8.3).

The top and bottom surfaces are defined for an initial zero point as the following:

$$U(i, j, 0) = L(i, j, 0) = g(i, j); \tag{8.4}$$

where $g(i, j)$ is an initial image.

Generally we have:

$$U(i, j, \varepsilon + 1) = \max \left\{ U(i, j, \varepsilon) + 1, \max_{k,m \in \eta} [U(k, m, \varepsilon)] \right\}; \tag{8.5}$$

$$L(i, j, \varepsilon - 1) = \min \left\{ U(i, j, \varepsilon) - 1, \max_{k,m \in \eta} [L(k, m, \varepsilon)] \right\}; \tag{8.6}$$

$$\eta = \{(k, m) | d[(k, m), (i, j)] \leq 1\}; \tag{8.7}$$

where d is a distance function.

The designed covering, formed by the two specified functions, has a thickness of 2ε. For the two-dimensional signal the area of a surface is the volume occupied by a covering, and divided by size 2ε. The "surface" area of intensity $A(\varepsilon)$ within the limits of the window of observation (R) is calculated

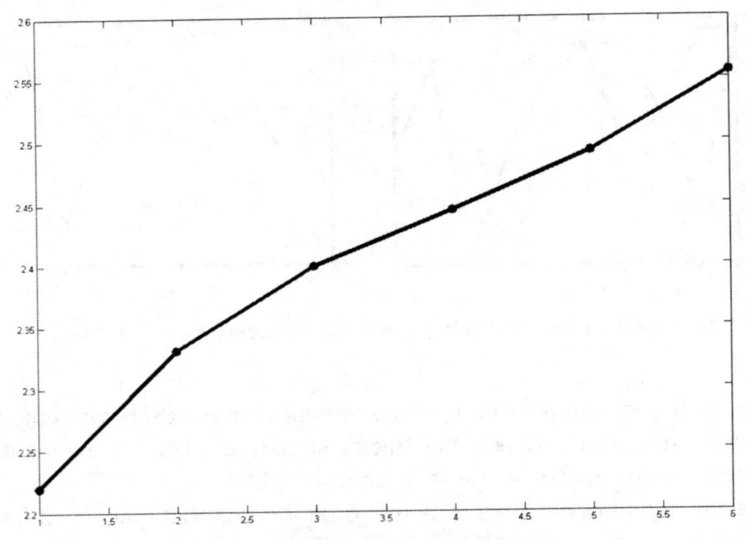

Figure 8.4 Image part fractal signature.

by the subtraction of the bottom "surface" points from top one with a further summation over the whole window:

$$A(\varepsilon) = \frac{\sum_{i,j \in R} U(i, j, \varepsilon) - L(i, j, \varepsilon)}{2\varepsilon} = \frac{V(\varepsilon)}{2\varepsilon}. \tag{8.8}$$

The fractal dimension is determined by the slope of $\log A(\varepsilon)$ as a function of $\log \varepsilon$. The example of the fractal signature is represented in Figure 8.4.

Fractal dimension $D(i, j)$ at point (i, j) for all scales is defined as the weighted sum of the local fractal dimension $F\varepsilon(i, j)$:

$$D(i, j) = \frac{\sum_{\varepsilon} C_{\varepsilon} F_{\varepsilon}(i, j)}{\sum_{\varepsilon} C_{\varepsilon}}; \tag{8.9}$$

where

$$C_{\varepsilon} = \frac{\log \varepsilon - \log(\varepsilon - 1)}{\log 2}; \tag{8.10}$$

$$F_{\varepsilon} = \frac{\log A(i, j, \varepsilon) - \log A(i, j, \varepsilon - 1)}{\log \varepsilon - \log(\varepsilon - 1)}. \tag{8.11}$$

Size $F_{\varepsilon}(i, j)$ is calculated by dividing $A(i, j, \varepsilon)$ by $A(i, j, \varepsilon - 1)$:

$$\frac{A(i, j, \varepsilon)}{A(i, j, \varepsilon - 1)} = \frac{K\varepsilon^{(2-D)}}{K(\varepsilon - 1)^{(2-D)}} = \left(\frac{\varepsilon}{\varepsilon - 1}\right)^{(2-D)}; \tag{8.12}$$

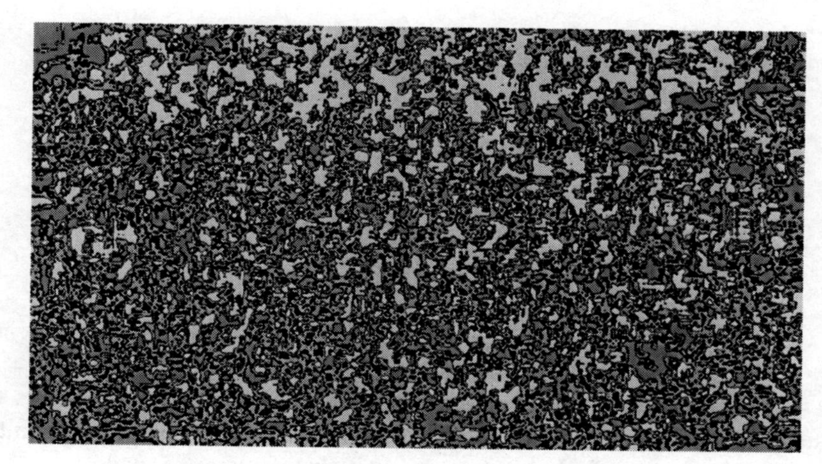

Figure 8.5 Fractal signatures of various image areas.

after finding the logarithm, we obtain:

$$\frac{\log A(i, j, \varepsilon) - \log A(i, j, \varepsilon - 1)}{\log \varepsilon - \log(\varepsilon - 1)} = 2 - D = F_{\varepsilon}(i, j). \tag{8.13}$$

Having substituted the values obtained in (8.11) and (8.12) in expression (8.13) we obtain the fractal dimensions:

$$D(i, j) = \frac{\log A(i, j, \varepsilon) - \log A(i, j, 1)}{\log \varepsilon - \log 1}. \tag{8.14}$$

Results of the fractal signature calculation are presented in Figure 8.5 (the visualization of the calculated values of the fractal signatures is displayed). Fractal dimension D approaches 2 at high irregularly and heterogeneity of the initial data and is reduced to 1 for homogeneous data.

8.2.3 Processing Algorithm Based on the Joint

The essence of the segmentation algorithm is the co-processing of the source images and their fractal and textural features (i.e., an initial multispectral image is complemented by images of texture and fractal features).

The calculation of the fractal signature and textural features of the images is carried out for the individual channels and the subsequent association performed using coefficients whose values depend on the type and condition of the vegetation.

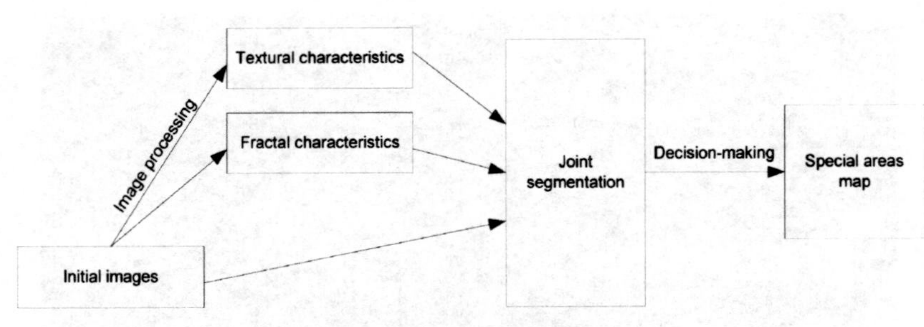

Figure 8.6 Processing algorithm scheme.

A weighed matrix of color features of the initial image, as well as the textural and the fractal features computed for each color channel of the initial image are used as the feature space over which a decision is made.

The color ranges of the corresponding healthy and sick parts of fields obtained from an expert are used as color features.

The algorithm is intended for segmentation of two-dimensional data representing matrices of various features of the initial image such as color channels, textural and fractal characteristics. Thus, the segmentation algorithm is executed in an N-dimensional space of attributes (where N is the number of characteristics used) where each dimension can be taken with a certain weight coefficient.

Thus, the algorithm consists of the following steps (Figure 8.6).

Step 1. Processing of the initial images for the extraction of additional information of channels representing matrices of textural and fractal characteristics of each initial image color channel separately;

Step 2. Joint segmentation of the textural and fractal characteristics matrices and the initial color channel images (using the K-means or ISOMAD algorithm);

Step 3. Special area map building on the basis of the results of the joint segmentation.

An example of the execution of the processing algorithm based on the joint segmentation applied to an image is shown in Figure 8.7. The segmentation results obtained allow detecting areas where there is disease development in an automatic manner. The knowledge of the location of such areas allows determining the requirements of those or other agricultural field areas for fertilizers and other chemicals. It allows making agricultural work more effective and less expensive.

Figure 8.7 Joint segmentation result example.

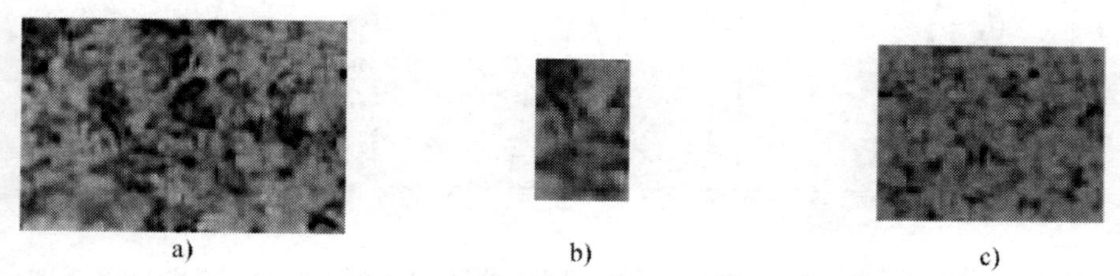

a) b) c)

Figure 8.8 Image of diseased plant areas with a size of 97 × 66 pixels (8.8a) and 20 × 32 pixels (8.8b), area of healthy plants with a size of 62 × 50 pixels (8.8c).

8.3 Image Recognition

8.3.1 RGB

Analysis of color features of various object types in the images showed that they differ slightly within the same type, and they are independent both from the height from which the images are taken, and the time of shooting. These color differences for each color channel (R, G, B) can be used for monitoring agricultural field diseases.

It should be noted that a histogram with a maximum global peak depends on the size and color of the images over which it is constructed – for small images the size of the individual elements of brightness can be set to zero, and the number of such elements may also decrease leading to histogram distortions. To reduce the influence of these factors, a histogram of brightness ranges is proposed for use here, i.e., the histogram based on a set of elements of brightness in each range. Such histogram is referred to as a reduced one

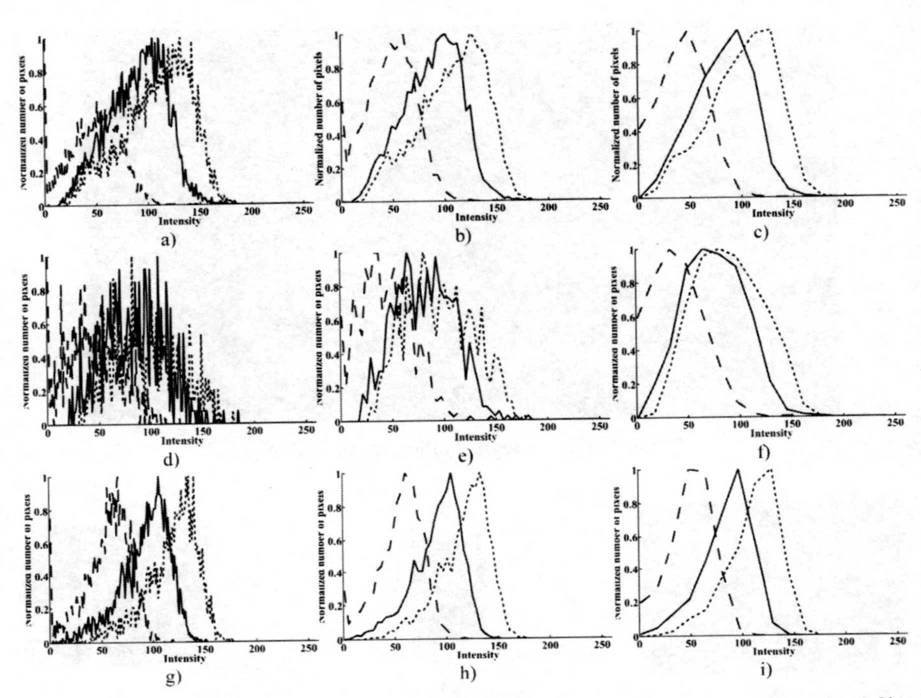

Figure 8.9 Histograms: original (8.9a, 8.9d, 8.9g), reduced to 64 segments (8.9b, 8.9e, 8.9h), up to 16 segments (8.9c, 8.9f, 8.9i) for objects 8.9a, 8.9b, 8.9c, respectively.

(or a histogram with the reduced number of readings along axis X). To ensure interoperability between the histograms of various sizes of the images we use a normalization procedure, after which the data takes values in the range [0, 1].

Figure 8.9 illustrates the influence of the number of ranges on the reduced and normalized histograms for the images of different classes of objects shown in Figure 8.8, where solid, dashed and dotted lines indicate the histograms of the red, green and blue channels respectively. Histogram distortions are seen in Figures 8.9d–8.9f: the smoothness decreases and a large number of gaps, i.e., regions with zero values in the histogram, appears. A loss of data can also be appreciated by comparing Figures 8.9b and 8.9c or Figures 8.9e and 8.9f (detail is lost if the number of segments is reduced from 64 to 16). In this case the difference of the color features of the objects of various types may be insufficient for classification (Figures 8.9c, 8.9f, 8.9i). To minimize the data loss produced by the variability in the number of the ranges, it should be chosen so that the reduced histogram becomes smoother than the original one, but contains enough data about the variability. A partition into 64 segments of the X axis was selected during the experiments.

A normalized histogram for one color channel of an image of size $M \times N$ pixels is formed by the following algorithm:

Step 1. Calculate histogram (*hist*) for the selected image areas. The histogram is an array of numbers in the range [0, 255], each number represents a number of elements of specified brightness on a halftone image.

Step 2. Calculate the reduced histogram (*hist*) with 256 points to 64 values – the amount calculated for each segment contains four values of the original histogram:

$$res(i) = \sum_{k=(i-1)*4+1}^{j*4} hist(k), \quad \text{for } i = 1, \ldots, 64, \tag{8.15}$$

where *res* is the histogram array with reduced number of elements.

Step 3. Calculate the maximum value of the histogram (*res*):

$$mx = \max(res(i)), \quad \text{for } i = 1, \ldots, 64, \tag{8.16}$$

Step 4. Normalize the histogram (*res*) to the range [0, 1] by dividing values res of the histogram array by mx:

$$res(i) = res(i)/mx, \quad \text{for } i = 1, \ldots, 64, \tag{8.17}$$

This algorithm is used for each color channel of the original image. As a result, three normalized reduced histograms are obtained, that together make up an array of 192 values, which is used for classification.

8.3.2 HSV

The HSV (Hue, Saturation, Value) color space can be used in addition to the RGB color space. The following transformations are performed to obtain HSV representation of given RGB image.

1. Convert the RGB channels in the range [0, 255] to the range [0, 1] by dividing the color values of the pixels of the image by 255.

2. Calculate HSV color values using the following formulas:

$$H = \begin{cases} 0, \text{ if } \max(R,G,B) = \min(R,G,B); \\ 60 \times \frac{G-B}{\max(R,G,B)-\min(R,G,B)} + 0, \text{ if } \max(R,G,B) = R, G \geq B; \\ 60 \times \frac{G-B}{\max(R,G,B)-\min(R,G,B)} + 360, \text{ if } \max(R,G,B) = R, G < B; \\ 60 \times \frac{B-R}{\max(R,G,B)-\min(R,G,B)} + 120, \text{ if } \max(R,G,B) = G; \\ 60 \times \frac{R-G}{\max(R,G,B)-\min(R,G,B)} + 240, \text{ if } \max(R,G,B) = B; \end{cases} \quad (8.18)$$

$$S = \begin{cases} 0, \text{ if } \max(R,G,B) = 0; \\ \text{else } 1 - \frac{\min(R,G,B)}{\max(R,G,B)}; \end{cases} \quad (8.19)$$

$$V = \max(R,G,B). \quad (8.20)$$

where H possesses the values in the range [0, 360), and S, V, R, G, B – in the range [0, 1].

3. Transform values H, S and V to the range [0, 255]:

$$\begin{aligned} H &= H/360 \times 255; \\ S &= S \times 255; \\ V &= V \times 255. \end{aligned} \quad (8.21)$$

The proposed image processing algorithms are applied to HSV data to construct normalized reduced histograms. Figure 8.10 shows the normalized histograms for reduced HSV color space, where solid, dashed and dotted lines show the values of the Hue, Saturation and Value channels respectively.

8.3.3 Perceptron as a Classifier

The task of object classification over images is partially solved by using the Contrast of quantity measure s texture feature corresponding to local variations over the images. Depending on the degree of image variability the Contras textural feature t takes higher values for images having high divergent objects and lower values in the case of low divergent objects. The analysis of Contrast texture feature showed that the soil plots are present low divergence and the areas of vegetation a high divergence [45]. Thus, Contrast can be used to separate image plots of soil. This will reduce the amount of computation, which may be important when processing large amounts of data. In contrast to soil, the leaves of vegetation equally present a low divergence over images

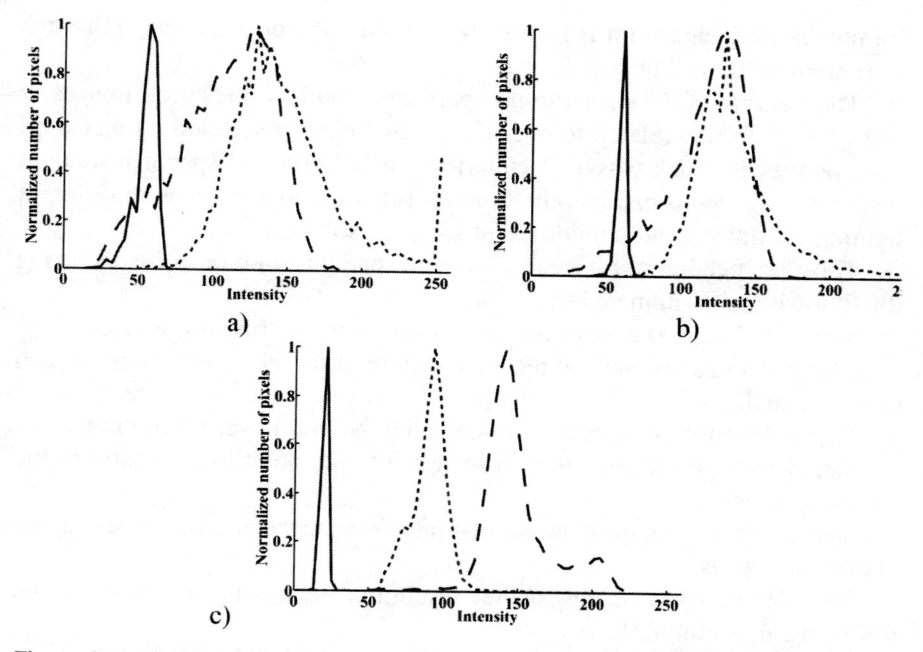

Figure 8.10 Normalized reduced histogram constructed for the HSV-representation: (a) "diseased plants"; (b) "healthy plants"; (c) "soil".

with high spatial resolution. A high variability is preserved only at edges of leaves, which does not correctly separate "soil" and "vegetation" classes. To use the Contrast texture feature for mapping image areas containing soil, some transformations should be performed. It is proposed not to use feature arrays from channel images but halftone images to visualize the arrays. This allows an expert to monitor and adjust the process as well as the widespread use of image processing algorithms for conversion.

To classify image areas a multilayer perceptron [46] with $N \times L$ inputs (where N is the number of segments of the reduced histogram, L is the number of channels), one hidden layer, containing 32×3 neurons (the number of the neurons is chosen experimentally), and an output layer containing three neurons corresponding to object types of the images is proposed. All neurons have a logistic activation function in a sigmoid form.

The back-propagation algorithm is used to adjust the weights of the perceptron. In this case, the input of the perceptron are fed normalized histograms obtained from images of objects selected by an operator. A data sample for the learning algorithm is formed by scanning the original image through a "running-window" of size $K \times K$.

The training of the perceptron is performed on low resolution images of one type of objects related to one of indicated classes selected by an expert (100 images for each class). Peculiarities of lighting and spatial resolution were not considered because the training set contains images with different lighting conditions and with different spatial resolution.

The classification of images with a high spatial resolution is carried using the following algorithm (Alg 1a):

Step 1. Select next area of a source image using a "running-window".

Step 2. Build a normalized reduced histogram for the chosen area for each color channel.

Step 3. Perform pixel classification using the multilayer perceptron.

Step 4. Assign a class obtained in step 3 to the point in the center of the "running-window".

Step 5. Form a map of morbidity rate from the obtained values of the classes of objects.

The classification of low spatial resolution images is carried using the following algorithm (Alg 1b):

Step 1. Construct a mask of vegetation using feature Contrast.

Step 2. If the image is processed completely, then go to step 7, otherwise choose an element from the source image using a "running-window".

Step 3. If the mask of vegetation in the center of the "running window" is not zero, then go to step 4, otherwise assign a point in the center of the "running window" to the "soil" class and go to step 2.

Step 4. Build the normalized reduced histogram for each color channel of the selected "running-window" element.

Step 5. Perform pixel classification using the multilayer perceptron.

Step 6. Assign a point in the center of the "running-window" class derived in step 5.

Step 7. Form a map of morbidity rate from the obtained values of the classes of objects.

The selection of image area and the corresponding vegetation mask area are carried out by means of "running-windows".

Figure 8.11 shows results from testing the algorithm. The selection of image areas was carried out using a running-window of size $K \times K$ pixels without a mask (Figures 8.11b, 8.11d) and with the mask (Figures 8.11c, 8.11e) (in the experiments a value of $K = 10$ is used). The mask is formed from vegetation maps obtained using feature Contrast. In Figures 8.11b,

Figure 8.11 Example disease map for $K = 10$, using RGB- and HSV-submission: (a) original image of a field; (b) disease map, "running-windows" without the mask (RGB); (c) disease map, "running-windows" with the mask (RGB); (d) disease map, "running-windows" without the mask (HSV); (e) disease map, "running-windows" with the mask (HSV).

8.11c, 8.11d and 8.11e non-classified boundaries are black, soil areas are dark gray, areas with healthy plants are light gray and areas of diseased plants are white.

The results of the experiments show that the algorithms applied to RGB images are more sensitive to details that create more shallow areas that are classified as "diseased plants". At the same time the details are not lost by using the HSV image representation. This distinction allows obtaining maps of incidence with varying detail levels, and thus more flexibility is provided for the recognition of the extent of diseased plants.

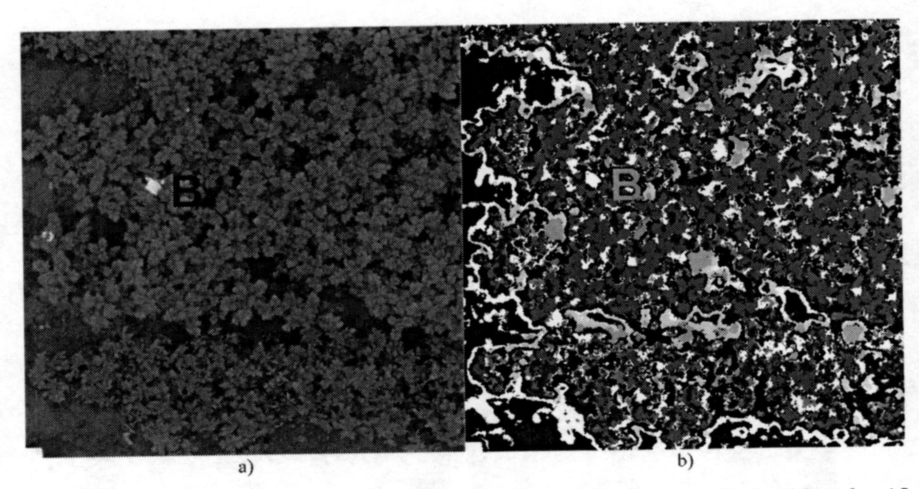

Figure 8.12 Initial image (a) and result of clustering using the GGC-algorithm for 15 clusters (b).

8.4 Improving Recognition Accuracy

8.4.1 Cluster Recognition

A significant disadvantage of the algorithm described above is its not sufficiently high recognition accuracy, due to the heterogeneity of vegetation, patches, to sunlight on the leaves, extraneous objects, etc. Masks of object classes can be used to increase recognition accuracy. Each mask is a binary image where areas of a particular class are marked using white color.

Recognition of clusters obtained as a result of the joint segmentation can be used for the construction of the masks of the classes. The algorithm includes the following steps (Alg 2):

Step 1. Conduct a clustering of the initial image for the joint segmentation algorithm, using color, texture and fractal features of images (see Figure 8.12).

Step 2. Obtain a binary mask for each of the obtained clusters, where 0 corresponds to areas not belonging to this cluster, and 1 in any other case (see Figure 8.13).

Step 3. Construct a reduced normalized histogram for each cluster in accordance with the obtained binary masks (for both RGB and HSV data representation).

Step 4. Recognize each of the resulting reduced normalized histograms using the trained neural network.

Figure 8.13 The binary mask of one of the clusters obtained by segmenting of the original image in Figure 8.12a.

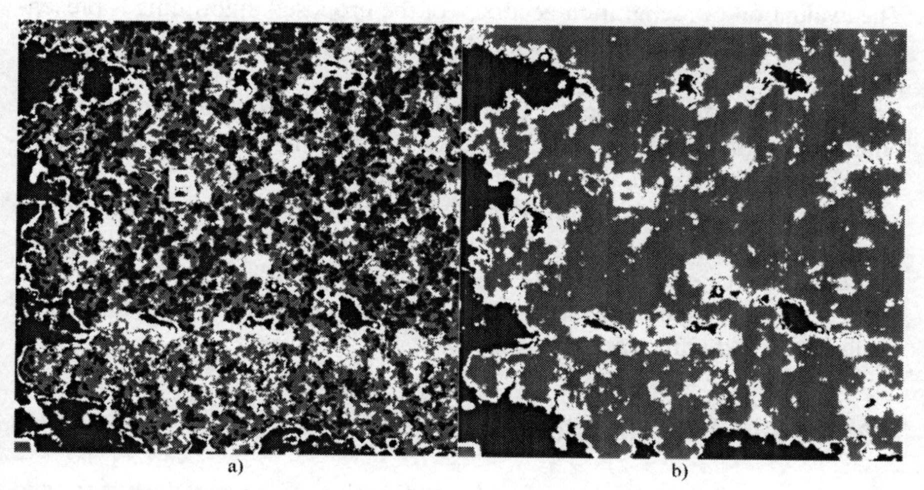

a) b)

Figure 8.14 Recognition result for the cluster RGB-representation of (a) and for HSV (b).

Step 5. Construct the cluster binary masks and the colored images in accordance with the recognition results (see Figure 8.14).

Thus, the image is formed as a result of the algorithm. The classes of area objects on the agricultural field image ("soil", "healthy plants" and "diseased plants") are marked with different colors. An example of recognition results

Table 8.1 Recognition errors.

Algorithm	**With mask**		**Without mask**	
	RGB	HSV	RGB	HSV
By recognition of initial image areas selected by "running window" (K=10)	4.835	4.957	20.028	16.888
By recognition of initial image areas selected by "running window" (K=20)	5.541	4.835	23.155	21.65
By recognition of initial image areas selected by "running window" (K=30)	5.591	6.132	24.15	24.996

is presented in Figure 8.13, where the dark gray areas denote "soil", light gray areas "healthy plants" and the white ones "diseased plants".

8.4.2 Comparison of Recognition Results

The evaluation of recognition accuracy of the proposed algorithms is presented in Table 8.1 (the error is calculated as the average recognition error obtained for several images).

It is evident from Table 8.1 that using the "diseased plants" class mask, obtained by the clustering process, allows improving recognition accuracy significantly for the initial agricultural field images.

Thus, the algorithm for cluster recognition (Alg 2) can be used to increasing recognition accuracy by forming a class binary mask. The algorithm for image area recognition (Alg 1) is required for the clusters recognition procedure. At the same time it can be used without masking (e.g., for mapping of diseases) and with masking (for statistics).

8.5 Practical Applications

A hardware-software complex for the application of mineral fertilizers and other chemicals over agricultural fields has been proposed on the basis of the developed algorithm (Figure 8.15). The technique based on the complex is the following:

1. Special area maps (for example, sites with developing disease of plants) are calculated using the proposed algorithms.
2. The built maps receive a geographical binding and they are kept in the GIS DB for further use.

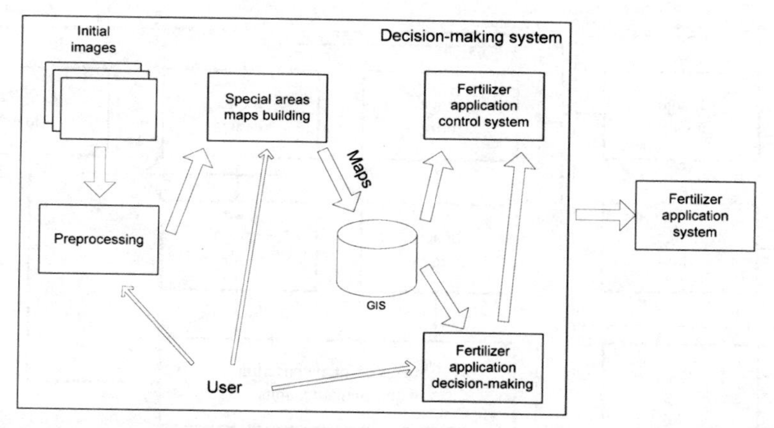

Figure 8.15 Scheme of the hardware-software complex for mineral fertilizers and other chemicals application on agricultural fields.

3. The obtained maps are used for decision-making on the necessity of application of this or that amount of fertilizers on a given farmland site.
4. The chemicals application control system supervises amount of chemicals put into the soil and directs a corresponding command to chemicals application system on the basis of available maps and real-time data.

The following real-time data can be used:

- data from a global navigating satellite system. In this case the control system, determines with the help from the navigating system, on what site of a field the application of a chemical is necessary, calculates the amount of chemicals, processing the corresponding special area map;
- data from color cameras in the visual range. In this case the control system can correct in real time the given special area maps, thus, leading to more exact decisions that increase efficiency in the application of chemicals.

The general structural scheme of the software is shown in Figure 8.16.

The software system contains three main units:

1. Image processing unit. The tools of the unit allow storing, processing and preserving the integrity of geographical data, represented as raster images. The unit implements the following functions:

   1.1. Image import. It is destined for the translation of raster images stored in the file system to the system's internal format.

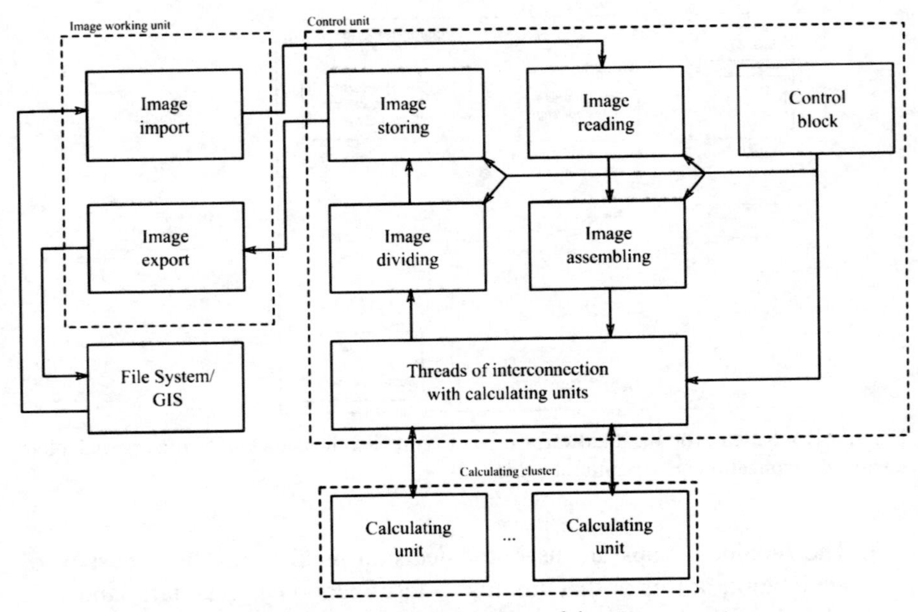

Figure 8.16 General structural scheme of the program system.

1.2. Image export. It is aimed at the translation from the system's internal format to raster images.

2. Control unit. This unit coordinates the calculation process, dividing the source image into parts, assembling the results of image processing from the parts. The unit contains three elements.
   2.1. Calculation control. This unit controls the image processing stages.
   2.2. Loading of data/Storage of result.
   2.3. Division/Assembly of image.
3. Calculating cluster. Parallel processing of source image parts is carried out here.
   3.1. Processing unit. This unit directly performs data processing.

Image processing is carried out as follows. The source image is divided into parts. Each part contains some sections of the original multispectral image. An example of a simple division is shown in Figure 8.17. Some image processing algorithms require taking into consideration the environment of the image element. In this case the image element environment should be added to this element, as shown in Figure 8.18.

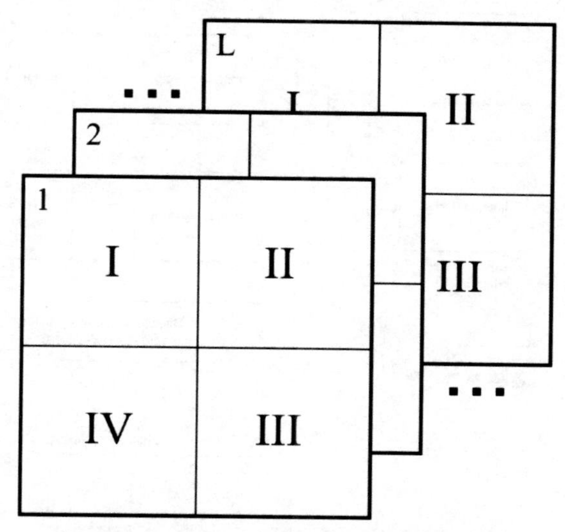

Figure 8.17 Example of simple image dividing.

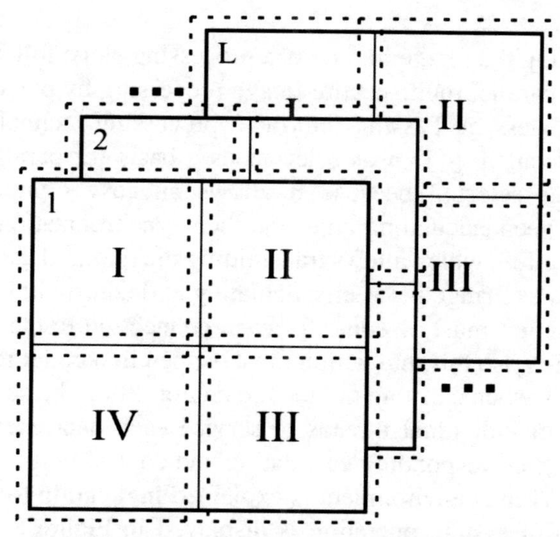

Figure 8.18 Example of dividing with taking into consideration environment of image element.

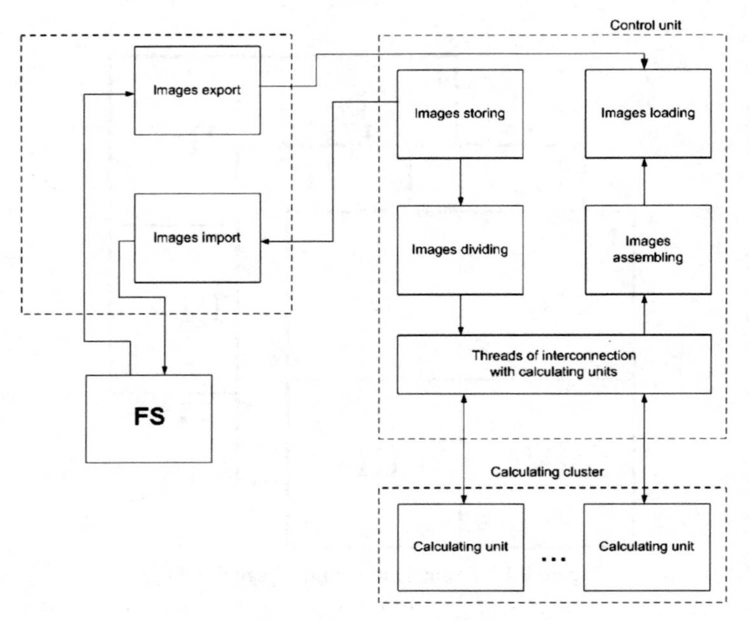

Figure 8.19 Schematic of system operation.

After dividing the image into parts a processing stage follows. This stage carries out the parallel multispectral image processing by parts.

The MPI (Message Passing Interface) processing technology, which is explained in detail in [47], was selected as a basis for parallelizaton. This technology was selected because it allows an easy organization of the interaction between calculating units and their synchronization.

TCP connection, which allows transmitting the required data without loss, is used for data exchange between calculating and control units. At the same time the interaction must be equal for each connection to the corresponding calculating unit. Also this interaction must happen in parallel to decrease performance losses when carrying out the interaction. It can be achieved through the allocation of individual threads to service each data exchange connection between the corresponding calculating and control units. Multithreading using the GNU/Linux environment is explained in detail in [48].

A diagram of system operation is displayed in Figure 8.19. The image processing system tests were carried out on a "SKIF K-1000" supercomputer [49, 50]. The experiments involved from 1 up to 64 computing nodes for color images of sizes 2000 × 2000 and 1000 × 1000 pixels. Dependence

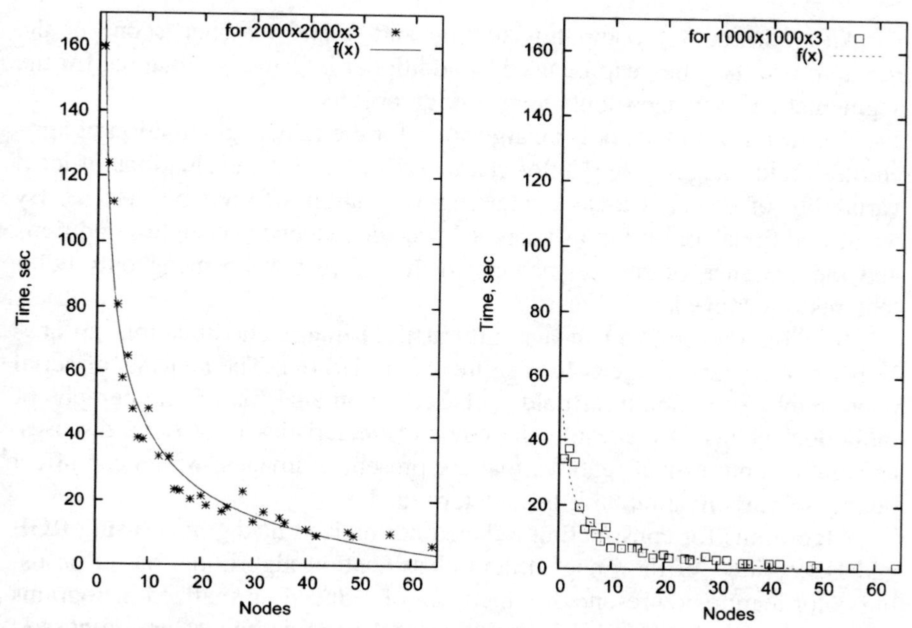

Figure 8.20 Dependence of the operation time of the image on the number of nodes for color images of sizes 2000 × 2000 (a) and 1000 × 1000 (b) pixels.

of system operation time on the number of computing nodes is represented in the graphs shown in Figure 8.20.

8.6 Conclusions

The analyses of the problem of special area detection and recognition over agricultural field images reveals a lack of methods and algorithms for the selection and classification of aerial objects in multi-temporal images of agricultural fields with different spatial resolution. To solve this problem, the analysis of subject area is carried out. As a result it was found that direct application of any of the existing methods do not provide a full solution to the detection and land cover classification of agricultural fields problem using multi-temporal data.

An algorithm for computing texture and fractal features is one of the research results. They can be used as additional informative features for the segmentation of images containing foreign objects.

The result of this work is an algorithm for the joint segmentation of agricultural field images, using color characteristics and the evaluations of local variability of detected areas containing vegetation affected by disease. By using additional information signs a lower dependence on light conditions and the presence of foreign objects in image than when using only color features is achieved.

To solve the problem of agricultural field image classification, an analysis of crop plant image color features is carried out. The analysis of aerial photographs of agricultural fields is based on an analysis of photographs of individual plants. As a result, the color characteristics of various diseases, as well as a number of features that are present in images, which can affect quality of classification have been determined.

Algorithms for constructing reduced normalized histograms (using RGB and HSV views of the images), and classification algorithms based on using color features represented in the form of reduced normalized histograms taking into account the spatial resolution of the source images are proposed.

Thus, as a result of the work a number of algorithms that permit solving the problem of monitoring agricultural plants state on the basis of the classification of color characteristics and using texture and fractal features to improve classification accuracy are obtained.

The scientific importance of the results obtained consists in the capability of creating new highly effective methods and highly efficient algorithms for the processing of images of objects of nonanthropogenic origin and the determination of the fractal nature of objects.

The practical importance consists in the application of the developed methods and algorithms to the classification of objects of natural origin objects that essentially permits increasing the accuracy and reliability of the operation of computer vision, monitoring and decision-making systems.

A possible area of application of the system is remote sensing of the Earth (in forestry, geology, agriculture).

References

[1] R. Bonnefon, P. Dherete, and J. Desachy. Geographic information system updating using remote sensing images. *Pattern Recognition Letters*, 23:1073–1083, 2002.

[2] W.J.D. van Leeuwen et al. Multi-sensor NDVI data continuity: Uncertainties and implications for vegetation monitoring applications. *Remote Sensing of Environment*, 3:67–81, 2006.

[3] A. Sofou, G. Evangelopoulos, and P. Maragos. Soil image segmentation and texture analysis: A computer vision approach. *IEEE Geoscience and Remote Sensing Letters*, 2:394–398, 2005.

[4] R.J. Shrivastava and J.L. Gebelein. Land cover classification and economic assessment of citrus groves using remote sensing. *Journal of Photogrammetry and Remote Sensing*, 61:341–353, 2007.

[5] M.M. Verstraete and B. Pinty. Designing optimal spectral indices for remote sensing applications. *IEEE Transactions on Geoscience and Remote Sensing*, 34(5):1254–1265, 1996.

[6] G.B. Senay et al. Using high spatial resolution nniltispectral data to classify corn and soybean crops. *Photogrammetric Engineering and Remote Sensing*, 66(3):319–327, 2000.

[7] S.A. Rubtsov, I.N. Golovanev, and A.N. Kashtanov. *Aerospace Equipment and Technologies for Precision Farming*, 330 pp. 2008 [in Russian].

[8] G. Laneve, M.M. Castronuovo, and E.G. Cadau. Continuous monitoring of forest fires in mediterranean area using MSG. *IEEE Transactions on Geoscience and Remote Sensing*, 44(10):2761–2767, 2006.

[9] F. Maselli and M. Chiesi. Evaluation of statistical methods to estimate forest volume in a Mediterranean region. *IEEE Transactions on Geoscience and Remote Sensing* 44(8):2239–2250, 2006.

[10] C. de Wasseige and P. Defourny. Retrieval of tropical forest structure characteristics from bi-directional reflectance of SPOT images. *Remote Sensing of Environment*, 83:362–375, 2002.

[11] D.P. Roy, L. Boschetti, and S. Trigg. Remote sensing of fire severity: Assessing the performance of the Normalized Burn Ratio. *IEEE Geoscience and Remote Sensing Letters*, 3(1):112–116, 2006.

[12] J.C. Ritchie, P.V. Zimba, and J.H. Everitt. Remote sensing techniques to assess water quality. *Photogrammetric Engineering and Remote Sensing*, 69(6):695–704, 2003.

[13] M.G. Turner, R.H. Gardner, and R.V. O'Neill. *Landscape Ecology in Theory and Practice*, 417 pp. Springer-Verlag, 2001.

[14] B.I. Belyaev and L.V. Katkovsky. *Optical Remote Sensing*, 455 pp., BSU, Minsk, 2006 [in Russian].

[15] N. Kumar, S. Pandey, A. Bhattacharya, and P.S. Ahuja. Do leaf surface characteristics affect agrobacterium infection in tea [Camellia Sinensis (L.) O Kuntze]? *J. Biosci.*, 29(3):309–317, 2004.

[16] Lanlan Wu, Youxian Wen, Xiaoyan Deng, Hui Peng. Identification of weedcorn using BP network based on wavelet features and fractal dimension. *Scientific Research and Essay*, 4(11):1194–1200, 2009.

[17] Zhihao Qin and Minghua Zhang. Detection of rice sheath blight for in-season disease management using multispectral remote sensing. *International Journal of Applied Earth Observation and Geoinformation*, 7(2):115–128, 2005.

[18] S. Aksoy, H.G. Akcay, and T. Wassenaar. Automatic mapping of linear woody vegetation features in agricultural landscapes using very high-resolution imagery. *IEEE Transactions on Geoscience and Remote Sensing*, 48(1,2):511–522, 2010.

[19] M. Torre and P. Radeva. Agricultural field extraction on aerial images by region competition algorithm. In *Proceedin gsof International Conference on Pattern Recognition (ICPR'00)*, Barcelona, Spain, September 3–8, pp. 137–139, 2000.

[20] A.V. Inyutin. The algorithm of image segmentation by grayscale pseudoskeleton. In *Proceedings of the III International Conferences on Neural Networks and Artificial Intelligence (ICNNAI'2003)*, Minsk, Belarus, November 12–14, pp. 263–265. Publishing Center of BSU, Minsk, 2003.

[21] P. Soille. Morphological image analysis applied to crop field mapping. *Image and Vision Computing*, 18(13):1025–1032, 1973.

[22] P. Tzionas, S.E. Papadakis, and D. Manolakis. Plant leaves classification based on morphological features and a fuzzy surface selection technique. In *Proceedings of 5th International Conference on Technology and Automation (ICTA'05)*, Thessaloniki, Greece, 15–16 October, pp. 365–370, 2005.

[23] A. Apan et al. Spectral discrimination and separability analysis of agricultural crops and soil attributes using aster imagery. In *Proceedings of 11th Australasian Remote Sensing and Photogrammetry Conference*, Brisbane, Queensland, 2–6 September, pp. 396–411, 2002.

[24] T.F. Burks, S.A. Shearer, and F.A. Payne. Classification of weed species using color texture features and discriminant analysis. *Transactions of ASAE*, 43(2):441–448, 2000.

[25] S.V. Ablameyko et al. Detection and analysis of stochastic data and digital images. *Bulletin of the Foundation for Fundamental Research*, 4:101–106, 2003 [in Russian].

[26] Ya.A. Furman et al. *Introduction to Contour Analysis: Applications to Image Processing and Signal*, 592 pp., 2003 [in Russian].

[27] I.B. Kerfoot and Y. Bresler. Theoretical analysis of multispectral image segmentation criteria. *IEEE Trans. Image Processing*, 8(6):768–820, 1999.

[28] G.B. Coleman and H.C. Andrews. Image segmentation by clustering. *Proc IEEE*, 67:773–785, 1979.

[29] Z. Zhang. A survey on evaluation methods for image segmentation. *Pattern Recognition*, 29(8):1335–1346, 1996.

[30] G.A. Carpenter, S. Gopal, and C.E. Woodcock. A neural network method for efficient vegetation mapping. *Remote Sensing Environment*, 70(9):326–338, 1999.

[31] N. Kussul et al. Remote sensing of vegetation using modular neural networks. In *Proceedings of the III International Conferences on Neural Networks and Artificial Intelligence (ICNNAI'2003)*, Minsk, Belarus, November 12–14, pp. 232–234, Publishing Center of BSU, Minsk, 2003.

[32] C.C. Yang et al. A neural network method for efficient vegetation mapping. Recognition of weeds with image processing and their use with fuzzy logic for precision farming. *Canadian Agricultural Engineering*, 42(4):195–200, 2000.

[33] Y.H. Tseng, P.H. Hsu, and Y.H. Chen. Automatic detecting rice fields by using multispectral satellite images, land-parcel data and domain knowledge. In *Proceedings of the 19th Asian Conference on Remote Sensing*, Manila, Philippines, 16–20 November, pp. R-1-1–R-1-7, Minsk, 1998.

[34] T.C. Haas. Kriging and automated variogram modeling within a moving window. *Atmospheric Environment – Part A. General Topics*, 24(7):1759–1769, 1990.

[35] B.M. Whelan, A.B. McBratney, and R.A. Viscarra-Rossel. Spatial prediction for precision agriculture. In *Proceedings of the 3rd International Conference on Precision Agriculture*, Minneapolis, Minnesota, June 23–26, pp. 331–342, 1996.

[36] C. Walter et al. Spatial prediction of topsoil salinity in the chelif valley, algeria, using local kriging with local variograms versus local kriging with whole-area variogram. *Australian Journal of Soil Research*, 39:259–272, 2001.

[37] A. Lucieer, P. Fisher, and A. Stein. Texture-based segmentation of high-resolution remotely sensed imagery for identification of fuzzy objects. *International Journal of Remote Sensing*, 26(14):2917–2936, 2005.

[38] D. Popescu and R. Dobrescu. Carriage road pursuit based on statistical and fractal analysis of the texture. *International Journal of Education and Information Technologies*, 2(1):62–70, 2008.

[39] C.W. Emerson, N. Siu-Ngan Lam, and D.A. Quattrochi. Multi-scale fractal analysis of image texture and pattern. *Photogrammetric Engineering & Remote Sensing*, 65(1):51–61, 1999.

[40] V.V. Starovoytov. *Local Geometric Methods of Digital Image Processing and Analysis*, 284 pp., Institute of Technical Cybernetics of BAS, Minsk, 1997 [in Russian].

[41] R.M. Haralick, K. Shanmugam, and I. Dinstein. Textural features for image classification. *IEEE Transactions on Systems, Man and Cybernetics*, 6:610–621, 1973.

[42] J. Feder. *Fractals*, 283 pp. New York: Plenum Press, 1988.

[43] A.A. Potapov and M. Logos. *Fractals in Radiophysics and Radiolocation*, 664 pp., 2002 [in Russian].

[44] R.R. Nigmatullin and A.A. Potapov. Fractals, fractional operators and fractional kinetics in dielectric spectroscopy and wave processes. *Physics of Wave Processes and Radio System*, 10(3):30–49, 2007 [in Russian].

[45] V. Ganchenko, A. Petrovsky, and B. Sobkoviak. Joint segmentation of aerial photographs with the various resolution. In *Proceedings of the 5th International Conference on Neural Networks and Artificial Intelligence (ICNNAI)*, Minsk, Belarus, May 27–30, pp. 177–181, Minsk, 2008.

[46] S. Haykin. *Neural Networks: A Comprehensive Foundation* (2nd ed.), 823 pp., Pearson Education, 2005.

[47] P.S. Pacheco. *Parallel Programming with MPI*, 418 pp., Morgan Kaufmann, San Francisco, CA, 1997.

[48] M. Mitchell, J. Oldham and A. Samuel. *Advanced Linux Programming*, 368 pp., New Riders Publishing, Indianapolis, 2001.

[49] The SKIF K-1000 Supercomputer 2.5 Tflops [Electronic document] – Access mode: http:skif.pereslavl.rupsi-inforcmsrcms-leaflets.engskif-k1000-leaflet-engl.pdf, 2004.

[50] Joint Belarussian-Russian program "SKIF" [Electronic document] – Access mode: http:skif.bas-net.byindex_en.htm, 2004.

9

Towards Real-Time Image Processing: A GPGPU Implementation of Target Identification

D.B. Heras[1], F. Argüello[1], J. López Gómez[1],
B. Priego[2] and J.A. Becerra[2]

[1] *Grupo de Arquitectura de Computadores, Universidade de Santiago de Compostela, Spain; e-mail: dora.blanco@usc.es*
[2] *Grupo Integrado de Ingeniería, Universidade da Coruña, Spain*

Abstract

In the quest for real time processing of hyperspectral images, this chapter presents three artificial intelligence algorithms for target detection specially developed for their implementation over GPU and applied to a search-and-rescue scenario. All the algorithms are based on the application of artificial neural networks to the hyperspectral data. In the first algorithm the neural networks are applied at the level of individual pixels of the image. The second algorithm is a multiresolution based approach to scale invariant target identification using a hierarchical artificial neural network architecture. The third algorithm is a refinement of the previous one but including also the ability to detect the orientation of the targets in cases for which this information is relevant. We have studied the main issues for the efficient implementation of the algorithms in GPU: the exploitation of thousands of threads that are available in this architecture and the adequate use of bandwidth of the device. The tests we have performed show both the effectiveness of detection of the algorithms and the efficiency of the GPU implementation in terms of execution times and bandwidth usage. These results bear out that the GPU

Richard J. Duro and Fernando López Peña (Eds.), Digital Image and Signal Processing for Measurement Systems, 235–265.

is an adequate computing platform for on-board processing of hyperspectral information.

Keywords: GPU, CUDA, target detection, hyperspectral images, image processing.

9.1 Introduction

The growth in airborne and satellite based hyperspectral sensors, as well as the appearance of more affordable hyperspectral cameras is fostering research both in new applications and in new problems arising from the ever increasing amount of hyperspectral data. Applications range from conventional remote sensing monitoring and exploration of the landscape, to manufacturing plant material surface inspection, art authentication or search and rescue operations among others [1–4]. A hyperspectral image is basically an image where for each pixel we have a set of m values that correspond to the spectral components in the visible or infrared areas as shown in Figure 9.1.

Real-time processing of hyperspectral images has led to algorithm implementations based on direct projections over clusters and networks of workstations [5]. But these systems are generally expensive, bulky and difficult to adapt to on-board data processing. Consequently, low-weight and low-power components are necessary to achieve real-time analysis. The real-time requirement is specially relevant for some applications such as for a search-and-rescue scenario, that it is our case, or for waterway control and security [6].

Nowadays GPUs (Graphics Processing Units) are cheap, high-performance, many-core processors that can be used to accelerate a wide range of

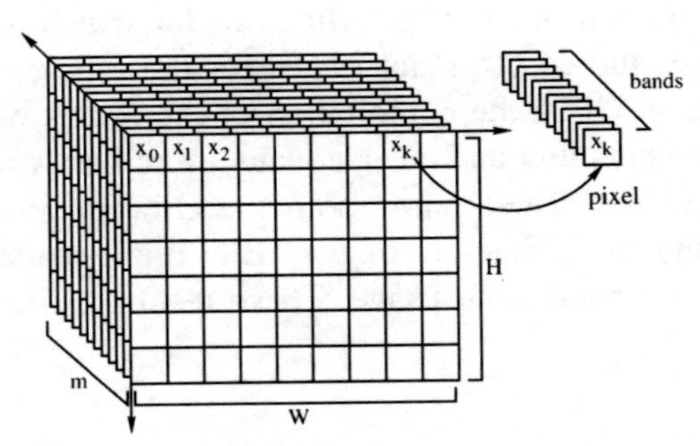

Figure 9.1 Hyperspectral image.

applications, not only the classical graphics processing ones [7]. CUDA for NVIDIA GPUs, is a language binding to the C/C++ language for general purpose parallel implementation oriented towards a fine grained level of parallelism [8]. However, the direct implementations of parallel algorithms in GPU offer irregular results depending mainly on the computation-communications balance and on the pattern of the memory accesses. Consequently, optimizations are frequently performed [7].

The automatic detection of targets that we address in this work is a very relevant problem in the field of hyperspectral image processing [9]. Most of the published work in target detection over hyperspectral images focuses in images taken at a large distance from the targets, typically images obtained by satellite sensors so each pixel of the image represents hundreds of square meters of land and contains a mixture of different materials. A target detection requires in this case a subpixel extraction of information. Different implementations in clusters or FPGAs have been developed for some algorithms performing target detection over hyperspectral images and comparing their performance to that obtained in GPU implementations [10–12]. The good results obtained indicate that efficient implementations in GPU may offer important advantages in applications that demand a real-time response.

The main objective of this work is to further explore the efficient implementation of multiresolution target tracking algorithms using GPU processing. In fact, the idea here is not to directly map already known algorithms, but, rather, to create computational intelligence based algorithms that are intrinsically adapted to the hardware architecture and the parallel processing characteristics that GPUs offer.

The artificial intelligence algorithms developed have in common that they compute independent ANNs over the image in the same way it was explained in [13]. The main benefit of this approach is that there are no write data dependencies among the calculations of different ANNs, which fits the requirements for an adequate exploitation of the SIMD parallelism paradigm exploited by GPUs. We introduce three different algorithms for target detection of full pixel targets applying them to a search and rescue scenario and working with sea images where the targets are shipwrecked persons and ships.

The outline of this chapter is the following. Section 9.2 describes the target detection algorithms developed for their efficient execution over GPUs and based on the use of ANNs. Section 9.3 outlines the main characteristics of GPU and CUDA that have been considered for the parallel implementations and Section 9.4 analyzes the parallel implementations of the algorithms for a GPU. The results are shown and discussed in Section 9.5. Finally the main contributions are explained in Section 9.6.

9.2 Methods for Target Detection

Hyperspectral sensors usually record the images scanning a scene line by line, in our case from left to right. For each line they obtain all the spectral components of the line. This way a hyperspectral image can be considered as a set of pixel vectors as shown in Figure 9.1 where m denotes the spectral bands.

In our detection algorithms the target can be a pixel of the image (a vector of m spectral components) or it can be spread over a larger area. We have developed one approach that searches for the target at pixel level identifying all the pixels containing parts of the target. The other two approaches perform a multiresolution search based on exploring the image by inspecting windows (volumes of the hyperspectral image) of decreasing sizes until focusing on the appropriate scale that fits the target size. The last of these multiresolution approaches also finds out the orientation of the targets in cases for which this information is relevant.

This section describes the artificial intelligence target detection algorithms that have been designed for their implementation on the GPU. We call them pixel level algorithm for target detection, multiresolution target detection algorithm and multiresolution rotation-invariant target detection algorithm.

9.2.1 Pixel Level Algorithm for Target Detection

This algorithm is based on inspecting each pixel (vector of m spectral components) of the image separately producing a detection index that corresponds to the probability of the pixel being a target or part of one. A radial basis function (RBF) network is applied to each pixel of the image. In our case the ANN is trained offline and only once for a variety of images of the same type with a varying number of different size targets, in our case shipwrecked persons. The training process should be repeated if the type of targets were changed. Figure 9.2 shows the execution flow for the detector at pixel level.

The input to the RBF is, as mentioned above, one pixel, i.e., a vector of m spectral components $x_1, x_2, \ldots, x_m$. The number of neurons in the hidden layer is in general k: $h_1, h_2, \ldots, h_k$. Each neuron i in the hidden layer has a set of associated values: m centers $u_{1i}, u_{2i}, \ldots, u_{mi}$, a weight w_i, and a

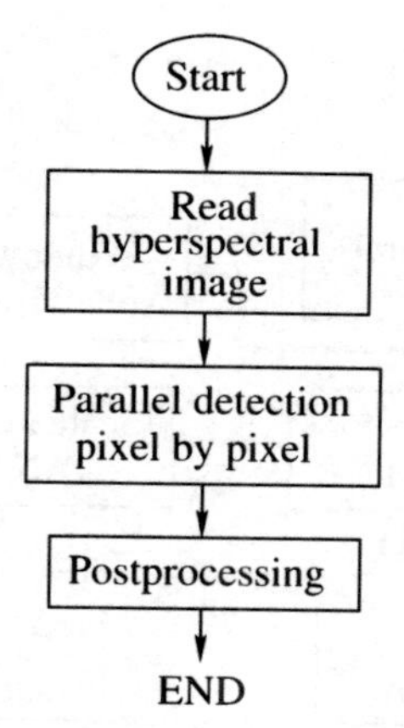

Figure 9.2 Target detection algorithm at pixel level.

deviation σ_i. Given these values the output of a neuron i, denoted as h_i, is calculated as:

$$h_i = e^{-\frac{\sum_{j=1}^{m}(x_i - u_{ji})^2}{2\sigma_i^2}}, \quad \text{being} \quad i = 1, 2, \ldots, k. \tag{9.1}$$

In our case the RBF network has four neurons in the hidden layer ($k = 4$) and only one output that is denoted by y. It has an associated threshold th and it is calculated as:

$$y = \sum_{i=1}^{k} w_i h_i + th. \tag{9.2}$$

Both Eqs. (9.1) and (9.2) are reduction operations. Note that the application of the RBF to different pixels (vectors of spectral components) does not present dependencies. The output value indicates if the pixel is part of the target or not. If we want to evaluate the size of the detected targets and therefore the distance at which the images were taken, a simple postprocessing of the outputs obtained for the whole image would be required.

9.2.2 Multiresolution Target Detection Algorithm

The second algorithm focuses on searching for the target at the right scale so the postprocessing step described for the previous algorithm is not required. This is a multiresolution based approach to scale invariant target identification using a hierarchical ANN based architecture consisting on two ANNs. The

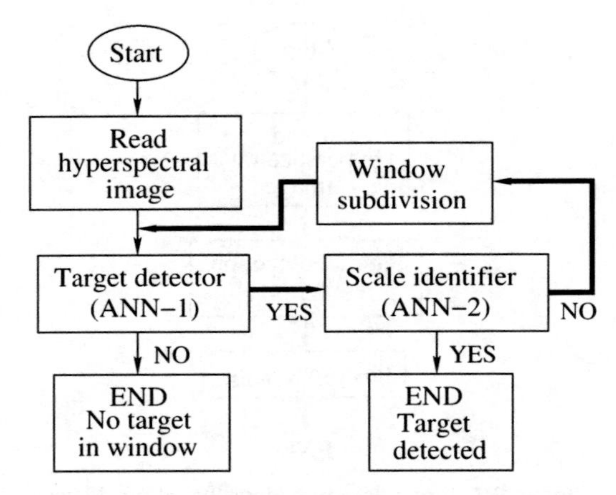

Figure 9.3 Multiresolution target detection algorithm.

neural network architecture is trained offline, once for each ANN in a similar manner to the previous algorithm.

The block diagram for the algorithm is shown in Figure 9.3. The 2-ANN detector is iteratively applied over decreasing size windows. In each iteration the window size for all the windows is reduced by a quarter with respect to the previous one, repeating the detection process in the resulting windows. The operations that the algorithm performs at each iteration and for each one of the selected functions are the following:

- *Target detection.* If the first ANN (ANN-1), that performs target detection, provides a positive detection over a window, it indicates that one or more instances of a target are somewhere inside the window.
- *Scale identification.* The second ANN (ANN-2) performs scale invariant identification. It answers if the target matches the window and the process over the window stops or if the scale must be reduced and the 2-ANN detection process must continue over smaller windows in subsequent iterations.

The process continues until all the windows have been tested. Finally the hierarchical structure decides how many instances of the target are present in the image, where they are, and their size within the image.

Both, ANN-1 and ANN-2, are feed forward back propagation networks of 2 layers and 5 neurons in the hidden layer. In order to apply one of these ANN to a window (a cube of information) it is necessary to calculate the deviation of all the spectral bands for all the pixels in the window. The deviation for each band is calculated as:

$$\sigma_i = \sqrt{\frac{\sum_{l=1}^{W_w H_w} (x_{li} - \overline{x_i})^2}{WH}}. \tag{9.3}$$

being W_w the width of the window, H_w its the height, and $W \times H$ the size of the entire image. Index i runs over the spectral bands, l denotes the index of each one of the windows for the i band, and $\overline{x_i}$ is the mean of the values in the window for band i. Therefore for a given window a vector of the deviations $\sigma_1, \sigma_2, \ldots, \sigma_m$ is calculated where m is the number of spectral bands. This vector, once normalized to the range $[-1, 1]$, is the input for each one of the ANNs when they are applied to that window.

When calculating the neurons of the hidden layer, each neuron i has an associated set of values: centers $u_{1i}, u_{2i}, \ldots, u_{mi}$ (being m the number of spectral bands considered), a bias b_i and an output h_i. The output of a neuron i is:

$$h_i = \sum_{j=1}^{m} (\sigma_j u_{ji}) + b_i, \quad \text{where} \quad i = 1, 2, \ldots, k, \tag{9.4}$$

with k being the number of neurons in the hidden layer. The output of each neuron h_i is modified by a sigmoid function and calculated as:

$$l_i = \frac{2}{1 + e^{-2.h_i}} - 1. \tag{9.5}$$

Considering that the hidden layer consists of 5 neurons ($k = 5$) and the output layer of only one neuron with a a set of associated weights $w_1, w_2, \ldots, w_5$ and a bias b, the output y is calculated as:

$$y = \sum_{j=1}^{5} (k_j.\omega_j) + b. \tag{9.6}$$

This value goes through a sigmoid activation function that is calculated in a similar way as in Eq. (9.5) and finally an output y_{norm} is obtained. The label of the output indicates that its value must be denormalized in a final step.

Note that (9.4) and (9.6) in a similar way as Eqs. (9.1) and (9.2) involve mainly reduction operations. The main difference with the pixel-level case is in the calculations for computing the entries for the ANNs required in the multiresolution algorithm.

From the explanation on this algorithm we conclude that the number of times the ANNs are computed for a specified image size is unknown a priory,

Figure 9.4 Working example for the multiresolution level target detection algorithm over an image with two targets.

unlike for the pixel level algorithm. This algorithm saves computations but at the cost of increasing the complexity of the algorithm flow. Figure 9.4 illustrates the application of the algorithm for an image with two targets.

9.2.3 Multiresolution Rotation-Invariant Target Detection Algorithm

This target detection algorithm follows a multiresolution strategy similar to that of the previous one adding the capability to estimate the rotation angle of the target. The process, as the previous one, is iterative over decreasing size windows using a hierarchical ANN based architecture but in this case consisting of four ANNs, each of which performs a specific operation on the selected window. All ANNs are feed-forward backpropagation networks of two layers that compute Eqs. (9.4), (9.5) and (9.6) with normalized inputs in the range $[-1, 1]$. The number of inputs is dependent on each ANN, the number of neurons in the hidden layer is 16 except for ANN-2 which has 8 neurons, and in all cases there is only one neuron in the output layer.

The block diagram for the algorithm is shown in Figure 9.5. The operations that the algorithm performs at each iteration and for each one of the selected windows are the following:

- *Target detection.* This first operation determines the presence or absence of a target in the selected area. It uses a feed-forward backpropagation network (ANN-1) that takes as input the vector of m deviations σ_i calculated on the corresponding window of the image using Eq. (9.3). The output value indicates whether the area being evaluated contains the

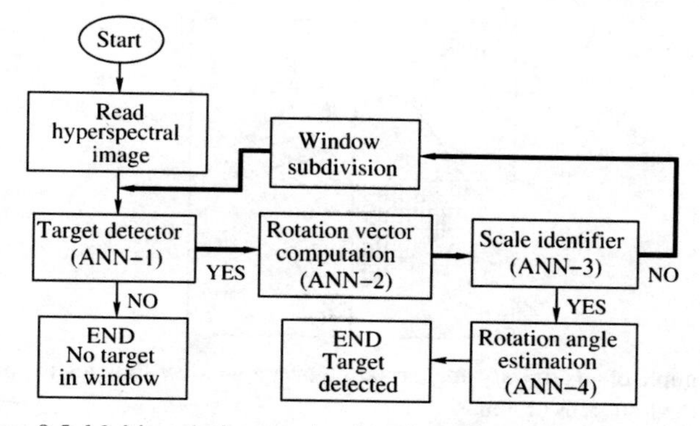

Figure 9.5 Multiresolution rotation-invariant target detection algorithm.

target (positive output), or if instead there is no target in that region (negative output).

- *Rotation vector computation.* The processes of scale identification and rotation angle estimation are executed on those windows for which the target detection has provided a positive output. The first step is the computation of the rotation vector. It uses as reference a mask that divides the window into p cells (in our case the four quadrants given that $p = 4$). The window to be tested is rotated q times (4 in our case) and for each one of the angles considered, $2\pi i/q, i = 0, 1, \ldots, q-1$, the average spectrum of the pixels contained in each cell of the mask is computed. This process is illustrated in Figure 9.6 for one arbitrary angle and considering a mask of four cells. We use a feedforward backpropagation network (ANN-2) with $p \times m$ inputs (where m is the number of spectral bands) and only one output. The network is applied as many times as rotation angles are considered, and it provides one output value per rotation angle that is calculated in a similar way as Eq. (9.6), so finally we have as output a rotation vector of q components.
- *Scale identification.* The next step determines from the rotation vector if the target is present in the right scale. To do this, we use a new network (ANN-3) whose input is the rotation vector obtained in the previous step and that provides as output, calculated as indicated by Eq. (9.6), a value greater than zero (if the identification is positive) or less than zero (if the identification is negative). If the identification is positive a rotation angle estimation is performed. Otherwise the window is divided by four and the process from the target detection step is repeated over each one of the resulting subwindows.

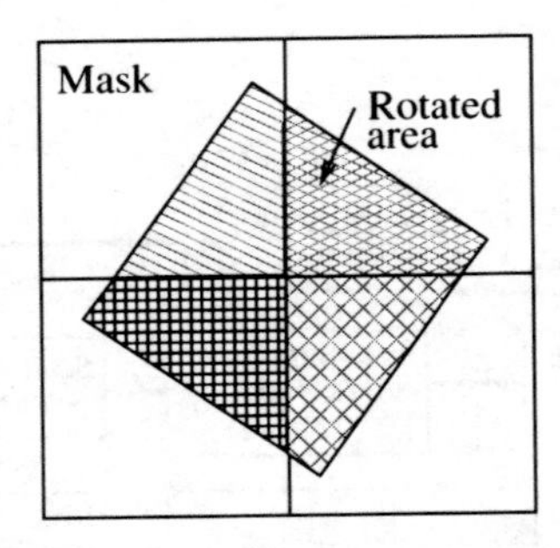

Figure 9.6 Example of an arbitrary angle rotation over a window showing the pixels contained in each one of the four cells of a mask.

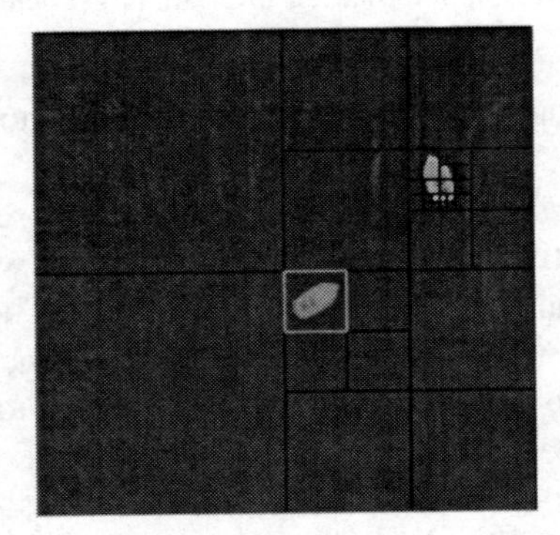

Figure 9.7 Working example with one valid target and one invalid target for the multiresolution rotation-invariant target detection algorithm.

- *Rotation angle estimation*. Finally, we estimate the rotation angle of the target using an additional feed-fordward network (ANN-4). The inputs are the same as in the previous network and the output corresponds to the rotation angle of the target.

This way the algorithm estimates for each target the right scale and the rotation angle. As for the previous algorithm, the number of windows of the image that must be processed is unknown a priori as well as the number of ANNs that must be applied. Figure 9.7 shows the required subdivision of the original image when only one valid target is present. Observe that the search over the area of the image containing the invalid target is iteratively performed but it finishes without a positive detection.

9.3 GPU Overview

GPUs provide massively parallel processing capabilities based on a data parallel architecture. CUDA [8] is one of the APIs that are available for exploiting the architecture. The NVIDIA CUDA architecture is organized into a set of streaming multiprocessors (SMs) each one with many cores or streaming processors (SPs). The number of cores per SM depends on the device architecture, e.g., NVIDIA Fermi GTX580 series has a total of 512 cores distributed into 16 SMs, each one with 32 SPs. These cores can manage hundreds of threads in a simple program multiple data programming model.

A CUDA program, which is called a kernel, is executed by thousands of threads grouped into blocks [14]. The blocks are arranged into a grid and scheduled to any of the available cores which enables scalability for future GPU models. If there are not enough processing units for the blocks assigned to a SM, the code is sequentially executed.

The smallest number of threads that are scheduled is known as a warp so this is the minimum size of SIMD (Single Instruction Multiple Data) processing. All the threads within a warp execute the same command at the same time. The size of a warp is implementation defined [14] and for the GTX580 GPU its size is 32 [15]. In the Fermi architecture [16] each SM is equipped with two warp schedulers and two instruction dispatch units allowing two warps to be issued and executed concurrently. Figure 9.8 shows a diagram of a NVIDIA's Fermi architecture which is the used in this work.

The memory hierarchy is organized, as shown in Figure 9.9, into a global memory, a read-only texture memory and a constant memory, with special features such as caching or prefetching data. These memories are available for all the threads. A two-level cache hierarchy per SM is available. The L1 cache is configurable to support both shared memory and caching of local and global memory operations. The L2 cache is unified for data and instructions and shared by al the SMs. In the case of the GTX580 architecture the L1 cache can be of 16 KB or 48 KB and the size of the L2 cache is 768 KB. There is an on-chip memory space called shared memory that is available per block. It enables a very fast read/write access to the data (is as fast as registers if there are no bank conflicts) but with the lifetime of the block. Each thread has also its own local memory and registers.

Threads within a block can be synchronized but not the threads that belong to different blocks. Due to this restriction it is not possible to share data among blocks so this type of communication must be through the global memory. In this situation achieving a low rate of communications

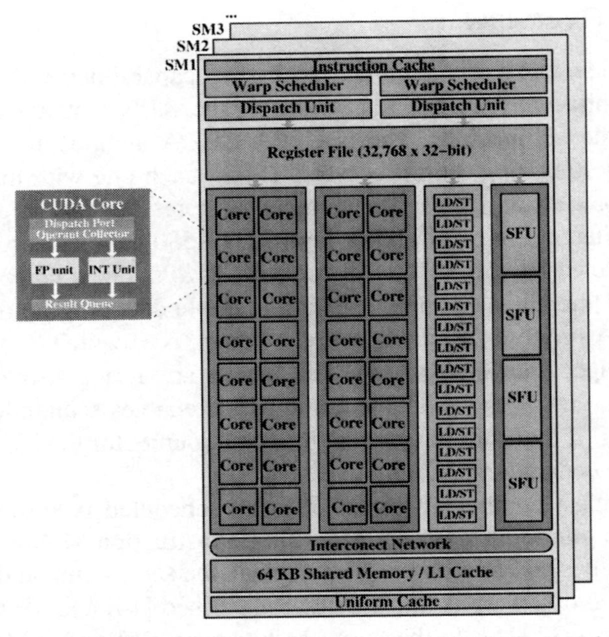

Figure 9.8 Simplified multiprocessor architecture for a NVIDIA's Fermi GPU [16].

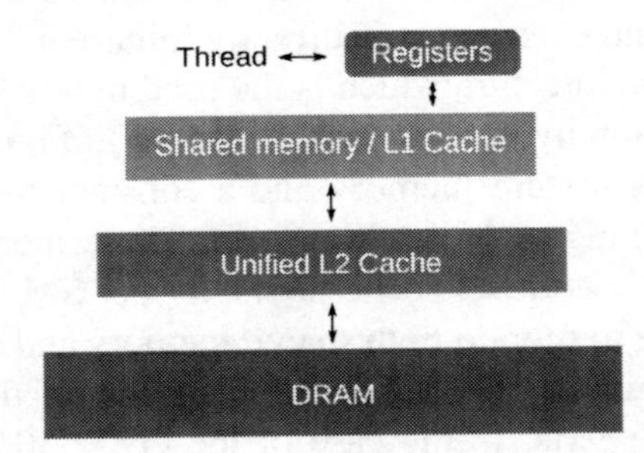

Figure 9.9 Memory hierarchy for the Fermi architecture.

through the global memory becomes a challenge, although in the most recent architectures the cache memory alleviates this problem.

9.4 Parallel Implementation

The core computations in the algorithms developed perform memory accesses that present spatial data locality. The amount of data is so high that in most cases it exceeds the capacity of the shared memory. There are two possible strategies for partitioning the image among thread blocks [12]: spatial-domain partitioning, that divides the image into cubes of pixels with their spectral bands, and spectral-domain partitioning, that divides the image into slices made up of one or several contiguous spectral bands.

In our case, the hyperspectral image processing is done with ANNs that take as inputs all the spectral bands, so it is advantageous for efficiently exploiting the memory hierarchy to use the first one of these approaches. Data are transferred from the global memory to the shared memory in blocks (cubes of data) where each pixel is stored in shared memory followed by its spectral components, processed by blocks of threads in the CUDA cores and rewritten again on the original positions of the global memory.

The hyperspectral data are stored in the global memory because if cached, like in the Fermi architecture, its bandwidth if higher than for the texture memory. The constant memory, that is optimized for broadcasting values, is used for the parameters of the ANNs (weights, thresholds, etc.) because these are accessed by multiple threads.

9.4.1 Target Detection Algorithm at Pixel Level

In this type of detection each pixel of the hyperspectral image is applied as input to an ANN. The implementation of this detector implies, therefore, the computation of a large number of ANN instances on GPU.

The operation performed by each neuron of an ANN, given by Eqs. (9.1) and (9.2), is basically a reduction operation. It can be implemented in an optimized way using sequential addressing, multiple elements per thread, unroll and other optimization techniques such as those described in [17]. This first strategy has been applied in [18], that focuses in the GPU implementation of large-scale neural networks in the context of biological model simulation. The alternative to the reduction operation is to organize the ANN operation in the form of matrix products and take advantage of the matrix algebra functions available in the libraries [19]. This approach has been analyzed in a similar context by [20, 21]. In this work we use the first option because it is the one that offers more opportunities for optimization.

In applications that require the computation of a large number of ANNs on GPU, how to perform the projection should be studied carefully. The alternatives for mapping ANNs onto the thread blocks are basically two [18]: (1) *Neuronal parallelism.* In this approach each one of the threads of a block computes a neuron as indicated in Eq. (9.1). Thus, the different synapses of the neuron are computed sequentially by the same thread and the results are stored in shared memory. In our implementation the threads are synchronized after the computation of the hidden layer and one of them computes the output of Eq. (9.2) that is stored in global memory. With a RBF network as described by Eqs. (9.1)–(9.2), of 64 inputs, one hidden layer of 4 neurons and one output, each block can compute in parallel $B/4$ ANN instances (B is the block size). If $B = 256$, 64 ANN instances will be computed by each block. This is shown in Figure 9.10(a).

Given the large number of data required the entries are loaded sequentially from the global memory when they are needed, and the accumulation of products is stored in shared memory.

(2) *Synaptic parallelism.* In this approach each one of the threads of a block computes the synapses of one neuron. All the threads involved in the computation of the same ANN cooperate because a reduction operation, calculated in a similar way as in [22], is required for computing Eq. (9.1). Thus, a set of threads collaborate in the computation of each neuron and, considering 256-thread blocks (the best value for our implementations) each block of threads can compute 4 complete ANNs. Input data and other initial values are loaded from the global memory and the constant memory in parallel and stored in shared memory as for the neuronal parallelization strategy. In the same way, when data load is complete, the computation of the neurons takes place entirely in shared memory. Finally one of the 64 neurons involved in the computation of each ANN calculates the ANN output as indicated by Eq. (9.2). The example for 256 threads per block is shown in Figure 9.10(b).

9.4.2 Multiresolution Target Detection Algorithm

This is an iterative algorithm, as explained in Section 9.2.2, that performs searches over decreasing size windows of the image. The operations that the algorithm performs at each iteration and for each one of the selected windows are *target detection* and *scale identification*. These operations are performed by ANNs whose computation requires *ANN input calculation* and *ANN application*. In this algorithm the application of the ANNs takes a secondary role, since the main bottlenecks are the reduction operations for calculating the inputs to the ANNs and the selection of windows of interest.

With respect to the reduction operations we employ a multi-pass approach. The first pass performs a block-level reduction. The hyperspectral image is divided into areas of size 8×8 pixels that are assigned to blocks of 512 threads. For each area, we compute two parameters per spectral band: the

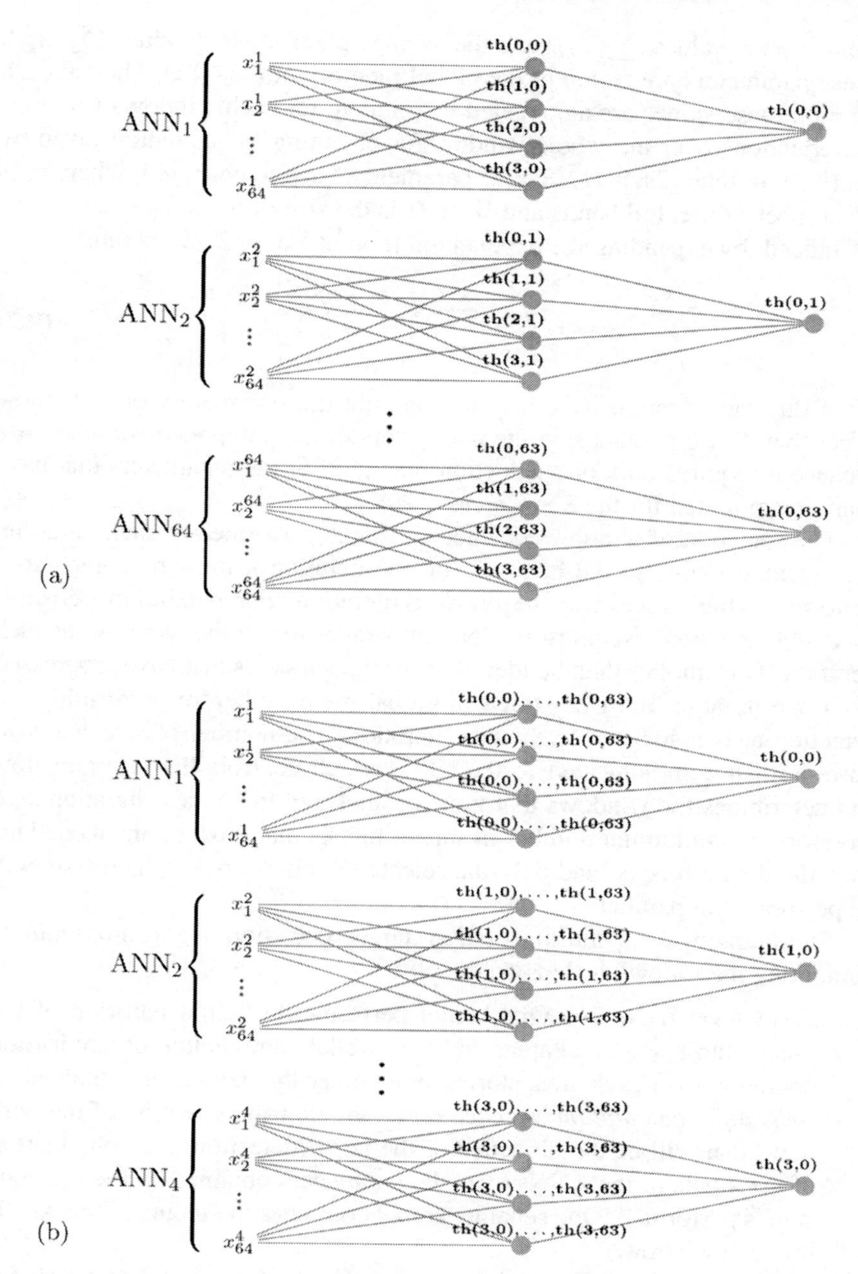

Figure 9.10 Different kinds of parallelism in the computation of ANNs on thread blocks considering 256 threads per block. $th(x, y)$ values denote the threads of the block. Neuronal parallelism (a), and synaptic parallelism (b).

average pixel values ($\sum_k x_{ki}$) and the average pixel squared values ($\sum_k x_{ki}^2$). These parameters are stored in the global memory of the GPU. Then at each iteration these values are aggregated to compute for each window of interest the reductions over the whole window for obtaining the deviation given by Eq. (9.3). In total, $2mWH/(8 \times 8)$ parameters have to be stored, where m is the number of spectral bands and $W \times H$ is the size of the image.

Indeed, by expanding the exponential term of Eq. (9.3) we obtain:

$$\sum_{k=1}^{N}(x_{ki} - \overline{x_i})^2 = \sum_{k=1}^{N} x_{ki}^2 - \frac{1}{N}\left(\sum_{k=1}^{N} x_{ki}\right)^2. \qquad (9.7)$$

From this factorization it is easy to compute the parameters of a window larger than 8×8 pixels using the parameters of the component subwindows because it requires only one reduction operation of the parameters that have been precomputed for the 8×8 areas.

The selection of windows of interest, that is performed at each iteration, is sequential owing to the fact that for one iteration it must be compulsory performed after processing the previous iteration. For maximum performance, the detection tests must be done in parallel for all the windows at each iteration. This implies that the identifiers of the windows that have progressed from one iteration must be stored in global memory before performing the detection tests associated to the next iteration. The number of selected windows is then communicated to the CPU, which controls the program flow and determines the windows that must be analyzed in the next iteration and, therefore, the minimum number of thread blocks that must be invoked. This way, the thread blocks load only the selected windows over which tests will be performed in parallel.

Summarizing, the multiresolution target detection algorithm requires computing the following steps on the GPU:

- *Block-level reduction.* One kernel performs the initial partition of the image into 8×8 pixel areas and the parallel computation of a reduction operation over each area, storing in memory the parameters obtained.
- *ANN input calculation.* At each iteration, after the selection of the windows that will be the objective of the target detection, the computation of the inputs of the ANNs from the parameters obtained by the previous step is performed. One separate kernel computes the inputs of the ANNs for each window.
- *ANN application.* At each iteration, application in parallel of ANNs for detection (ANN-1) and, if necessary, for scale identification (ANN-2) to

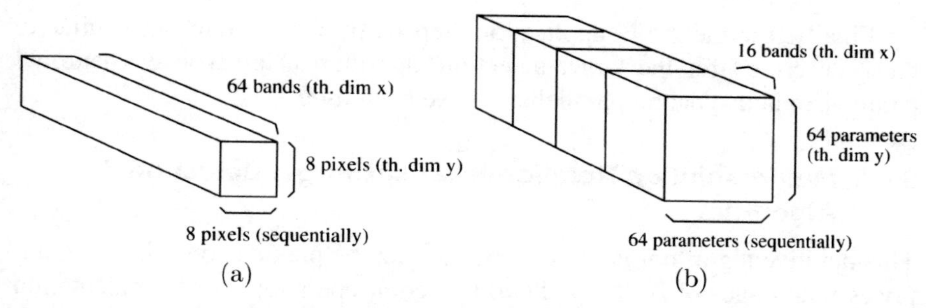

Figure 9.11 Mapping of computations on thread blocks for block-level reduction (a) and computation of the ANN inputs (b).

all the selected windows. The ANN application is computed by another separate kernel.

In order to maintain the coalescence of memory accesses, we use a spatial-domain partitioning which divides the image into cubes and bands in a similar way as [3] for computing the initial block reduction and the aggregation of results at each iteration of the algorithm. Figure 9.11 shows the mapping of computations to thread blocks in steps 1 (block-level reduction) and 2 (ANN input calculation). In this figure, "th. dim x" and "th. dim y" refer to the values that index the group of threads in a block.

In the block-level reduction step, shown in Figure 9.11(a), each block of 64×8 threads performs the reduction of the 64 bands of an area of 8×8 pixels. Since the number of pixels is larger than the number of available threads, some operations are performed sequentially over 8 pixels and 8 threads collaborate to perform a reduction operation for data in the same band. This is illustrated in the figure using two-dimensional thread blocks (of size $X \times Y$ threads) and one additional temporal dimension. These three dimensions of processing are mapped on the data cube as indicated in the figure.

For the ANN input calculation, shown in Figure 9.11(b), the thread blocks are 64×16 and process a data cube of $64 \times 64 \times 16$ parameters where each thread sequentially performs computations over 64 parameters and 64 threads collaborate in the reduction operation for the parameters of the same band. Note that in this case a block of threads does not compute all the spectral components of the image area considered, so different blocks must cooperate in calculating each ANN input.

Finally, for the ANN application step partitioning strategies similar to those described for the target detection algorithm at pixel level (neuronal parallelism and synaptic parallelism) have been used.

9.4.3 Multiresolution Rotation-Invariant Target Detection Algorithm

This iterative algorithm is more complex than the previous one since it comprises four stages of ANNs and two reduction operations for the calculation of the ANN inputs. These reduction operations compute the deviations which are inputs to ANN-1, and the average spectra of rotated areas which are inputs to ANN-2 (see Section 9.2.3). The algorithm also involves the selection of windows on which reduction operations have to be performed. Due to the iterative nature of this algorithm, the main bottlenecks are the selection of windows and the reduction operations for calculating the inputs to the ANNs.

As for the previous algorithm, the algorithm initially computes the average pixel values ($\sum_k x_{ki}$) and the average pixel squared values ($\sum_k x_{ki}^2$) using a *block level reduction* in order to accelerate the reduction operations. The whole image is divided into 8×8 pixel blocks and the results obtained for each spectral band are stored in the global memory of the GPU. These precomputed values will be used in the *target detection* and *rotation vector* computation stages. In the first case, they accelerate the computation of the deviations which are used as inputs to ANN-1 while in the second case the acceleration is over the calculation of the average spectra in rotations used as inputs to ANN-2. In our case, the mask is made up of 4 cells, and 4 rotation angles are considered, so that we can estimate rotations over targets with a minimum size of 16×16 (since it is the smallest area that contains 4 cells of size 8×8 pixels).

Similarly to the previous algorithms, the computation of the hierarchical ANN structure (composed by 4 ANNs in this case), can be performed using the partitioning strategies called neural parallelism and synaptic parallelism, as described in Section 9.4.1.

9.5 Experimental Results

The proposed GPU implementations have been tested on a NVIDIA Fermi-based GPU, a GeForce GTX 580 [15], which features 512 processor cores grouped into 16 SMs of 32 SPs each. The GPU global memory consists of 1.5GB of GDDR5 memory at 2004 MHz whose bandwidth with the CPU is

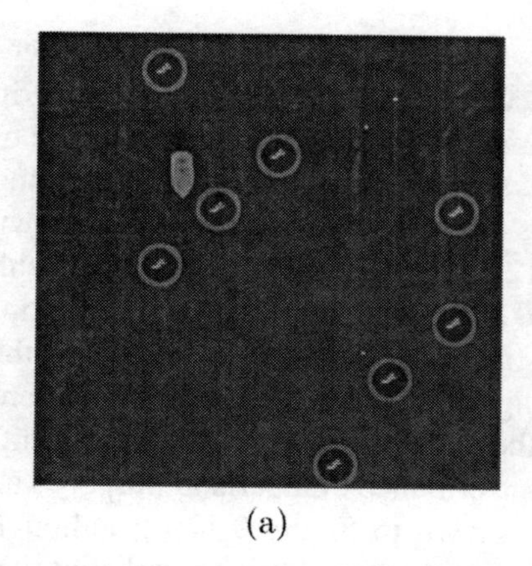

(a)

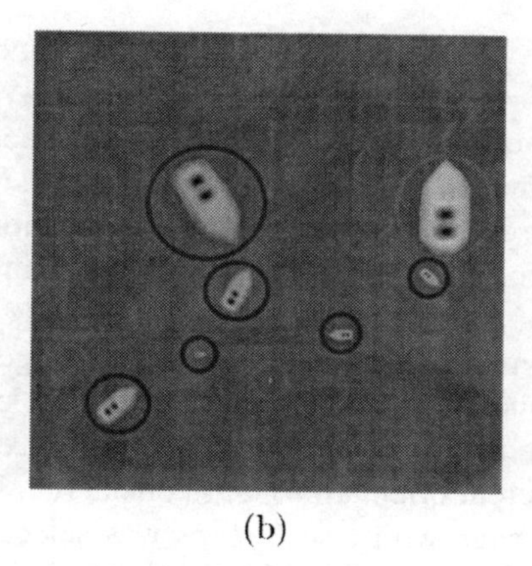

(b)

Figure 9.12 1024 × 1024 × 64 hyperspectral images used for the experiments with the target detection algorithms: pixel-level and the multiresolution (8 shipwrecked persons) (a), and rotation-invariant multiresolution (8 ships) (b).

through the PCI Express 2.0 with a peak bandwidth (counting both directions) of up to 20 GB/s. The bandwidth of the global memory inside the GPU has a peak value of 223.8 GB/s. In the case of the shared memory given that the clock rate is 1544 MHz, that 4 bytes per cycle can be read from each bank of the shared memory, that there are 32 banks per SM and 16 SMs, a peak value of 1581 GB/s is obtained. The 64 KB of on-chip memory per SM can be configured as 48 KB of shared memory and 16 KB of L1 cache or vice versa being the first one the configuration selected for this work. There is also a 768 KB L2 cache that is shared among all the SMs. Regarding the computation rate, the GTX580 card is made up of 512 cores, each one can deliver two floating point instructions per cycle and each one works at 1544 MHz, so a peak rate of 1581.1 GFLOPS is obtained [24]. The CUDA code has been compiled using the `nvcc` compiler provided by the CUDA 4.0 toolkit, also under Linux.

The CPU code was optimized through data access reordering in order to increase the memory hierarchy performance. It was evaluated in an Intel Quad Core Xeon E5440 at 2.83 GHz, with one 32 KB L1 cache per core for instructions and another one for data. There is also a 6 MB L2 cache per core pair and a unified 8 MB L3 cache shared by all cores. The code was compiled using the `gcc` version 4.4.3. In both platforms the algorithms were compiled with the -O2 flag.

The images used in the experiments were obtained using the hyperespectrometer developed by our group. We focus on a search and rescue scenario working on pre-recorded hyperspectral images. Although a large number of images with different number of targets were required for the ANN training process, the performance evaluation has been done for the images shown in Figure 9.12. The first one (Figure 9.12(a)) is a 1024 $\times$ 1024 image with 64 spectral bands (the size is 1024 $\times$ 1024 $\times$ 64 floats) that contains shipwrecked persons in the sea as targets. This image is used for analyzing the pixel level and the multiresolution level algorithms. For the rotation-invariant multiresolution algorithm an image of the same size but with ships in different orientations was considered. For performance comparison purposes an image with also 8 ships was selected as shown in Figure 9.12(b); although only 6 out of the 8 ships in the image are valid targets. It is remarkable that the images include not only the targets but also different objects of similar characteristics for proving the ability of the developed algorithms to identify targets in complex scenarios.

For the speedup measures we compare the execution time in the GPU to the execution time on the CPU, obtaining each one as an average of one hundred executions. The time measurement is, in general, started right after the hyperspectral image is in the GPU global memory and stopped after the detection results are obtained and stored in the CPU memory. Nevertheless, we have also studied the time cost of data transfer between the CPU and the GPU because it is a crucial factor when searching for real-time processing.

The bandwidth utilization and the GFLOPS rate obtained for each algorithm are also analyzed. For this purpose we have estimated the execution time for global memory accesses, shared memory accesses, and computation separately. To estimate global memory access time, we replace all global memory accesses with register accesses, and calculate the global memory access time as the time difference between this program and the original one. The same approach is used to measure shared memory access time. This method was previously used in [23]. We must consider that this method performs only an estimation of time breakdown and it does not provide accurate time measures.

The ANNs used in this work are two-layer networks, so that all arithmetic operations can be performed in single precision because the accuracy loss does not alter the final results. This reduces to approximately half the number of bytes transferred and simplifies the arithmetic operations. In the case of ANNs with a high number of layers, double precision could be required.

Regarding effectiveness it is important to emphasize that the designed algorithms detect all the instances of the target in the image and that parallel versions of the designed detection algorithms provide exactly the same detection results as those implemented in CPU.

9.5.1 Target Detection Algorithm at Pixel Level

For this algorithm the computational work consists mainly on applying the RBF ANN to each pixel of the image (a vector of hyperspectral information). Two different partitioning strategies, as was explained in Section 9.4.1, were used: neuronal parallelism and synaptic parallelism.

The performance results for the $1024 \times 1024 \times 64$ sample image that contains 8 instances of the target (Figure 9.12(a)) are shown in Figure 9.13. The execution times for both GPU implementations as well as the sequential execution times in CPU are displayed in Figure 9.13(a). Different measurements were obtained varying the number of pixels per chunk, i.e., the maximum number of pixels that can be simultaneously processed. We observe that increasing this parameter (X-axis) the execution time in CPU remains constant. Nevertheless for the GPU implementations if the X-axis increases, the amount of work that can be performed in parallel also increases and, as a consequence, the execution time decreases. This effect can be more clearly observed in Figure 9.13(b) where the corresponding speedups for the same configurations are displayed. All the executions reflected in these graphs consider 256 threads per block, being this number of threads the best configuration obtained from the experiments.

We can observe that for the synaptic parallelism the maximum possible chunk size (maximum number of ANNs that will be executed in parallel) is 2^{16}. The reason is that computing 2^{18} ANNs, the next value in the X-axis, requires 65536 active blocks ($2^{18} \times 64 = 2^{26}$ threads, owing that each ANN requires 64 threads to be computed, and $2^{26}/256 = 65536$ blocks) being 65535 the maximum for this architecture [8].

Regarding the optimum chunk size we observe in Figure 9.13(b) that it depends on the partitioning strategy, being 2^{12} or bigger for neuronal parallelism and 2^{8} or bigger for synaptic parallelism. These limits can be explained by the GPU *occupancy*. This parameter indicates the degree of utilization of the hardware, takes values in the interval [0, 1] and it can be measured by using the *CUDA Visual Profiler* tool [25].

Table 9.1 shows the occupancy and the number of active blocks per SM for different chunk sizes for both parallelizing strategies. For the neuronal

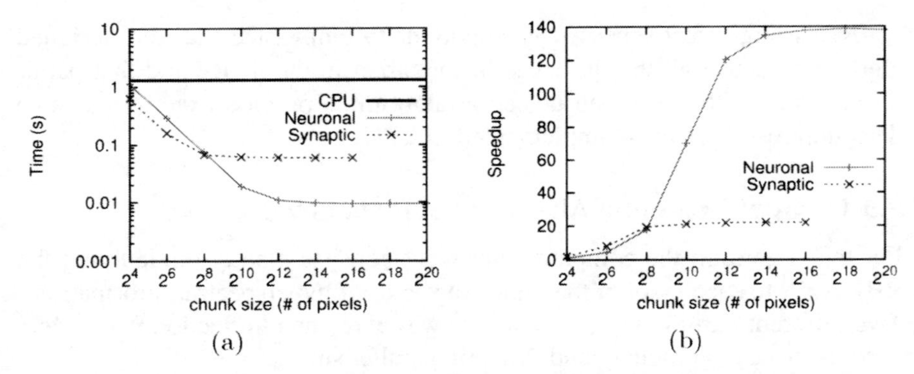

Figure 9.13 Execution times (a) and speedups (b) of the CPU and GPU implementations of the target detection algorithm at pixel level for a $1024 \times 1024 \times 64$ image.

Table 9.1 Target detection at pixel level. GPU occupancy and active blocks per SM for neuronal parallelism and synaptic parallelism considering 256-thread blocks.

	Neuronal parallelism		Synaptic parallelism	
Chunk	**Occupancy**	**Blocks per SM**	**Occupancy**	**Blocks per SM**
2^4	0.042	1	0.667	1
2^6	0.167	1	1.000	1
2^8	0.667	1	1.000	4
2^{10}	1.000	1	1.000	16
2^{12}	1.000	4	1.000	64
2^{14}	1.000	16	1.000	256
2^{16}	1.000	64	1.000	1024
2^{18}	1.000	256	–	–
2^{20}	1.000	1024	–	–

parallelism a chunk size of 2^{10} already presents occupancy 1. Indeed, considering that in the neuronal parallelism 4 threads are necessary for computing each ANN and also considering 256-thread blocks, a chunk of 2^{10} requires 16 blocks that are assigned to the 16 SMs in the GPU. Although in this case all the SMs are being used, the limit in hardware use has not been reached yet because each SM can keep 1536 threads (48 warps) active simultaneously [16]. If the chunk size is increased to 2^{12} all the SMs continue to be busy but with 4 blocks per SM (1024 threads), that approaches the limit of 1536 active threads. This is the reason for the maximum speedup for chunk sizes above 2^{12}. The same can be argued for the synaptic parallelism but considering that 64 threads are required for computing each ANN, achieving the maximum speedup for a chunk size of 2^8 and above.

In Figure 9.13 we observe that the implementation based on neuronal parallelism is the most efficient one reaching for 2^{20} pixels per chunk a value of 138.5×, being 0.0094 seconds the execution time, while for synaptic parallelism the best value was a speedup of 22.0× for 2^{16} pixels per chunk with 0.0592 seconds. The CPU time for the optimized detection algorithm at pixel level is 1.3018 seconds. In the neuronal parallelism only 4 threads are necessary for computing each ANN. For the case of synaptic parallelism 64 threads are necessary to compute one ANN. It is clear that each thread performs more computations in the case of neuronal parallelism thus exploiting the GPU architecture better and obtaining larger speedups.

Although no previous author we are aware of tested the same ANN-based target detection algorithms, we can compare these results to the target detection algorithms studied in [12] where the best result was a speedup of 70.1× for a hyperspectral image of 140 MB. In our case the best result is a speedup of 138.5× for an image nearly twice this size, 256 MB.

Figure 9.14 shows the time consumed in the target detection at pixel level for the two parallelizing strategies for a chunk size of 2^{16}, broken down into memory accesses and computation times. As we can see, the memory access is the major overhead in the execution of the two versions of the algorithm. The memory access for the neuronal parallelism consumes 91% of the time, being nearly 82% for the synaptic parallelism. For the neuronal parallelism 59% of the time is for global memory accesses because in this implementation each thread computes one neuron in the hidden layer thus not being necessary to store intermediate results in the shared memory. Note that, owing to the fact that the method for measuring the memory access time is not accurate, we obtain a global memory bandwidth for this algorithm that is larger than the theoretical peak of 223.8 GB/s.

For the synaptic parallelism the situation is the opposite because the number of accesses to shared memory that are required to store the intermediate results computed by each thread for each neuron is very large. The shared memory consumes 67% of the time so its bandwidth is 134.25 GB/s, still far from the theoretical peak of 1581 GB/s for this GPU. This is due to a better utilization of the memory transactions.

With respect to the computation time, it is a smaller fraction of the total for the neuronal parallelism, 9% corresponding to 0.0008 seconds, than for the synaptic case, 18.8% corresponding to 0.0132 seconds. The reason is the better GPU architecture exploitation for the neuronal case each thread performs more computations than for the synaptic case. As the computation time is smaller for the neuronal case, and the number of operations remains

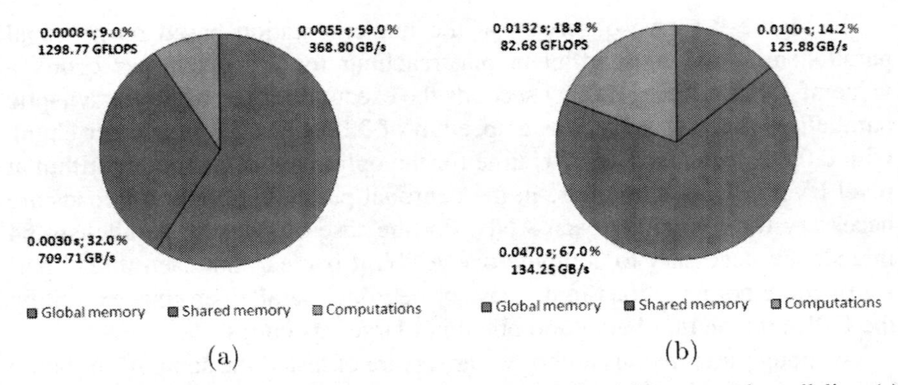

(a) (b)

Figure 9.14 Time breakdown for the pixel level algorithm applying neuronal parallelism (a) and synaptic parallelism (b).

constant, the GFLOPS rate is bigger. The value for the neuronal case is 1299 GFLOPS while for the synaptic one 83 GFLOPS were obtained.

9.5.2 Multiresolution Target Detection Algorithm

The multiresolution target detection algorithm searches iteratively over the image in decreasing size windows for consecutive iterations. In Table 9.2 the performance results are displayed in terms of execution time and speedup of the GPU implementation over the CPU one for the $1024 \times 1024 \times 64$ image of 8 shipwrecked persons (Figure 9.12(a)). The GPU implementation performs initially a block level reduction in shared memory over 8×8 pixel blocks. This stage of the algorithm is very efficient given that the large number of reduction operations are performed in parallel by different blocks of threads and each thread performs the reduction of not only 2 elements but a group of 8 elements. The speedup obtained and denoted in the table as “block level reduction” is $65.5\times$.

Afterwards an iterative process is executed. The ANN detector application is performed at each iteration in the same way as the detector at pixel level in the previous section but joining in the same block the application of the ANNs to different active windows of the image. For this task there are two possible approaches, as explained for the pixel by pixel algorithm: neuronal parallelism and synaptic parallelism. Given that at each iteration the ANN detector is applied only to the windows marked as promising in the previous iteration, the number of ANNs per iteration is low so both approaches present similar speedups of around $25\times$. For the sample image considered, the total

Table 9.2 Comparison of execution times and speedups for the GPU and CPU implementations of the multiresolution target algorithm for a 1024 × 1024 × 64 image of 8 shipwrecked persons.

	CPU (s)	GPU (s)	Speedup
CPU-GPU transfer time	–	0.0871	–
neuronal parallelism			
block level reduction	0.1113	0.0017	65.5×
ANN computation	0.0231	0.0010	22.8×
TOTAL	0.1344	0.0027	48.9×
synaptic parallelism			
block level reduction	0.1113	0.0017	65.5×
ANN computation	0.0231	0.0009	25.6×
TOTAL	0.1344	0.0026	**51.7x**

number of windows searched by the algorithm is 129 and the number of active windows in one iteration is at most 36. With a maximum of 36 windows being simultaneously processed we are working in the region of values below 2^6 for the X-axis in Figure 9.13(b), which corresponds to a situation when both parallelizing strategies give similar performance results. The first result for the input calculation and the ANN application ("ANN computation" in the table) is a speedup of 22.8× that is obtained for neuronal parallelism. A larger value of 25.6× is obtained for synaptic parallelism as shown in the table.

The "TOTAL" rows of Table 9.2 represent the sums of times for the "block level reduction" and "ANN computation" stages. The initial CPU-GPU transfer time that is carried out at the beginning of each algorithm is not included in the speedup calculations. This time, as detailed in the table, is 0.0871 seconds for the image considered. Given that the size of the image is 256 MB, we obtain a CPU-GPU bandwidth of 2.9 GB/s being 20 GB/s the theoretical peak for the GTX580 GPU.

The table also shows that for the whole multiresolution algorithm a speedup of 51.7× is the best value achieved. Although this technique obtains lower speedups than the target detection at pixel level (that obtained around 140×), we must consider that the execution time for the multiresolution algorithm (0.0026 seconds) is lower than for the approach at pixel level (0.0094 seconds). The time for the multiresolution approach depends not only on the size of the image but also on the number of targets.

Figure 9.15 shows separately the time consumed in accesses to memory and computation when computing the multiresolution algorithm for the case with best performance, the version with synaptic parallelism. The time is nearly equally divided among computation (46.4%) and memory accesses

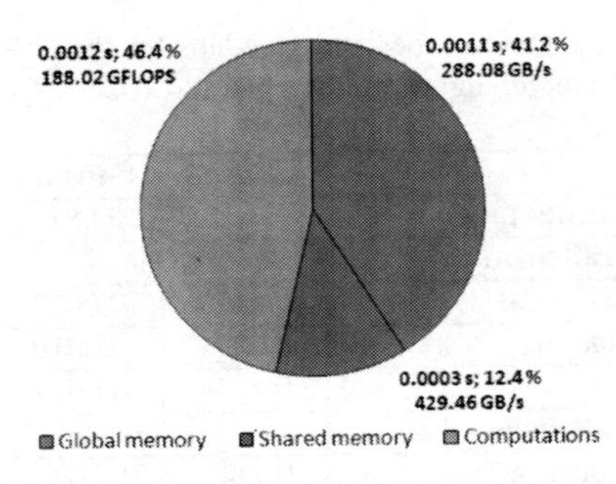

Figure 9.15 Time breakdown for the multiresolution target algorithm with synaptic parallelism for the 1024 × 1024 × 64 image of shipwrecked persons.

(53.6%). The global memory is responsible for most of the memory access time (41.2%) because the results for each iteration of the algorithm must be returned to global memory and after this to the CPU memory in order to decide the number of thread blocks for the subsequent iterations. The bandwidth measured is around 288 GB/s, hence the theoretical peak is achieved. For the shared memory the value obtained is 429.46 GB/s, far from the theoretical peak of 1544 GB/s.

9.5.3 Multiresolution Rotation-Invariant Target Detection Algorithm

This algorithm also performs an iterative search of targets over decreasing size windows and the computational stages are similar to those of the previous algorithm. The main difference is that it detects not only the targets but also gives their orientation but at the cost of increasing the complexity of the ANN detector and, therefore, increasing the computational time. For the rotation-invariant algorithm the ANN detector consists of 4 ANNs that must be sequentially applied to each window at each iteration. The input calculation for ANN-1 and ANN-2, as it was explained in Section 9.4.3, requires reduction operations based on the results of a first stage called block level reduction that is computed over 8 × 8 areas of the whole image.

The results for this algorithm have been obtained considering the sample image in Figure 9.12(b). This is a 1024 × 1024 × 64 image (64 spectral bands) containing 8 ships although only 6 are valid targets. Given that the minimum considered size of a target is 16 × 16 pixels, 7 iterations of the algorithm are required. A total of 173 windows are computed along the 7 iterations, being 64 the maximum number of active windows in one iteration. Note that the

Table 9.3 Comparison of execution times and speedups for the GPU and CPU implementations of the rotation-invariant multiresolution target algorithm for a 1024 × 1024 × 64 image of ships.

	CPU (s)	**GPU (s)**	**Speedup**
CPU-GPU transfer time	–	0.0871	–
neuronal parallelism			
block level reduction	0.1113	0.0017	65.5×
ANN computation	0.0405	0.0022	18.4×
TOTAL	0.1518	0.0039	38.9×
synaptic parallelism			
block level reduction	0.1113	0.0017	65.5×
ANN computation	0.0405	0.0021	19.3×
TOTAL	0.1518	0.0038	**39.9x**

number of windows that must be analyzed, and therefore the execution time, depend on the size of the image, the size of the targets and the number of targets.

The performance results for the CPU and the GPU code are compared in terms of execution time and speedup in Table 9.3. The results have been divided into the same stages as for the previous algorithm. The "CPU-GPU transfer time" and the "block level reduction" are the same as for that algorithm. Nevertheless, we can observe that for both, the CPU and the GPU code, the time for the "ANN application" is higher here because the ANN hierarchical structure is more complex. We have 0.0009 seconds for "ANN application" in the multiresolution algorithm and twice this time (0.0021 seconds) for the rotation-invariant algorithm.

As for the previous algorithm, the "ANN application" can be partitioned through neuronal parallelism or synaptic parallelism. The results are also similar applying both strategies, obtaining 19.3× for synaptic parallelism and 18.4× for neuronal parallelism. The reason is the small number of active windows at each iteration, that is at most 64. The best speedup for the whole algorithm is 39.9× with 0.0038 seconds. It is worth remembering that for the multiresolution algorithm the best speedup was 51.7× and the execution time 0.0026 seconds.

The time consumed in memory accesses and computation for the case of the best speedup (synaptic parallelism) are displayed in Figure 9.16. As for the previous algorithm (Figure 9.15) the time is nearly equally divided among computation (47.7%) and memory accesses (52.3%) and the memory access time is split among global and shared memories in similar percentages as for that case (37.5% for global memory and 15% for shared memory).

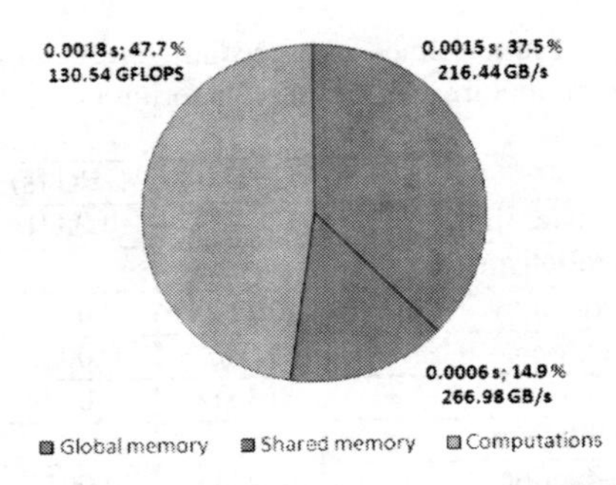

Figure 9.16 Time breakdown for the rotation-invariant multiresolution target algorithm with synaptic parallelism for the 1024 × 1024 × 64 image of ships.

Observe that the times are not exactly the same as for Table 9.3 because the method for measuring separately the time for computation and memory access perform some approximations, as it was explained at the beginning of this section. Nevertheless, there is an important difference to the previous case as the rotation-invariant algorithm presents a lower computational efficiency shown by lower values of bandwidth usage and GFLOPS.

9.6 Conclusions and Future Work

The objective of real-time processing has led us to propose three different artificial intelligence target detection algorithms for hyperspectral images. The target detection is applied to a search-and-rescue scenario where the targets are shipwrecked persons and ships of the size of one pixel or bigger, so no subpixel information extraction is required. The algorithms, based on the application of artificial neural networks (ANNs), have been developed for their efficient execution on GPUs, considering this a low cost, low-weight parallel computing platform suitable for on-board processing. These are: an algorithm for target detection at pixel level, a multiresolution target detection algorithm and a rotation-invariant multiresolution detection algorithm.

Given that this processing of hyperspectral information results in memory-bound applications the key is on reducing the memory requirements of our implementations and increasing the number of computations that each block performs, consequently hiding the latency of data movement. The memory requirements are reduced by storing intermediate results on the shared memory and by performing accesses exploiting the temporal and spatial locality of data in memory.

The first detection algorithm, detection algorithm at pixel level, searches pixel by pixel over the whole image determining if each pixel is part of a target or not. For this algorithm two different parallelizing approaches, neuronal and synaptic, were considered depending on how the computations are assigned to threads. The best speedup obtained over a 1024 × 1024 image with 64 spectral bands and 7 shipwrecked persons as targets is 138.5× and it was obtained for the neuronal approach.

The other two algorithms are called multiresolution algorithms because they perform and iterative search of the targets over decreasing size windows of the image. The size of the windows is divided by four in each iteration. For each window an ANN detector is applied that determines if a window must be subdivided or the detection over it finishes. The main computational cost of the multiresolution algorithms is associated to the computation of the inputs to the ANNs, that requires mainly reduction operations that should be repeated for each window of the image. With the objective of reusing data, the inputs are computed performing a multipass approach that increases the data reuse. Both multiresolution algorithms identify the targets without the need of a postprocessing stage for identifying the scale of the targets. The difference between the algorithms is that the rotation-invariant multiresolution algorithm not only detects the targets but also their rotation angle. Both implementations differs mainly on the complexity of the ANN-detector.

For the multiresolution target detection algorithm a 2-ANN detector is applied at each iteration and a maximum speedup of 51.7× was achieved for a 1024 × 1024 image with 64 spectral bands and 7 targets. For the rotation-invariant multiresolution algorithm a detector consisting of 4 ANNs is required and for an image of the same size but considering ships as targets, a speedup of 39.9× was obtained.

The developed algorithms are competitive in accomplishing the goal of real-time on-board processing of hyperspectral data because they are specially adapted to GPU processing, a flexible and portable computing platform. To reach the real-time objective depends on the rate at which images need to be analyzed. These algorithms could be applied with some modifications for a broad range of hyperspectral applications.

Acknowledgment

This work was supported in part by the Ministry of Science and Innovation, Government of Spain, and FEDER funds under contract TIN 2010-17541, and by the Xunta de Galicia under contracts 2010/28 and 08TIC001206PR.

References

[1] J. Oldeland, W. Dorigo, L. Lieckfeld, A. Lucieer, and N. Jurgens. Combining vegetation indices, constrained ordination and fuzzy classification for mapping semi-natural vegetation units from hyperspectral imagery. *Remote Sensing of Environment*, 114(6):1155–1166, 2010.

[2] F. López-Peña, J.L. Crespo, and R.J. Duro. Unmixing low-ratio endmembers in hyperspectral images through Gaussian synapse ANNs. *Instrum. Meas., IEEE Trans.*, 59(7):1834–1840, 2010.

[3] V. Fresse, D. Houzet, and C. Gravier. GPU architecture evaluation for multispectral and hyperspectral image analysis. *Society of Photo-Optical Instrumentation Engineers (SPIE) Conference Series*, 7872:121–127, 2010.

[4] J.E. Freeman, S. Panasyuk, A.E. Rogers, S. Yang, and R. Lew. Advantages of intraoperative medical hyperspectral imaging (MHSI) for the evaluation of the breast cancer resection bed for residual tumor. *Journal of Clinical Oncology*, 2005 ASCO Annual Meeting Proceedings, 23(16S), Part I of II (June 1 Supplement), pp. 709, 2005.

[5] J. Brazile, M.E. Schaepman, D. Schlapfer, J.W. Kaiser, J. Nieke and K.I. Itten. Cluster versus grid for large-volume hyperspectral image preprocessing. *Proceedings of SPIE*, 5542:480-491, 2003.

[6] B. Priego, D. Souto, F. López-Peña, and R.J. Duro. An ANN based hyperspectral waterway control and security system. In *Proceedings of IEEE International Conference on Computational Intelligence for Measurement Systems and Applications (CIMSA)*, Tianjim, pp. 1–6, 2011.

[7] J.D. Owens et al. A survey of general-purpose computation on graphics hardware. *Computer Graphics Forum*, 26(1):80–113, 2007.

[8] NVIDIA. NVIDIA CUDA C programming guide (version 4.0). NVIDIA Santa Clara, CA, 2011.

[9] D. Manolakis, D. Marden, and G.A. Shaw. Hyperspecral image processing for automatic target detection applications. *MIT Lincolm Laboratory Journal*, 14:79–116, 2003.

[10] A. Plaza, J. Plaza, and H. Vegas. Improving the performance of hyperspectral image and signal processing algorithms using parallel, distributed and specialized hardware-based systems. *Journal of Signal Processing Systems*, 61(3):293–315, 2009.

[11] S. Sánchez, G. Martin, A. Plaza, and Ch.-I. Chang. GPU implementation of fully constrained linear spectral unmixing for remotely sensed hyperspectral data exploitation. *Proceedings of the SPIE Optics and Photonics*, 7810:1–11, 2010.

[12] A. Paz and A. Plaza. Clusters versus GPUs for parallel target and anomaly detection in hyperspectral images. *EURASIP Journal on Advances in Signal Processing*, 2010:1–18, 2010.

[13] B. Priego, R.J. Duro, F. Bellas, and D. Souto. Neural based rotation and scale independent detection of targets in a hyperspectral waterway monitoring system. In *Proceedings of ICPRAM International Conference on Pattern Recognition Applications and Methods*, Vol. 1, pp. 419–425, 2012.

[14] D.B. Kirk, W.-M.W. Hwu. *Programming Massively Parallel Processors: A Hands-on Approach*. Elsevier, Burlington, MA, 2010.

[15] NVIDIA. NVIDIA GeForce GTX 580 GPU Datasheet. NVIDIA Santa Clara, CA, 2010.

[16] NVIDIA. Whitepaper. NVIDIAs Next Generation CUDA Compute Architecture: Fermi. USA, 2009.

[17] M. Harris, S. Sengupta, and J.D. Owens. Parallel prefix sum (scan) with CUDA. In *GPUGems 3*, H. Nguyen (Ed.), pp. 851-876. Addison Wesley, 2007,

[18] J.M. Nageswaran, N. Dutt, J.L. Krichmar, A. Nicolau, and A.V. Veidenbaum. A configurable simulation environment for the efficient simulation of large-scale spiking neural networks on graphics processors. *Neural Networks*, 22:791–800, 2009.

[19] NVIDIA. NVIDIA CUDA CUBLAS Library. March 2008.

[20] D.L. Ly, V. Paprotski, and D. Yen. Neural networks on GPUs: Restricted Boltzmann machines. Department of Electrical and Computer Engineering, University of Toronto, 2009.

[21] H. Jang, A. Park, and K. Jung. Neural network implementation using CUDA and OpenMP. In *Proceedings of the 2008 Digital Image Computing: Techniques and Applications*, pp. 155–161, 2008.

[22] Mark Harris. Optimizing parallel reduction in CUDA, 2008. Available at `http://developer.download.nvidia.com/compute/cuda/1_1/Website/projects/reduction/doc/reduction.pdf` [accessed on 1/25/2010].

[23] Y. Zhang, J. Cohen, and J.D. Owens. Fast tridiagonal solvers on the GPU. In *Proceedings of 15th ACM SIGPLAN Symposium on Principles and Practice of Parallel Computing*, Bangalore, India, January, Vol. 9-14, pp. 127-136, ACM, New York, 2010.

[24] G. Chen, L. Chacón, and D.C. Barnes. An efficient mixed-precision, hybrid CPU-GPU implementation of a fully implicit particle-in-cell algorithm. *J. Comput. Phys.*, submitted, 2011.

[25] NVIDIA. Compute Visual Profiler, User Guide. DU-05162-001_v02, October 2010.

10

A Temporal Based Process Classification Approach Using Hyperspectral Image Sequences

Blanca Priego, Daniel Souto, Francisco Bellas, Richard J. Duro and Fernando López-Peña

Grupo Integrado de Ingeniería, Universidade da Coruña, 15403 Ferrol (A Coruña), Spain; e-mail: blanca.priego@udc.es

Abstract

This chapter is concerned with the problem of classifying processes using temporal characteristics. To this end the approach followed employs the temporal information in sequences of hyperspectral images that are obtained as the processes take place. In other words, classification is achieved taking into account the temporal evolution of the process in the discrimination that must be made. To facilitate obtaining the appropriate classifiers, a particular type of artificial neural networks with trainable delays in their synapses as well as a training algorithm that permits adapting these temporal delays to the process are considered. The classification scheme is presented and applied to a real processing problem in the area of curing resins. Several experiments that considered different proportions of resin components as well as varying environmental parameters such as humidity were carried out and are described here.

Keywords: artificial neural networks, synaptic delays, hyperspectral images, process classification.

Richard J. Duro and Fernando López Peña (Eds.), Digital Image and Signal Processing for Measurement Systems, 267–287.

10.1 Hyperspectrometry

Hyperspectrometry or hyperspectral imaging is a relatively new technique that has been traditionally used in the field of remote sensing. The first hyperspectrometers were developed in order to produce images from high flying airborne platforms or satellites. They were usually large instruments with complex deployment and handling characteristics typically run by space agencies or other large providers of imaging resources [1]. Typical applications of remote hyperspectral sensing are related to vegetation monitoring [2–4], target detection [5, 6], and many others. Currently the situation has changed quite a bit. As a consequence of the popularization of imaging sensors and the advances in digital photography and video capture technology, new implementations of many smaller designs and platforms have been proposed and have led to the aperture of many new application domains. This has been especially relevant in the area of close up inspection tasks such as medical imaging [7, 8] or quality control in processing plants [9, 10], leading to a flurry of activity in research into new algorithms and strategies that would adapt to these new application areas.

A hyperspectral image is characterized by the fact that for each spatial resolution element (pixel), data is collected with high spectral resolution over the electromagnetic spectrum in the 400–2500 nm band (visible to near infrared). It is commonplace to use 50 to 250 spectral bands with bandwidths in the 5 to 20 nm range. The large amount of information that any hyperspectral image provides allows for a very detailed description of the spectral signature in each pixel in the image, and consequently, facilitates the detection and identification of individual materials or classes. However, these very large amounts of data are the source of some of the main challenges currently associated with hyperspectral imaging: efficient handling of high data rates and accurate and fast segmentation and classification of areas within the images.

Hyperspectral sensing systems represent today a mature technology, and, as indicated, have been used in many different fields. Nevertheless, there is presently a wide open field for new applications using hyperspectral imaging at close and mid-range that would make these instruments much more accessible and ubiquitous. However, addressing these applications does require an impulse in two main directions for the technology to become commonplace and popular. On one hand, hyperspectrometers must be made more affordable, small, light and rugged and, on the other, they must be made as autonomous as possible and very easy to use by non-experts. This obviously requires developments in the hardware and control structure and motivates the research into efficient and accurate methods to process this data. In particular, as autonomy and data processing resilience would be a very important aspect of this new generation of sensors, the introduction of more intensive computational intelligence techniques that allow the systems to be used in less specialized applications than those currently contemplated becomes a necessity.

From a software or data processing point of view, Manolakis et al. [11] have indicated that most algorithms used in hyperspectral applications can be grouped into four categories: change detection, target/anomaly detection, classification, and spectral unmixing. In the last twenty years, almost all research efforts have focused on the analysis of static hyperspectral images in terms of the last three categories. However, more recently and as a consequence of improvements in hyperspectrometers, it has become feasible to capture a continuous flow of images allowing for the processing of hyperspectral information from a dynamic point of view. Some algorithms have been designed using temporal information [12, 13], mostly for the detection of changes, directly in the images as in [14–16] or detecting changes in the abundances of endmembers as in [17]. Some other authors use the sequence of images to enhance resolution, this is the case in [18], but still without really using the temporal evolution of the frame sequence as a classification strategy. The objective of this chapter is to present a technique that permits addressing the problem of using temporal sequences in the classification process and not only to detect changes. In fact, there are many problems where information from a single image provides ambiguous classification results and it is the integration of the evolution of the subject in time that really provides an unambiguous classification.

This problem has been tackled in the image processing field dealing with one or three dimensional (RGB) images by using video sequences to obtain better classifications, to eliminate noise or to improve image resolution. Some authors have even used low resolution multispectrometers to this end. In fact, an example of this is the work by Murguia et al. [19] who characterize astronomical objects in terms of their periodic signatures to determine their rotation. Here the objective is to extend this work to the realm of high dimensional images, such as those obtained from hyperspectral sensors, through the use of new techniques and algorithms that are adapted to the high dimensionality involved. In particular, in this chapter we make use of specific types of neural network architectures and training procedures for the temporal processing of this type of images.

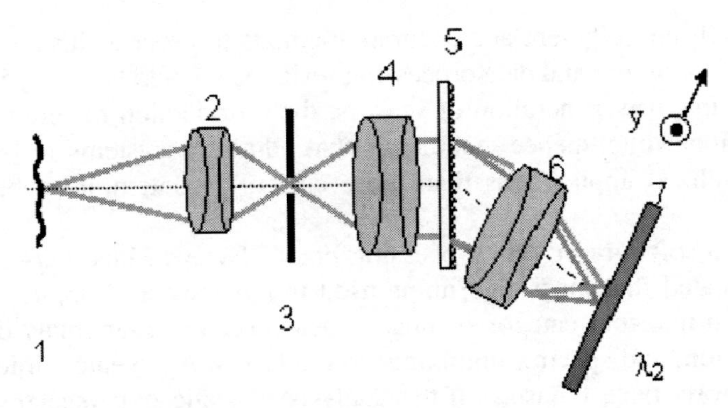

Figure 10.1 Hyperspectrometer schematics: (1) subject, (2), (4), (6) objectives, (3) slit, (5) grating, (7) CCD.

10.2 Sensing System

Hyperspectrometric sensors come in a variety of flavors and structures. The sensing system used here in the framework of a research project on the developing of a complete, automatic, flexible, reliable, fast and precise real time detection and representation system. At its core is a push-broom type light imaging hyperspectrometer which is an independent module developed to be used as a portable device on its own. Figure 10.1 displays its configuration.

It is constructed out of commercially available components allowing for a low-cost product. The slit allows just one line of the subject image which is decomposed into a spatial dimension plus a spectral dimension in the image plane. A 12 bit CCD camera acquires these images corresponding to a section of the hyperspectral cube. The instrument presents a good signal to noise ratio for wavelengths in the range between 430 and 1000 nm, which is the range it covers with a resolution of up to 1040 bands. The measured spatial resolution was about 30 um in the image plane. In terms of speed, this sensor is capable of capturing up to 47.2 hyperspectral 1040 pixel lines per second with the spectral depth indicated before.

This instrument was used in a processing line and Figure 10.2 displays the experimental setup that was used and a schematic view of the hyperspectrometer. In addition, a virtual instrument has been developed to implement the ANN based processing module plus a user friendly graphical interface for visualization, control and post-processing. For all the experiments presented here, all of the measurements were carried out in our laboratory in a simulated processing line as shown in the top image of Figure 10.2.

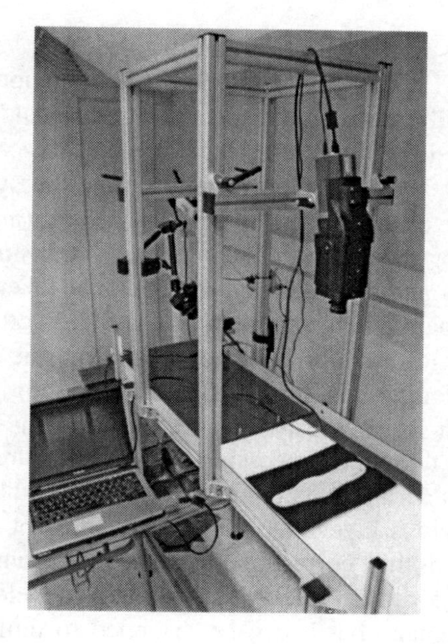

Figure 10.2 Experimental setup (top) and schematic view of the hyperspectrometer developed.

10.3 Temporal Based Pixel Classification

Now the problem that must be addressed is how to perform classifications taking into account the temporal evolution of the spectra of the pixels. To this end it was determined that a neural network architecture developed in our group and called Temporal Delay Based ANN (TDB-ANN) as well as its training algorithm (TDBP) could be very adequate for this purpose. The architecture and training algorithm of the artificial neural network we consider were introduced in [20] and studied with respect to its noise rejection characteristics in [21].

A TDB-ANN consists of several layers of neurons connected in a very similar way to a Multiple Layer Perceptron (MLP). Thus, every neuron of one layer is connected through a synapsis to every neuron of the next layer. Each neuron performs a sum of its inputs and passes these values through some non-linear function (in this case a sigmoid). It is obviously a feed-forward network. The only difference with respect to a traditional MLP is that the synapses are represented by two trainable parameters: the classical weight term and an additional delay term. Consequently, now the synaptic connections between neurons are characterized by a pair of values, (w_{ij}, τ_{ij}), where w_{ij} is the weight describing the ability of the synapsis to transmit information from neuron i to neuron j and τ_{ij} is a delay, which can be taken as an indication of the length of the synapsis between neurons i and j. In other words, it models the time it takes for information to traverse the synapsis and reach the target neuron, that is, the longer it is, it will take more time for information to reach the target neuron. It is, obviously, this second term the one that allows the network to perform temporal processing as a target neuron in a given instant of time will be considering at its inputs information chunks coming from the neurons in the previous layer that have taken different amounts of time to reach it, meaning that they correspond to different instants of time.

The training algorithm (TDBP) was initially developed for processing one-dimensional signals, but it can be extended to multidimensional ones. The main assumption that is made when training networks using this algorithm is that each neuron in a given layer can choose the delay it wishes to impose on its inputs. Time is discretized into instants, each one of which corresponds to the period of time between an input to the network and the next input. Every neuron of the network computes an output every instant of time.

A selection function is added to the processing of each neuron in order to allow it to choose from all of the possible previous inputs to a neuron the ones we are actually going to take as input it in a given instant of time (think of all of the inputs in time as a list of values out of which the neuron can choose one by pointing at it or, actually, by modifying a delay term). This selection function can be something as simple as:

$$\delta_{ij} = \begin{cases} 1 \rightarrow i = j \\ 0 \rightarrow i \neq j \end{cases} \tag{10.1}$$

When considering an output neuron, the output of neuron k in instant t can be expressed as:

$$O_{kt} = F\left(\sum_{i=0}^{N}\sum_{j=0}^{t} \delta_{j(t-\tau_{ik})} w_{ik} h_{ij}\right) \tag{10.2}$$

where F is the activation function of neuron k, h_{ij} is the output of neuron i of the previous layer in instant j and w_{ik} is the weight of the synapsis between neuron i and neuron k. The first sum is over all the neurons that reach neuron k (those of the previous layer) and the second one is over all the instants of time considered, that is, an interval as long as the maximum delay term we want to allow.

The result of this function is the sum of the outputs of the hidden neurons in times $t - \tau_{jk}$ (where τ_{jk} is the delay in each synapsis jk) weighed by the corresponding weight values. We thus obtain the output of every neuron as a function of the outputs of the neurons in the previous layer and the weights and delays in the synapses linking them.

If we now consider the output of a neuron in the hidden layer, the equation would look like:

$$h_{it} = F\left(\sum_{r=0}^{M}\sum_{s=0}^{t} \delta_{s,(t-\tau_{ri})} w_{ri} I_{rs}\right) \tag{10.3}$$

In this case I_{rs} corresponds to the output of an input neuron.

The key to this type of networks is that both the weights and delays can be trained. To train these weights and delays we have resorted to a modification of the basic gradient descent algorithm of traditional backpropagation, but taking into account the delay terms when computing the gradients of the error with respect to weights and delays.

Therefore, starting from an error term for the outputs consisting in the squared difference between what the network outputs and the target values we want it to produce summed for all outputs:

$$E_{total,t} = \sum_{k} (O_{kt} - T_{kt})^2 \tag{10.4}$$

We can now calculate the gradients of the error with respect to the weights and delays in the synapses of the different layers.

10.3.1 Weight Terms

The influence of modifying a weight term w_{ik} connecting a neuron in a hidden layer to one in the output layer over this error term can be written as:

$$\frac{\partial E_{total,t}}{\partial w_{jk}} = \frac{\partial E_{total,t}}{\partial O_{kt}} \frac{\partial O_{kt}}{\partial oNet_{kt}} \frac{\partial oNet_{kt}}{\partial w_{jk}} \tag{10.5}$$

with $oNet_{kt}$ being the weighed sum of the inputs to neuron k in time t, in other words, its combination function.

Now, much in the same way as in the case of the traditional backpropagation algorithm, we define

$$\Delta_{kt} = \frac{\partial E_{total,t}}{\partial O_{kt}} \frac{\partial O_{kt}}{\partial oNet_{kt}} = 2(O_{kt} - T_{kt})F'(oNet_{kt}) \tag{10.6}$$

And the last term of (10.5) results in

$$\frac{\partial oNet_{kt}}{\partial w_{jk}} = \frac{\partial \left[\sum_{i=0}^{N} \sum_{l=0}^{t} \delta_{l,(t-\tau_{ik})} w_{ik} h_{il}\right]}{\partial w_{jk}} = \sum_{l=0}^{t} \delta_{l,(t-\tau_{jk})} h_{jl} = h_{j,(t-\tau_{jk})} \tag{10.7}$$

Therefore, the gradient of the error in terms of the weights of the synapses to neurons in the output layer can be written as:

$$\frac{\partial E_{total,t}}{\partial w_{jk}} = \Delta_{kt} h_{j,(t-\tau_{jk})} \tag{10.8}$$

Now, for synapses to neurons in the hidden layer,

$$\frac{\partial E_{total,t}}{\partial w_{jk}} = \frac{\partial E_{total,t}}{\partial hNet_{kt}} \frac{\partial hNet_{kt}}{\partial w_{jk}} \tag{10.9}$$

and using the same reasoning as in the case of output neurons:

$$\frac{\partial hNet_{kt}}{\partial w_{jk}} = \sum_{l=0}^{t} \delta_{l,(t-\tau_{jk})} I_{jl} = I_{j,(t-\tau_{jk})} \tag{10.10}$$

being I_{jt} the output of input neuron j in time t. On the other hand,

$$\Delta_{kt} = \frac{\partial E_{total,t}}{\partial hNet_{kt}} = \frac{\partial E_{total,t}}{\partial h_{kt}} \frac{\partial h_{kt}}{\partial hNet_{kt}} = \frac{\partial E_{total,t}}{\partial h_{kt}} F'(hNet_{kt}) \tag{10.11}$$

and backpropagating the error through the output neurons

$$\frac{\partial E_{total,t}}{\partial h_{kt}} = \sum_{r} \frac{\partial E_{total,t}}{\partial oNet_{rt}} \frac{\partial oNet_{rt}}{\partial h_{kt}} = \sum_{r} \Delta_{rt} \frac{\partial oNet_{rt}}{\partial h_{kt}} = \sum_{r} \Delta_{rt} w_{kr} \tag{10.12}$$

where r goes through all the output neurons. Therefore

$$\Delta_{kt} = \frac{\partial E_{total,t}}{\partial hNet_{kt}} = F'(hNet_{kt}) \sum_{r} \Delta_{rt} w_{kr} \tag{10.13}$$

and, consequently, the gradients in this case are very similar to those corresponding to the output neurons.

$$\frac{\partial E_{total,t}}{\partial w_{jk}} = \frac{\partial E_{total,t}}{\partial hNet_{kt}} \frac{\partial hNet_{kt}}{\partial w_{jk}} = I_{j,(t-\tau_{jk})} F'(hNet_{kt}) \sum_r \Delta_{rt} w_{kr} \quad (10.14)$$

10.3.2 Delay Terms

Doing the same for the influence on the total error of a variation of a delay term τ_{jk}, that is, of which previous output is selected, we obtain:

$$\frac{\partial E_{total,t}}{\partial \tau_{jk}} = \frac{\partial E_{total,t}}{\partial O_{kt}} \frac{\partial O_{kt}}{\partial oNet_{kt}} \frac{\partial oNet_{kt}}{\partial \tau_{jk}} \quad (10.15)$$

where

$$\frac{\partial oNet_{kt}}{\partial \tau_{jk}} = \frac{\partial \left[\sum_{i=0}^{N} \sum_{l=0}^{t} \delta_{l,(t-\tau_{ik})} w_{ik} h_{il} \right]}{\partial \tau_{jk}} \quad (10.16)$$

Using the same arguments as in the case of the weights, we have

$$\frac{\partial oNet_{kt}}{\partial \tau_{jk}} = w_{jk} \frac{\partial h_{j,(t-\tau_{jk})}}{\partial \tau_{jk}} \quad (10.17)$$

The second term of (10.17), that is, the variation of the output of a neuron in the hidden layer due to a change in the delay term, which is the same as saying how the output of the hidden layer varies with time around $t - \tau_{jk}$ can be approximated to a first order by considering discrete time as:

$$\frac{\partial oNet_{kt}}{\partial \tau_{jk}} = w_{jk} \left(h_{j,(t-\tau_{jk})} - h_{j,(t-\tau_{jk}-1)} \right) \quad (10.18)$$

This approximation only implies assuming a degree of smoothness in the outputs of the neurons with time in the time scale of two discrete events, which in practice turns out to be a valid assumption. Obviously, this term could be made more precise by considering more points around $t - \tau_{jk}$, but this is usually not necessary.

Thus, the gradient terms can be written as:

$$\frac{\partial E_{total}}{\partial \tau_{jk}} = \Delta_k w_{jk} \left(h_{j(t-\tau_{jk})} - h_{j(t-\tau_{jk}-1)} \right) \quad (10.19)$$

in which E_{total}, as indicated above, is the total squared error for all the training vectors.

In the case of the hidden layer we have

$$\frac{\partial E_{total,t}}{\partial \tau_{jk}} = \frac{\partial E_{total,t}}{\partial hNet_{kt}} \frac{\partial hNet_{kt}}{\partial \tau_{jk}} \tag{10.20}$$

and, as in the case of output neurons,

$$\frac{\partial hNet_{kt}}{\partial \tau_{jk}} = w_{jk}\left(I_{j,(t-\tau_{jk})} - I_{j,(t-\tau_{jk}-1)}\right) \tag{10.21}$$

being I_{jt} the output of input neuron j in time t. Consequently,

$$\frac{\partial E_{total,t}}{\partial \tau_{jk}} = \frac{\partial E_{total,t}}{\partial hNet_{kt}} \frac{\partial hNet_{kt}}{\partial \tau_{jk}} = w_{jk}\left(I_{j,(t-\tau_{jk})} - I_{j,(t-\tau_{jk}-1)}\right)$$
$$F'(hNet_{kt}) \sum_r \Delta_{rt} w_{kr} \tag{10.22}$$

where index r represents the neuron of the next layer, whether output or hidden.

Taking all of the previous equations into account, now the algorithm is basically the same traditional gradient descent employed in standard backpropagation. That is, modify the weight term and the delay terms in the direction of the negative gradient until a minimum of the error for the whole system is obtained. Consequently, the equations for the increments of weights and time delays for the output nodes during the backpropagation phase result in:

$$\Delta w_{jk} = 2(O_{kt} - T_{kt})F'(oNet_{kt})h_{j,(t-\tau_{jk})} \tag{10.23}$$

$$\Delta \tau_{jk} = 2(O_{kt} - T_{kt})F'(oNet_{kt})w_{jk}\left(h_{j,(t-\tau_{jk})} - h_{j,(t-\tau_{jk}-1)}\right) \tag{10.24}$$

and for those in the hidden layers:

$$\Delta w_{jk} = I_{j,(t-\tau_{jk})} F'(hNet_{kt}) \sum_r \Delta_{rt} w_{kr} \tag{10.25}$$

$$\Delta \tau_{jk} = w_{jk}\left(I_{j,(t-\tau_{jk})} - I_{j,(t-\tau_{jk}-1)}\right) F'(hNet_{kt}) \sum_r \Delta_{rt} w_{kr} \tag{10.26}$$

To address the problem of obtaining fractional values for t after updating (which would not be valid due to the fact that time is discretized into integer valued instants) and as in practice we have seen that rounding off to the nearest integer is not a good solution as this obviously leads to a very discontinuous gradient and thus hinders the training process, we linearly interpolate during the training process values for the outputs of the neurons in between the values they produce for the two integer delays around the fractional one. This way, the gradient function is continuous and no additional assumptions to the one that was already pointed out regarding a degree of smoothness in the outputs of the neurons are made.

Summarizing, by discretizing the time derivative, simple expressions may be obtained for the modification of the weights and delays of the synapses in an algorithm that is basically a backpropagation algorithm where the activation function of the neuron has been modified to permit selecting delays or, in order words, choosing from the list of previous outputs of the neuron in the previous layer. By adding input neurons to the network, any dimensionality of the signals can be chosen.

10.4 An Application within an Industrial Process

The objective of this chapter is to present a technique whereby considering the temporal sequence of hyperspectral images taken during the evolution of a process, and appropriately processing them, we can obtain information that is hard or impossible to obtain using static images. A series of experiments related to controlling the quality of the drying or curing processes after applying different surface coatings were carried out to test this hypothesis and to demonstrate how the algorithm can be applied.

In terms of the industrial process, the aim is to determine whether a surface coating applied in an industrial line meets the quality levels required. This will be achieved by studying the temporal drying sequence of the coating by capturing a sequence of hyperspectral images of the products as time progresses. The coatings under consideration are resins. To produce appropriately cured resins it is necessary to mix certain components (in this case two) in the adequate proportions and let the resin cure for a period of time under controlled environmental conditions. Obviously, what is being considered here is really a spatial-temporal problem, as the classification may be different for different areas of the resin surface as shown in Figure 10.3. The final objective is to be able to determine whether the resin has cured appropriately in every point of the surface by analysing the temporal evolution of the process.

To test the validity of the approach, an experimental data set was constructed by making several different mixtures of resins under different humidity conditions and with different proportions of two components A and B. The different cases considered are displayed in Table 10.1. The first one of these cases (50% A and 50% B) was taken as the reference "correct curing process"

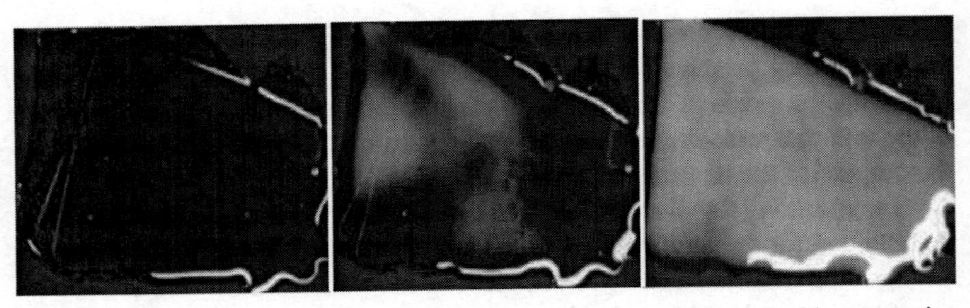

Figure 10.3 Three instants in the evolution of the drying process as obtained by composing bands from the visible spectra.

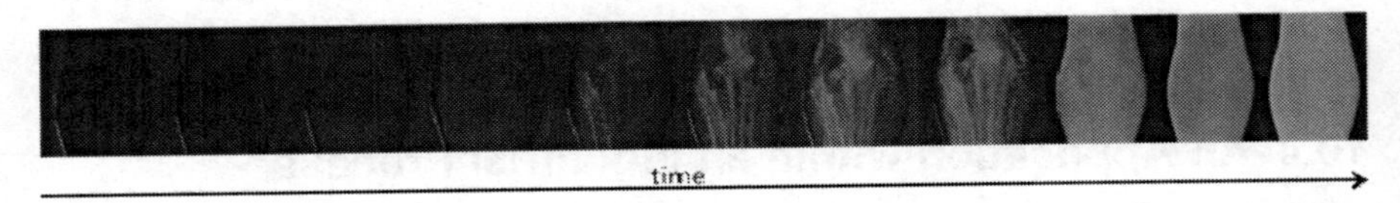

Figure 10.4 Whole curing process for the reference sample.

and as such it is labeled in the graphs and comments that will follow. This case contemplates the reference mixture and conditions for which the curing process is optimal.

As the objective is to analyse the curing process through the determination of how it evolves in time, the mixtures were made and left to cure for 500 seconds. Using the hyperspectrometric setup described before and shown in the top image of Figure 10.2, images were taken of a line of 1000 points traversing the mixtures at a rate of close to one image per second (485 images in 500 seconds). Figure 10.4 displays a sequence of images of this curing process for the reference sample and Figure 10.5 displays the visual result of the final material obtained for some of the experiments. It is hard to discriminate between the correct curing process and the others only by analyzing these final images without considering the temporal evolution of the material's mixture.

The spectra corresponding to the points obtained were normalized and corrected using as a reference the background area, where there was no mixture. The objective was for the processing system to discriminate the optimal curing from suboptimal ones as soon as possible.

As a first test in order to validate the need to perform a time based processing, we evaluated the possibility of discriminating the correctly cured resin from the rest once the whole curing process was finished. Figure 10.6

Table 10.1 Curing processes.

EXPERIMENT 1 Different component percentage. Mixing performed by person X	
1.1	50% A, 50% B
1.2	20% A, 80% B
1.3	30% A, 70% B
1.4	40% A, 60% B
1.5	45% A, 55% B
1.6	55% A, 45% B
1.7	60% A, 30% B
EXPERIMENT 2 Humid Conditions	
2.1	Very wet curing container
2.2	Slightly wet curing container
2.3	5%Water, 47.5% A, 47.5% B
2.4	1%Water, 49.5% A, 49.5% B
EXPERIMENT 3	
	Non-consistent, irregular stirring motion
EXPERIMENT 4 Different component percentage. Mixing performed by person Y	
4.1	40%A, 60%B
4.2	45%A, 55%B
4.3	55%A, 45%B
4.4	60%A, 40%B
EXPERIMENT 5	
	Two different mixtures mixed with slight stirring Mixing 1: 50%A, 50%B Mixing 2: 40%A, 60%B

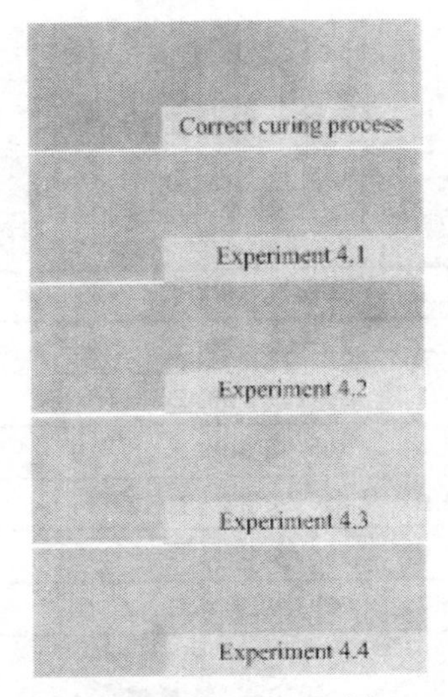

Figure 10.5 Final result for some of the experiments.

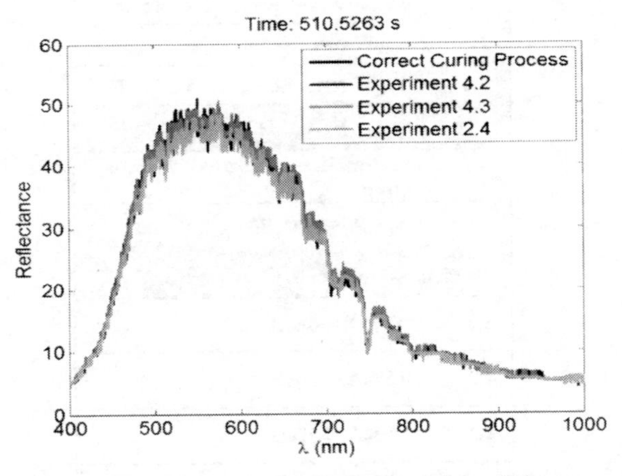

Figure 10.6 Spectra of the final resins after curing has ended for different cases (a correct and three incorrect curing processes).

displays the spectra corresponding to points on three different experiments once all of them had cured. As expected, the spectra are practically identical and it was impossible to obtain any kind of correct classification using only this information.

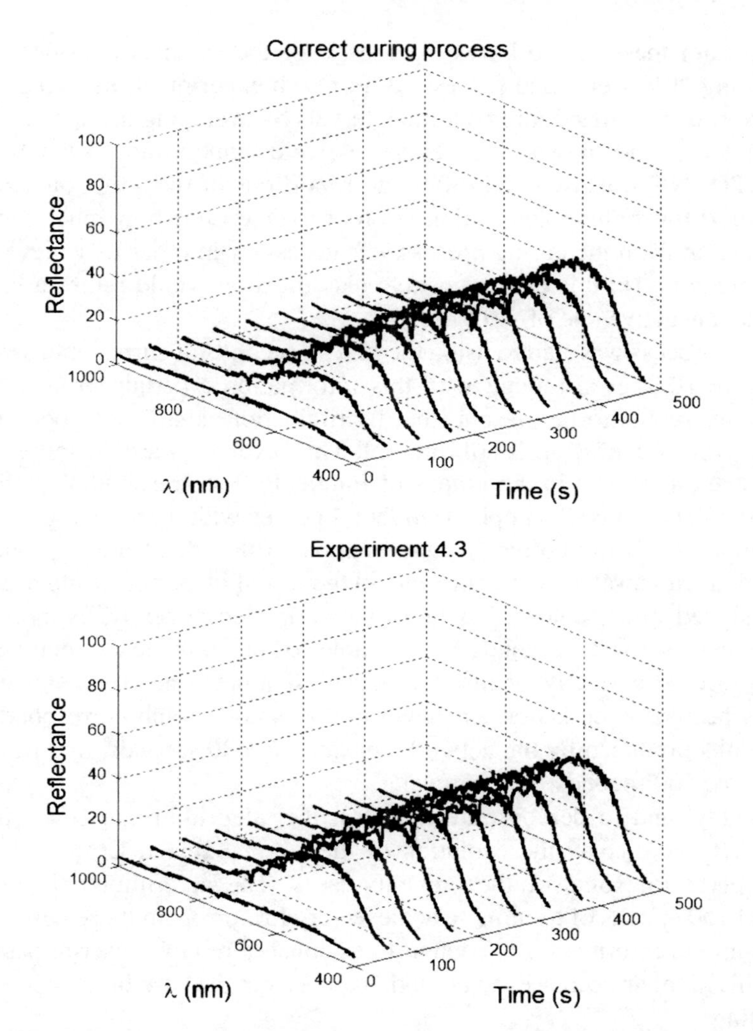

Figure 10.7 Evolution of the spectra during the curing process for the correct curing process (top) and that of experiment 4.3 (bottom).

As a consequence, we considered the temporal evolution and introduced the delay based neural networks explained above as classifiers over a sequence of images taken in time. Figure 10.7 displays a sequence of spectra for two cases, one the correct curing process and another one that is incorrect (4.3 in Table 10.1).

To train these networks we used 80% of the samples considered. The remaining 20% were used for testing. For each experiment, the temporal sequence was discretized into ten time intervals by averaging the spectra within each interval and the spectra themselves were binned into 16 bands. A 16 input TDANN was used and different proportions of the temporal sequence were used for training and testing in order to determine how much temporal information from the curing process was necessary in order to achieve 100% discrimination (in this case a positive classification would return a value of 0.5 and a negative one of −0.5).

This process was started using the first 36 seconds of curing and, as shown in Figure 10.8, even though with this information a certain order in terms of quality seems to be present, the discrimination is still not good enough and a level of confusion is still present. This level decreases if more time is considered as shown in the graphs of Figure 10.9, corresponding to 42, 47, 53 and 480 seconds of sampling. In fact, in cases with more than 47 seconds of sampling, all the correctly cured samples (blue dots) are appropriately classified with a value above zero and all the rest of incorrect curing processes are assigned values below zero by the network. Consequently, by monitoring the processes for 50 seconds we are able to discriminate perfectly if it is going to cure correctly or not. Obviously, if more time is considered, the results become even better, as shown in the bottom graph corresponding to the results produced by the network considering 480 seconds, that is, almost the whole 500 seconds time interval.

Finally, and to demonstrate how fast this algorithm can learn, in Figure 10.10 we display the evolution of the Mean Squared Error as training takes place for some of the previous cases. The algorithm only requires around 150 epochs of training to achieve very low errors in those cases where the temporal information is enough to establish a reliable discrimination. In fact, this number decreases to around 70 epochs in the case of the 480 second sampling.

10.5 Conclusions

This chapter has presented an approach to consider the temporal evolution of the spectra within hyperspectral images to classify the quality of processes even when the final spectra obtained (and therefore their visual appearance) are almost the same for correctly performed processes or processes with deficiencies.

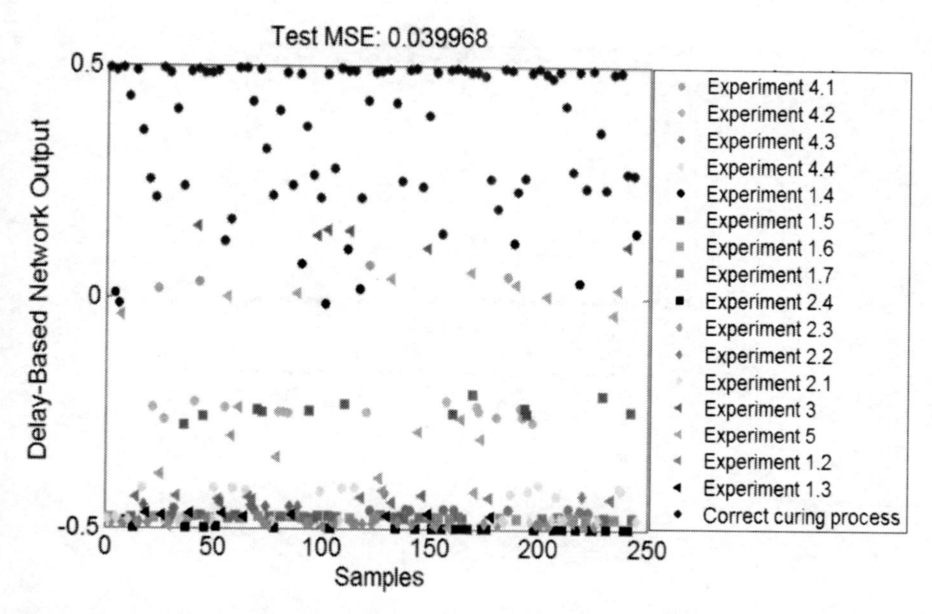

Figure 10.8 Results produced by the TDANN when using a sequence of 10 averaged spectra from the first 36 seconds of the curing process.

The approach involves using artificial neural networks that have been specifically developed to take into account temporal considerations. In this particular case we have extended the operation of Trainable Delay Based Artificial Neural Networks to multidimensional signals to achieve this end. Their application has been straightforward and in the industrial case to which it has been applied, that is, resin curing processes, the results obtained have been very satisfactory with quite short training periods.

This type of approach has been shown to be able to discriminate very clearly between correctly performed curing processes and processes that present problems due to wrong component mixtures or humidity. In addition, it is important to note that using this type of procedure, the discrimination can be performed using just the first few seconds of the curing process and this allows for an immediate rejection of suboptimal products without having to wait for the whole curing process to end.

Anyway, this is just one application within a field that is becoming very important in hyperspectral image processing: Temporal based classification. This field presents many opportunities to detect and classify events as a function of the temporal variations of the features (whether directly spectra or

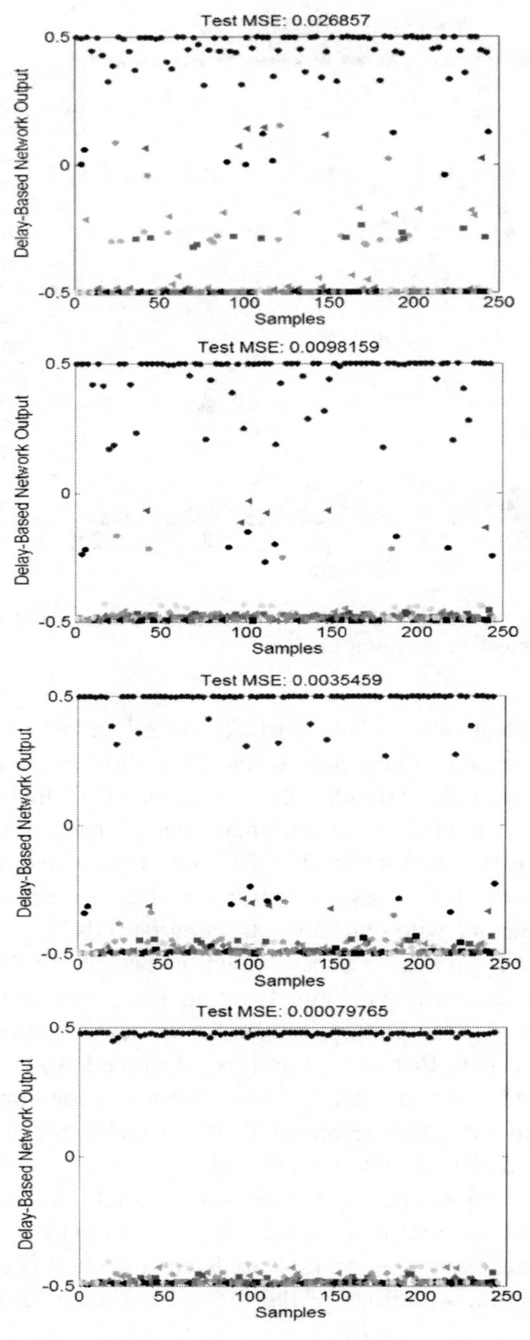

Figure 10.9 Same as Figure 10.5 (including process labels) but taking into account, from top to bottom, 42, 47, 53 and 480 seconds of curing process.

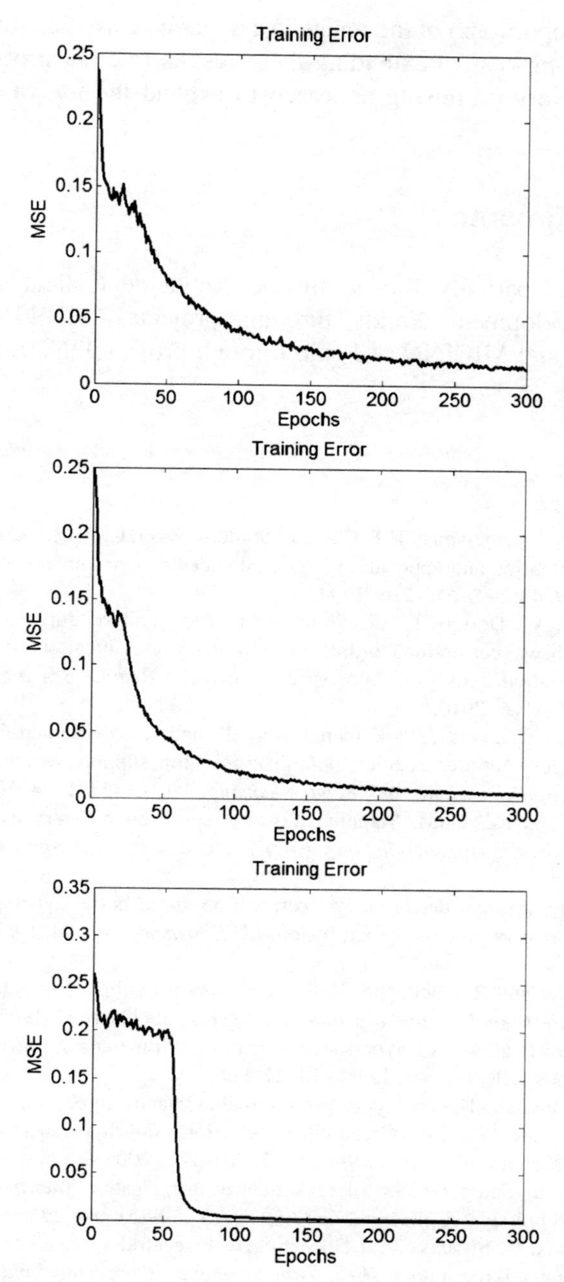

Figure 10.10 Training Error for the training processes using 47, 53 and 480 seconds of sampling.

endmember proportions) of the image in a dynamic classification process. We are now in the process of extending these results to other processes in what we think is a very promising approach to expand the use of hyperspectral imaging.

Acknowledgments

This work was partially funded by the Xunta de Galicia and European Regional Development Funds through projects 09DPI012166PR and 10DPI005CT, and MICINN of Spain through project TIN2011-28753-C02-01.

References

[1] K.L. Carder, P. Reinersman, R.F. Chen, F. Muller-Karger, C.O. Davis, and M. Hamilton. AVIRIS calibration and application in coastal oceanic environments *Remote Sensing of Environment*, 44(2–3):205–216, 1993.

[2] J. Oldeland, W. Dorigo, L. Lieckfeld, A. Lucieer, and N. Jürgens. Combining vegetation indices, constrained ordination and fuzzy classification for mapping semi-natural vegetation units from hyperspectral imagery. *Remote Sensing of Environment*, 114(6):1155–1166, 2010.

[3] V. Heikkinen, T. Tokola, J. Parkkinen, I. Korpela, and T. Jaaskelainen. Simulated multispectral imagery for tree species classification using support vector machines. *IEEE Transactions on Geoscience and Remote Sensing*, 48(3):1355–1364, 2010.

[4] G.H. Mitri and I.Z. Gitas. Mapping postfire vegetation recovery using EO-1 hyperion imagery. *IEEE Transactions on Geoscience and Remote Sensing*, 48(3):1613–1618, 2010.

[5] D.E. Bar et al. Target detection and verification via airborne hyperspectral and high-resolution imagery processing and fusion. *IEEE Sensors Journal*, 10(3):711315-1–711, 2010.

[6] J. Broadwater and R. Chellappa. Hybrid detectors for subpixel targets. *IEEE Transactions on Pattern Analysis and Machine Intelligence*, 29(11):1891–1903, 2007.

[7] A.M. Siddiqi et al. Use of hyperspectral imaging to distinguish normal, precancerous, and cancerous cells. *Cancer*, 114(1):13–21, Feb. 2008.

[8] Q. Li, Y. Wang, and H. Liu. Hyperspectral tongue images. In *Biosystems*, pp. 1–3, 2010.

[9] H. Okamoto and W.S. Lee. Green citrus detection using hyperspectral imaging. *Computers and Electronics in Agriculture*, 66(2):201–208, 2009.

[10] M.O.N.L. Liu. Categorization of pork quality using gabor filter-based hyperspectral imaging technologye. *Journal of Food Engineering*, 99(3):284–293, 2010.

[11] D. Manolakis, C. Siracusa, and G. Shaw. Hyperspectral subpixel target detection using the linear mixing model. *IEEE Transactions on Geoscience and Remote Sensing*, 39(7):1392–1409, 2001.

[12] M.V.A. Murzina and J.P. Farrel. Dynamic hyperspectral imaging. *Nondestructive Detection*, 5679:135–144, 2005.

[13] J. Blackburn, M. Mendenhall, A. Rice, P. Shelnutt, N. Soliman, and J. Vasquez. Feature aided tracking with hyperspectral imagery. In *Proceedings of the SPIE Conference on Signal and Data Processing of Small Targets*, p. 6699, 2007.

[14] A. Schaum. Local covariance equalization of hyperspectral imagery: Advantages and limitations for target detection. In *Proceedings of IEEE Aerospace Conference*, pp. 2001–2011, IEEE, 2005.

[15] J.M. Tirpak and W.M. Giuliano. Using multitemporal satellite imagery to characterize forest wildlife habitat: The case of ruffed grouse. *Forest Ecology and Management*, 260(9):1539–1547, September 2010.

[16] C. Ong, T.J. Cudahy, and G. Swayze. Predicting acid drainage related physicochemical measurements using hyperspectral data. *Chemical Analysis*, pp. 13–16, May 2003.

[17] Q. Du, L. Wasson, and R. King. Unsupervised linear unmixing for change detection in multitemporal airborne hyperspectral imagery. In *Proceedings of 2005 International Workshop on the Analysis of Multi-Temporal Remote Sensing Images*, pp. 136–140, 2005.

[18] J.C.-wai Chan, J. Ma, and F. Canters. Superresolution enhancement for temporal hyperspectral oriented data sets. In *Proceedings of 2009 IEEE International Geoscience and Remote Sensing Symposium (IGARSS 2009)*, Vol. 3, pp. 1003–1006, 2009.

[19] J. Murguia et al. Applications of multispectral video. *Proceedings of SPIE*, 7780:77800B–77800B-11, 2010.

[20] R.J. Duro and J.S. Reyes. Discrete-time backpropagation for training synaptic delay-based artificial neural networks. *IEEE Transactions on Neural Networks*, 10(4):779–789, 1999.

[21] J.S. Santos-Reyes and R.J. Duro. Influence of noise on discrete time backpropagation trained networks. *Neurocomputing*, 41(1–4):67–89, 2001.

Subject Index

About the Editors

Richard J. Duro is a Professor in the Department of Computer Science, Head of the Integrated Group for Engineering Research at the University of A Coruña, and Research Director at the Center for Marine Technology of Galicia. His research interests include autonomous and evolutionary robotics in highly unstructured and dynamic environments, higher order neural network structures and signal processing. He has authored more than 200 papers in peer reviewed journals and high level conferences and four books on these topics and has participated in the organization of several IEEE conferences as well as in the program and scientific committees of over 100 international conferences. He is a Senior member of the IEEE and has chaired the Task Force on Instrumentation and Measurement within the IEEE Computer society. He has been the principal researcher of more than 80 research projects.

Fernando López Peña received a Master degree from the Polytechnic University of Madrid, Madrid, Spain, in 1981, a Research Master from the von Karman Institute of Fluid Dynamics, Belgium, in 1987, and the Ph.D. degree from the University of Louvain, Louvain-le-Neuve, Belgium, in 1992. He is a Professor at the University of A Coruña, Ferrol, Spain and leads the Fluids Engineering Group that is part of the Integrated Group for Engineering Research at this university. From February 2005 until February 2012 he was the Head of the Maritime and Oceanic Department of this university. He was an Assistant Professor at the Polytechnic University of Madrid from January 1993 to September 1994 before incorporating to his current position. He has authored about 100 papers in peer reviewed journals and international conferences, holds four patents, and has led more than 80 research projects. His current research activities are related to intelligent hydrodynamic and aerodynamic design and optimization, flow measurement and diagnosis, and signal and image processing.

Lightning Source UK Ltd.
Milton Keynes UK
UKOW02n0842151214

243137UK00001B/28/P